Earth Slopes & Retaining Structures

Custom 8th Edition

Braja M. Das | Khaled Sobhan

CENGAGE
Learning·

Australia • Brazil • Japan • Korea • Mexico • Singapore • Spain • United Kingdom • United States

CENGAGE
Learning

Earth Slopes & Retaining Structures:
Custom 8th Edition

Principles of Foundation Engineering, Eighth Edition
Braja M. Das
© 2016, 2012 Cengage Learning. All rights reserved.
Library of Congress Control Number: 2014947850

Principles of Geotechnical Engineering, Eighth Edition
Braja M. Das | Khaled Sobhan
© 2014, 2010 Cengage Learning. All rights reserved.
Library of Congress Control Number: 2012942875

For product information and technology assistance, contact us at
Cengage Learning Customer & Sales Support, 1-800-354-9706

For permission to use material from this text or product,
submit all requests online at **cengage.com/permissions**
Further permissions questions can be emailed to
permissionrequest@cengage.com

This book contains select works from existing Cengage Learning resources and
was produced by Cengage Learning Custom Solutions for collegiate use. As
such, those adopting and/or contributing to this work are responsible for
editorial content accuracy, continuity and completeness.

Compilation © 2016 Cengage Learning

ISBN: 978-1-337-03992-5

Cengage Learning

20 Channel Center Street
Boston, MA 02210
USA

Cengage Learning is a leading provider of customized learning solutions with
office locations around the globe, including Singapore, the United Kingdom,
Australia, Mexico, Brazil, and Japan. Locate your local office at:
www.international.cengage.com/region.
Cengage Learning products are represented in Canada by Nelson Education, Ltd.

For your lifelong learning solutions, visit **www.cengage.com /custom.**

Visit our corporate website at **www.cengage.com.**

Printed at CLDPC, USA, 07-18

Brief Contents

1 Natural Soil Deposits and Subsoil Exploration

3.1 Introduction

To design a foundation that will support a structure, an engineer must understand the types of soil deposits that will support the foundation. Moreover, foundation engineers must remember that soil at any site frequently is nonhomogeneous; that is, the soil profile may vary. Soil mechanics theories involve idealized conditions, so the application of the theories to foundation engineering problems involves a judicious evaluation of site conditions and soil parameters. To do this requires some knowledge of the geological process by which the soil deposit at the site was formed, supplemented by subsurface exploration. Good professional judgment constitutes an essential part of geotechnical engineering—and it comes only with practice.

This chapter is divided into two parts. The first is a general overview of natural soil deposits generally encountered, and the second describes the general principles of subsoil exploration.

Natural Soil Deposits

3.2 Soil Origin

Most of the soils that cover the earth are formed by the weathering of various rocks. There are two general types of weathering: (1) mechanical weathering and (2) chemical weathering.

Mechanical weathering is a process by which rocks are broken down into smaller and smaller pieces by physical forces without any change in the chemical composition. Changes in temperature result in *expansion and contraction of rock* due to gain and loss of heat. Continuous expansion and contraction will result in the development of cracks in rocks. Flakes and large fragments of rocks are split. *Frost action* is another source of mechanical weathering of rocks. Water can enter the pores, cracks, and other openings in the rock. When the temperature drops, the water freezes, thereby increasing the volume by about 9%. This results in an outward pressure from inside the rock. Continuous freezing and thawing will result in the breakup of a rock mass. *Exfoliation* is another mechanical weathering process by which rock plates are peeled off from large rocks by physical forces. Mechanical weathering of rocks also takes place due to the action of *running water, glaciers, wind, ocean waves*, and so forth.

Chemical weathering is a process of decomposition or mineral alteration in which the original minerals are changed into something entirely different. For example, the common minerals in igneous rocks are quartz, feldspars, and ferromagnesian minerals. The decomposed products of these minerals due to chemical weathering are listed in Table 3.1.

Most rock weathering is a combination of mechanical and chemical weathering. Soil produced by the weathering of rocks can be transported by physical processes to other places. The resulting soil deposits are called *transported soils.* In contrast, some soils stay where they were formed and cover the rock surface from which they derive. These soils are referred to as *residual soils.*

Transported soils can be subdivided into five major categories based on the *transporting agent:*

1. *Gravity transported* soil
2. *Lacustrine* (lake) deposits
3. *Alluvial* or *fluvial* soil deposited by running water
4. *Glacial* deposited by glaciers
5. *Aeolian* deposited by the wind

In addition to transported and residual soils, there are *peats* and *organic soils,* which derive from the decomposition of organic materials.

Table 3.1 Some Decomposed Products of Minerals in Igneous Rock

Mineral	Decomposed Product
Quartz	Quartz (sand grains)
Potassium feldspar ($KAlSi_3O_8$) and Sodium feldspar ($NaAlSi_3O_8$)	Kaolinite (clay) Bauxite Illite (clay) Silica
Calcium feldspar ($CaAl_2Si_2O_8$)	Silica Calcite
Biotite	Clay Limonite Hematite Silica Calcite
Olivine ($Mg, Fe)_2SiO_4$	Limonite Serpentine Hematite Silica

3.3   Residual Soil

Residual soils are found in areas where the rate of weathering is more than the rate at which the weathered materials are carried away by transporting agents. The rate of weathering is higher in warm and humid regions compared to cooler and drier regions and, depending on the climatic conditions, the effect of weathering may vary widely.

Residual soil deposits are common in the tropics, on islands such as the Hawaiian Islands, and in the southeastern United States. The nature of a residual soil deposit will generally depend on the parent rock. When hard rocks such as granite and gneiss undergo weathering, most of the materials are likely to remain in place. These soil deposits generally have a top layer of clayey or silty clay material, below which are silty or sandy soil layers. These layers in turn are generally underlain by a partially weathered rock and then sound bedrock. The depth of the sound bedrock may vary widely, even within a distance of a few meters. Figure 3.1 shows the boring log of a residual soil deposit derived from the weathering of granite.

In contrast to hard rocks, there are some chemical rocks, such as limestone, that are chiefly made up of calcite ($CaCO_3$) mineral. Chalk and dolomite have large concentrations of dolomite minerals [$Ca\,Mg(CO_3)_2$]. These rocks have large amounts of soluble materials,

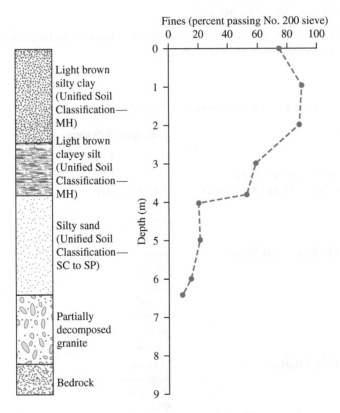

Figure 3.1 Boring log for a residual soil derived from granite

Table 3.2 Velocity Scale for Soil Movement on a Slope

Description	Velocity (mm/sec)
Very slow	5×10^{-5} to 5×10^{-7}
Slow	5×10^{-3} to 5×10^{-5}
Moderate	5×10^{-1} to 5×10^{-3}
Rapid	5×10^{1} to 5×10^{-1}

some of which are removed by groundwater, leaving behind the insoluble fraction of the rock. Residual soils that derive from chemical rocks do not possess a gradual transition zone to the bedrock, as seen in Figure 3.1. The residual soils derived from the weathering of limestone-like rocks are mostly red in color. Although uniform in kind, the depth of weathering may vary greatly. The residual soils immediately above the bedrock may be normally consolidated. Large foundations with heavy loads may be susceptible to large consolidation settlements on these soils.

3.4 Gravity Transported Soil

Residual soils on a natural slope can move downwards. Cruden and Varnes (1996) proposed a velocity scale for soil movement on a slope, which is summarized in Table 3.2. When residual soils move down a natural slope very slowly, the process is usually referred to as *creep*. When the downward movement of soil is sudden and rapid, it is called a *landslide*. The deposits formed by down-slope creep and landslides are *colluvium*.

Colluvium is a heterogeneous mixture of soils and rock fragments ranging from clay-sized particles to rocks having diameters of one meter or more. *Mudflows* are one type of gravity-transported soil. Flows are downward movements of earth that resemble a viscous fluid (Figure 3.2) and come to rest in a more dense condition. The soil deposits derived from past mudflows are highly heterogeneous in composition.

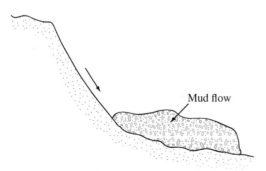

Mud flow

Figure 3.2 Mudflow

3.5   Alluvial Deposits

Alluvial soil deposits derive from the action of streams and rivers and can be divided into two major categories: (1) *braided-stream deposits* and (2) deposits caused by the *meandering belt of streams.*

Deposits from Braided Streams

Braided streams are high-gradient, rapidly flowing streams that are highly erosive and carry large amounts of sediment. Because of the high bed load, a minor change in the velocity of flow will cause sediments to deposit. By this process, these streams may build up a complex tangle of converging and diverging channels separated by sandbars and islands.

   The deposits formed from braided streams are highly irregular in stratification and have a wide range of grain sizes. Figure 3.3 shows a cross section of such a deposit. These deposits share several characteristics:

1. The grain sizes usually range from gravel to silt. Clay-sized particles are generally *not* found in deposits from braided streams.
2. Although grain size varies widely, the soil in a given pocket or lens is rather uniform.
3. At any given depth, the void ratio and unit weight may vary over a wide range within a lateral distance of only a few meters. This variation can be observed during soil exploration for the construction of a foundation for a structure. The standard penetration resistance at a given depth obtained from various boreholes will be highly irregular and variable.

   Alluvial deposits are present in several parts of the western United States, such as Southern California, Utah, and the basin and range sections of Nevada. Also, a large amount of sediment originally derived from the Rocky Mountain range was carried eastward to form the alluvial deposits of the Great Plains. On a smaller scale, this type of natural soil deposit, left by braided streams, can be encountered locally.

Meander Belt Deposits

The term *meander* is derived from the Greek word *maiandros,* after the Maiandros (now Menderes) River in Asia, famous for its winding course. Mature streams in a valley curve

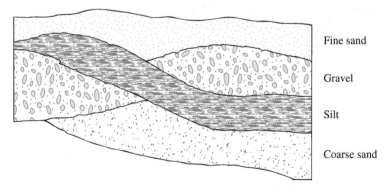

Fine sand

Gravel

Silt

Coarse sand

Figure 3.3 Cross section of a braided-stream deposit

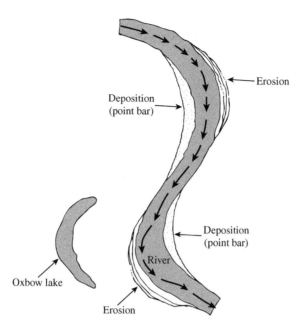

Figure 3.4 Formation of point bar deposits and oxbow lake in a meandering stream

back and forth. The valley floor in which a river meanders is referred to as the *meander belt*. In a meandering river, the soil from the bank is continually eroded from the points where it is concave in shape and is deposited at points where the bank is convex in shape, as shown in Figure 3.4. These deposits are called *point bar deposits*, and they usually consist of sand and silt-size particles. Sometimes, during the process of erosion and deposition, the river abandons a meander and cuts a shorter path. The abandoned meander, when filled with water, is called an *oxbow lake*. (See Figure 3.4.)

During floods, rivers overflow low-lying areas. The sand and silt-size particles carried by the river are deposited along the banks to form ridges known as *natural levees* (Figure 3.5).

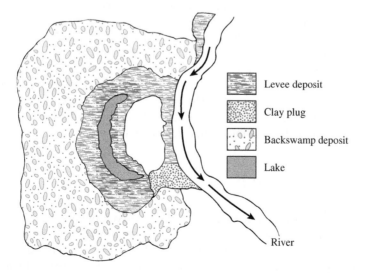

Figure 3.5 Levee and backswamp deposit

Table 3.3 Properties of Deposits within the Mississippi Alluvial Valley

Environment	Soil texture	Natural water content (%)	Liquid limit	Plasticity index
Natural levee	Clay (CL)	25–35	35–45	15–25
	Silt (ML)	15–35	NP–35	NP–5
Point bar	Silt (ML) and silty sand (SM)	25–45	30–55	10–25
Abandoned channel	Clay (CL, CH)	30–95	30–100	10–65
Backswamps	Clay (CH)	25–70	40–115	25–100
Swamp	Organic clay (OH)	100–265	135–300	100–165

(*Note:* NP—Nonplastic)

Finer soil particles consisting of silts and clays are carried by the water farther onto the floodplains. These particles settle at different rates to form what is referred to as *backswamp deposits* (Figure 3.5), often highly plastic clays.

Table 3.3 gives some properties of soil deposits found in natural levees, point bars, abandoned channels, backswamps and swamps within the alluvial Mississippi Valley (Kolb and Shockley, 1959).

3.6 Lacustrine Deposits

Water from rivers and springs flows into lakes. In arid regions, streams carry large amounts of suspended solids. Where the stream enters the lake, granular particles are deposited in the area forming a delta. Some coarser particles and the finer particles (that is, silt and clay) that are carried into the lake are deposited onto the lake bottom in alternate layers of coarse-grained and fine-grained particles. The deltas formed in humid regions usually have finer grained soil deposits compared to those in arid regions.

Varved clays are alternate layers of silt and silty clay with layer thicknesses rarely exceeding about 13 mm. (½ in.). The silt and silty clay that constitute the layers were carried into fresh water lakes by melt water at the end of the Ice Age. The hydraulic conductivity of varved clays exhibits a high degree of anisotropy.

3.7 Glacial Deposits

During the Pleistocene Ice Age, glaciers covered large areas of the earth. The glaciers advanced and retreated with time. During their advance, the glaciers carried large amounts of sand, silt, clay, gravel, and boulders. *Drift* is a general term usually applied to the deposits laid down by glaciers. The drifts can be broadly divided into two major

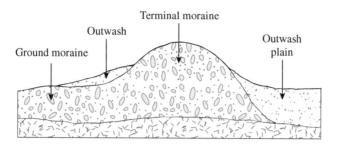

Figure 3.6 Terminal moraine, ground moraine, and outwash plain

categories: (a) unstratified drifts and (b) stratified drifts. A brief description of each category follows.

Unstratified Drifts

The *unstratified drifts* laid down by melting glaciers are referred to as *till*. The physical characteristics of till may vary from glacier to glacier. Till is called *clay till* because of the presence of the large amount of clay-sized particles in it. In some areas, tills constitute large amounts of boulders, and they are referred to as *boulder till*. The range of grain sizes in a given till varies greatly. The amount of clay-sized fractions present and the plasticity indices of tills also vary widely. During the field exploration program, erratic values of standard penetration resistance (Section 3.13) also may be expected.

The land forms that developed from the till deposits are called *moraines*. A *terminal moraine* (Figure 3.6) is a ridge of till that marks the maximum limit of a glacier's advance. *Recessional moraines* are ridges of till developed behind the terminal moraine at varying distances apart. They are the result of temporary stabilization of the glacier during the recessional period. The till deposited by the glacier between the moraines is referred to as *ground moraine* (Figure 3.6). Ground moraines constitute large areas of the central United States and are called *till plains*.

Stratified Drifts

The sand, silt, and gravel that are carried by the melting water from the front of a glacier are called *outwash*. The melted water sorts out the particles by the grain size and forms stratified deposits. In a pattern similar to that of braided-stream deposits, the melted water also deposits the outwash, forming *outwash plains* (Figure 3.6), also called *glaciofluvial deposits*.

3.8 Aeolian Soil Deposits

Wind is also a major transporting agent leading to the formation of soil deposits. When large areas of sand lie exposed, wind can blow the sand away and redeposit it elsewhere. Deposits of windblown sand generally take the shape of *dunes* (Figure 3.7). As dunes are formed, the sand is blown over the crest by the wind. Beyond the crest, the sand particles

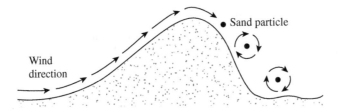

Figure 3.7 Sand dune

roll down the slope. The process tends to form a *compact sand deposit* on the *windward side,* and a rather *loose deposit* on the *leeward side,* of the dune.

Dunes exist along the southern and eastern shores of Lake Michigan, the Atlantic Coast, the southern coast of California, and at various places along the coasts of Oregon and Washington. Sand dunes can also be found in the alluvial and rocky plains of the western United States. Following are some of the typical properties of *dune sand:*

1. The grain-size distribution of the sand at any particular location is surprisingly uniform. This uniformity can be attributed to the sorting action of the wind.
2. The general grain size decreases with distance from the source, because the wind carries the small particles farther than the large ones.
3. The relative density of sand deposited on the windward side of dunes may be as high as 50 to 65%, decreasing to about 0 to 15% on the leeward side.

Figure 3.8 shows some sand dunes in the Sahara desert in Egypt.

Loess is an aeolian deposit consisting of silt and silt-sized particles. The grain-size distribution of loess is rather uniform. The cohesion of loess is generally derived from a clay coating over the silt-sized particles, which contributes to a stable soil structure in an unsaturated state. The cohesion may also be the result of the precipitation of chemicals leached by rainwater. Loess is a *collapsing* soil, because when the soil becomes saturated,

Figure 3.8 Sand dunes in the Sahara desert in Egypt (*Courtesy of Janice Das*)

it loses its binding strength between particles. Special precautions need to be taken for the construction of foundations over loessial deposits. There are extensive deposits of loess in the United States, mostly in the midwestern states of Iowa, Missouri, Illinois, and Nebraska and for some distance along the Mississippi River in Tennessee and Mississippi.

Volcanic ash (with grain sizes between 0.25 to 4 mm) and volcanic dust (with grain sizes less than 0.25 mm) may be classified as wind-transported soil. Volcanic ash is a lightweight sand or sandy gravel. Decomposition of volcanic ash results in highly plastic and compressible clays.

3.9 Organic Soil

Organic soils are usually found in low-lying areas where the water table is near or above the ground surface. The presence of a high water table helps in the growth of aquatic plants that, when decomposed, form organic soil. This type of soil deposit is usually encountered in coastal areas and in glaciated regions. Organic soils show the following characteristics:

1. Their natural moisture content may range from 200 to 300%.
2. They are highly compressible.
3. Laboratory tests have shown that, under loads, a large amount of settlement is derived from secondary consolidation.

3.10 Some Local Terms for Soils

Soils are sometimes referred to by local terms. The following are a few of these terms with a brief description of each.

1. *Caliche*: a Spanish word derived from the Latin word *calix*, meaning *lime*. It is mostly found in the desert southwest of the United States. It is a mixture of sand, silt, and gravel bonded together by *calcareous deposits*. The calcareous deposits are brought to the surface by a net upward migration of water. The water evaporates in the high local temperature. Because of the sparse rainfall, the carbonates are not washed out of the top layer of soil.
2. *Gumbo*: a highly plastic, clayey soil.
3. *Adobe*: a highly plastic, clayey soil found in the southwestern United States.
4. *Terra Rossa*: residual soil deposits that are red in color and derive from limestone and dolomite.
5. *Muck*: organic soil with a very high moisture content.
6. *Muskeg*: organic soil deposit.
7. *Saprolite*: residual soil deposit derived from mostly insoluble rock.
8. *Loam*: a mixture of soil grains of various sizes, such as sand, silt, and clay.
9. *Laterite*: characterized by the accumulation of iron oxide ($Fe_2 O_3$) and aluminum oxide ($Al_2 O_3$) near the surface, and the leaching of silica. Lateritic soils in Central America contain about 80 to 90% of clay and silt-size particles. In the United States, lateritic soils can be found in the southeastern states, such as Alabama, Georgia, and the Carolinas.

Subsurface Exploration

3.11   Purpose of Subsurface Exploration

The process of identifying the layers of deposits that underlie a proposed structure and their physical characteristics is generally referred to as *subsurface exploration*. The purpose of subsurface exploration is to obtain information that will aid the geotechnical engineer in

1. Selecting the type and depth of foundation suitable for a given structure.
2. Evaluating the load-bearing capacity of the foundation.
3. Estimating the probable settlement of a structure.
4. Determining potential foundation problems (e.g., expansive soil, collapsible soil, sanitary landfill, and so on).
5. Determining the location of the water table.
6. Predicting the lateral earth pressure for structures such as retaining walls, sheet pile bulkheads, and braced cuts.
7. Establishing construction methods for changing subsoil conditions.

Subsurface exploration may also be necessary when additions and alterations to existing structures are contemplated.

3.12   Subsurface Exploration Program

Subsurface exploration comprises several steps, including the collection of preliminary information, reconnaissance, and site investigation.

Collection of Preliminary Information

This step involves obtaining information regarding the type of structure to be built and its general use. For the construction of buildings, the approximate column loads and their spacing and the local building-code and basement requirements should be known. The construction of bridges requires determining the lengths of their spans and the loading on piers and abutments.

A general idea of the topography and the type of soil to be encountered near and around the proposed site can be obtained from the following sources:

1. United States Geological Survey maps.
2. State government geological survey maps.
3. United States Department of Agriculture's Soil Conservation Service county soil reports.
4. Agronomy maps published by the agriculture departments of various states.
5. Hydrological information published by the United States Corps of Engineers, including records of stream flow, information on high flood levels, tidal records, and so on.
6. Highway department soil manuals published by several states.

The information collected from these sources can be extremely helpful in planning a site investigation. In some cases, substantial savings may be realized by anticipating problems that may be encountered later in the exploration program.

Reconnaissance

The engineer should always make a visual inspection of the site to obtain information about

1. The general topography of the site, the possible existence of drainage ditches, abandoned dumps of debris, and other materials present at the site. Also, evidence of creep of slopes and deep, wide shrinkage cracks at regularly spaced intervals may be indicative of expansive soils.
2. Soil stratification from deep cuts, such as those made for the construction of nearby highways and railroads.
3. The type of vegetation at the site, which may indicate the nature of the soil. For example, a mesquite cover in central Texas may indicate the existence of expansive clays that can cause foundation problems.
4. High-water marks on nearby buildings and bridge abutments.
5. Groundwater levels, which can be determined by checking nearby wells.
6. The types of construction nearby and the existence of any cracks in walls or other problems.

The nature of the stratification and physical properties of the soil nearby also can be obtained from any available soil-exploration reports on existing structures.

Site Investigation

The site investigation phase of the exploration program consists of planning, making test boreholes, and collecting soil samples at desired intervals for subsequent observation and laboratory tests. The approximate required minimum depth of the borings should be predetermined. The depth can be changed during the drilling operation, depending on the subsoil encountered. To determine the approximate minimum depth of boring, engineers may use the rules established by the American Society of Civil Engineers (1972):

1. Determine the net increase in the effective stress, $\Delta \sigma'$, under a foundation with depth as shown in Figure 3.9. (The general equations for estimating increases in stress are given in Chapter 6.)
2. Estimate the variation of the vertical effective stress, σ_o', with depth.

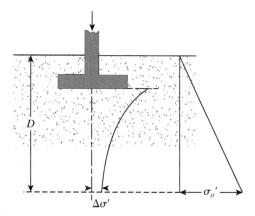

Figure 3.9 Determination of the minimum depth of boring

3. Determine the depth, $D = D_1$, at which the effective stress increase $\Delta\sigma'$ is equal to $(\frac{1}{10})q$ (q = estimated net stress on the foundation).
4. Determine the depth, $D = D_2$, at which $\Delta\sigma'/\sigma_o' = 0.05$.
5. Choose the smaller of the two depths, D_1 and D_2, just determined as the approximate minimum depth of boring required, unless bedrock is encountered.

If the preceding rules are used, the depths of boring for a building with a width of 30 m (100 ft) will be approximately the following, according to Sowers and Sowers (1970):

No. of stories	Boring depth	
1	3.5 m	(11 ft)
2	6 m	(20 ft)
3	10 m	(33 ft)
4	16 m	(53 ft)
5	24 m	(79 ft)

To determine the boring depth for hospitals and office buildings, Sowers and Sowers (1970) also used the following rules.

- For light steel or narrow concrete buildings,

$$\frac{D_b}{S^{0.7}} = a \tag{3.1}$$

where

D_b = depth of boring
S = number of stories
$$a = \begin{cases} \approx 3 \text{ if } D_b \text{ is in meters} \\ \approx 10 \text{ if } D_b \text{ is in feet} \end{cases}$$

- For heavy steel or wide concrete buildings,

$$\frac{D_b}{S^{0.7}} = b \tag{3.2}$$

where

$$b = \begin{cases} \approx 6 \text{ if } D_b \text{ is in meters} \\ \approx 20 \text{ if } D_b \text{ is in feet} \end{cases}$$

When deep excavations are anticipated, the depth of boring should be at least 1.5 times the depth of excavation.

Sometimes, subsoil conditions require that the foundation load be transmitted to bedrock. The minimum depth of core boring into the bedrock is about 3 m (10 ft). If the bedrock is irregular or weathered, the core borings may have to be deeper.

Table 3.4 Approximate Spacing of Boreholes

	Spacing	
Type of project	**(m)**	**(ft)**
Multistory building	10–30	30–100
One-story industrial plants	20–60	60–200
Highways	250–500	800–1600
Residential subdivision	250–500	800–1600
Dams and dikes	40–80	130–260

There are no hard-and-fast rules for borehole spacing. Table 3.4 gives some general guidelines. Spacing can be increased or decreased, depending on the condition of the sub-soil. If various soil strata are more or less uniform and predictable, fewer boreholes are needed than in nonhomogeneous soil strata.

The engineer should also take into account the ultimate cost of the structure when making decisions regarding the extent of field exploration. The exploration cost generally should be 0.1 to 0.5% of the cost of the structure. Soil borings can be made by several methods, including auger boring, wash boring, percussion drilling, and rotary drilling.

3.13 Exploratory Borings in the Field

Auger boring is the simplest method of making exploratory boreholes. Figure 3.10 shows two types of hand auger: the *posthole auger* and the *helical auger.* Hand augers cannot be used for advancing holes to depths exceeding 3 to 5 m (10 to 16 ft). However, they

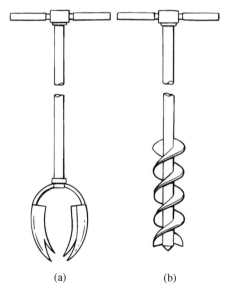

(a)   (b)

Figure 3.10 Hand tools: (a) posthole auger; (b) helical auger

can be used for soil exploration work on some highways and small structures. *Portable power-driven helical augers* (76 mm to 305 mm in diameter) are available for making deeper boreholes. The soil samples obtained from such borings are highly disturbed. In some noncohesive soils or soils having low cohesion, the walls of the boreholes will not stand unsupported. In such circumstances, a metal pipe is used as a *casing* to prevent the soil from caving in.

When power is available, *continuous-flight augers* are probably the most common method used for advancing a borehole. The power for drilling is delivered by truck- or tractor-mounted drilling rigs. Boreholes up to about 60 to 70 m (200 to 230 ft) can easily be made by this method. Continuous-flight augers are available in sections of about 1 to 2 m (3 to 6 ft) with either a solid or hollow stem. Some of the commonly used solid-stem augers have outside diameters of 66.68 mm ($2\frac{5}{8}$ in.), 82.55 mm ($3\frac{1}{4}$ in.), 101.6 mm (4 in.), and 114.3 mm ($4\frac{1}{2}$ in.). Common commercially available hollow-stem augers have dimensions of 63.5 mm ID and 158.75 mm OD (2.5 in. × 6.25 in.), 69.85 mm ID and 177.8 OD (2.75 in. × 7 in.), 76.2 mm ID and 203.2 OD (3 in. × 8 in.), and 82.55 mm ID and 228.6 mm OD (3.25 in. × 9 in.).

The tip of the auger is attached to a cutter head (Figure 3.11). During the drilling operation (Figure 3.12), section after section of auger can be added and the hole extended downward. The flights of the augers bring the loose soil from the bottom of the hole to the

Figure 3.11 Carbide-tipped cutting head on auger flight (*Courtesy of Braja M. Das, Henderson, Nevada*)

Figure 3.12 Drilling with continuous-flight augers (Danny R. Anderson, PE of Professional Service Industries, Inc, El Paso, Texas.)

surface. The driller can detect changes in the type of soil by noting changes in the speed and sound of drilling. When solid-stem augers are used, the auger must be withdrawn at regular intervals to obtain soil samples and also to conduct other operations such as standard penetration tests. Hollow-stem augers have a distinct advantage over solid-stem augers in that they do not have to be removed frequently for sampling or other tests. As shown schematically in Figure 3.13, the outside of the hollow-stem auger acts as a casing.

The hollow-stem auger system includes the following components:

Outer component:	(a) hollow auger sections, (b) hollow auger cap, and (c) drive cap
Inner component:	(a) pilot assembly, (b) center rod column, and (c) rod-to-cap adapter

The auger head contains replaceable carbide teeth. During drilling, if soil samples are to be collected at a certain depth, the pilot assembly and the center rod are removed. The soil sampler is then inserted through the hollow stem of the auger column.

Wash boring is another method of advancing boreholes. In this method, a casing about 2 to 3 m (6 to 10 ft) long is driven into the ground. The soil inside the casing is then

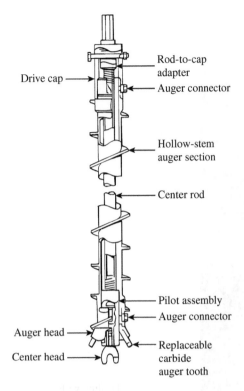

Figure 3.13 Hollow-stem auger components (After ASTM, 2001) (Based on ASTM D4700-91: Standard Guide for Soil Sampling from the Vadose Zone.)

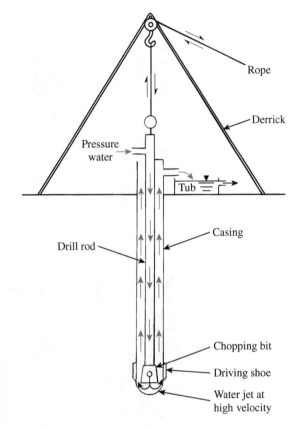

Figure 3.14 Wash boring

removed by means of a chopping bit attached to a drilling rod. Water is forced through the drilling rod and exits at a very high velocity through the holes at the bottom of the chopping bit (Figure 3.14). The water and the chopped soil particles rise in the drill hole and overflow at the top of the casing through a T connection. The washwater is collected in a container. The casing can be extended with additional pieces as the borehole progresses; however, that is not required if the borehole will stay open and not cave in. Wash borings are rarely used now in the United States and other developed countries.

Rotary drilling is a procedure by which rapidly rotating drilling bits attached to the bottom of drilling rods cut and grind the soil and advance the borehole. There are several types of drilling bit. Rotary drilling can be used in sand, clay, and rocks (unless they are badly fissured). Water or *drilling mud* is forced down the drilling rods to the bits, and the return flow forces the cuttings to the surface. Boreholes with diameters of 50 to 203 mm (2 to 8 in.) can easily be made by this technique. The drilling mud is a slurry of water and bentonite. Generally, it is used when the soil that is encountered is likely to cave in. When soil samples are needed, the drilling rod is raised and the drilling bit is replaced by a sampler. With the environmental drilling applications, rotary drilling with air is becoming more common.

Percussion drilling is an alternative method of advancing a borehole, particularly through hard soil and rock. A heavy drilling bit is raised and lowered to chop the hard soil. The chopped soil particles are brought up by the circulation of water. Percussion drilling may require casing.

3.14 Procedures for Sampling Soil

Two types of soil samples can be obtained during subsurface exploration: *disturbed* and *undisturbed.* Disturbed, but representative, samples can generally be used for the following types of laboratory test:

1. Grain-size analysis
2. Determination of liquid and plastic limits
3. Specific gravity of soil solids
4. Determination of organic content
5. Classification of soil

Disturbed soil samples, however, cannot be used for consolidation, hydraulic conductivity, or shear strength tests. Undisturbed soil samples must be obtained for these types of laboratory tests. Sections 3.15 through 3.18 describe some procedures for obtaining soil samples during field exploration.

3.15 Split-Spoon Sampling

Split-spoon samplers can be used in the field to obtain soil samples that are generally disturbed, but still representative. A section of a *standard split-spoon sampler* is shown in Figure 3.15a. The tool consists of a steel driving shoe, a steel tube that is split longitudinally in half, and a coupling at the top. The coupling connects the sampler to the drill rod. The standard split tube has an inside diameter of 34.93 mm ($1\frac{3}{8}$ in.) and an outside diameter of 50.8 mm (2 in.); however, samplers having inside and outside diameters up to 63.5 mm ($2\frac{1}{2}$ in.) and 76.2 mm (3 in.), respectively, are also available. When a borehole is extended to a predetermined depth, the drill tools are removed and the sampler is lowered to the bottom of the hole. The sampler is driven into the soil by hammer blows to the top of the drill rod. The standard weight of the hammer is 622.72 N (140 lb), and for each blow, the hammer drops a distance of 0.762 m (30 in.). The number of blows required for a spoon penetration of three 152.4-mm (6-in.) intervals are recorded. The number of blows required for the last two intervals are added to give the *standard penetration number, N,* at that depth. This number is generally referred to as the *N value* (American Society for Testing and Materials, 2014, Designation D-1586-11). The sampler is then withdrawn, and the shoe and coupling are removed. Finally, the soil sample recovered from the tube is placed in a glass bottle and transported to the laboratory. This field test is called the standard penetration test (SPT). Figure 3.16a and b show a split-spoon sampler unassembled before and after sampling.

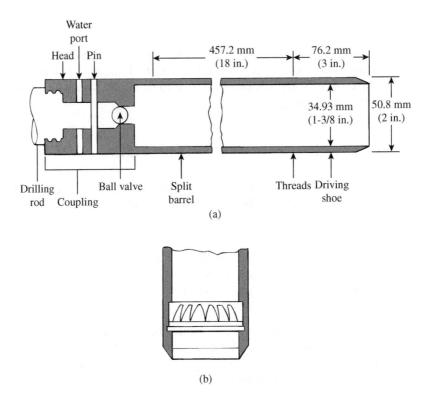

Figure 3.15 (a) Standard split-spoon sampler; (b) spring core catcher

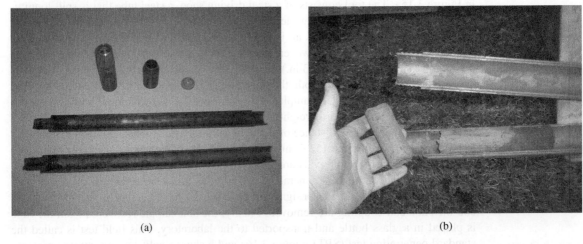

Figure 3.16 (a) Unassembled split-spoon sampler; (b) after sampling (*Courtesy of Professional Service Industries, Inc. (PSI), Waukesha, Wisconsin*)

The degree of disturbance for a soil sample is usually expressed as

$$A_R(\%) = \frac{D_o^2 - D_i^2}{D_i^2}(100) \tag{3.3}$$

where

A_R = area ratio (ratio of disturbed area to total area of soil)
D_o = outside diameter of the sampling tube
D_i = inside diameter of the sampling tube

When the area ratio is 10% or less, the sample generally is considered to be undisturbed. For a standard split-spoon sampler,

$$A_R(\%) = \frac{(50.8)^2 - (34.93)^2}{(34.93)^2}(100) = 111.5\%$$

Hence, these samples are highly disturbed. Split-spoon samples generally are taken at intervals of about 1.5 m (5 ft). When the material encountered in the field is sand (particularly fine sand below the water table), recovery of the sample by a split-spoon sampler may be difficult. In that case, a device such as a *spring core catcher* may have to be placed inside the split spoon (Figure 3.15b).

At this juncture, it is important to point out that several factors contribute to the variation of the standard penetration number N at a given depth for similar soil profiles. Among these factors are the SPT hammer efficiency, borehole diameter, sampling method, and rod length (Skempton, 1986; Seed, et al., 1985). The SPT hammer energy efficiency can be expressed as

$$E_r(\%) = \frac{\text{actual hammer energy to the sampler}}{\text{input energy}} \times 100 \tag{3.4}$$

$$\text{Theoretical input energy} = Wh \tag{3.5}$$

where

W = weight of the hammer $\approx$ 0.623 kN (140 lb)
h = height of drop $\approx$ 0.76 mm (30 in.)

So,

$$Wh = (0.623)(0.76) = 0.474 \text{ kN-m } (4200 \text{ in.-lb})$$

In the field, the magnitude of E_r can vary from 30 to 90%. The standard practice now in the U.S. is to express the N-value to an average energy ratio of 60% ($\approx N_{60}$). Thus, correcting for field procedures and on the basis of field observations, it appears reasonable to standardize the field penetration number as a function of the input driving energy and its dissipation around the sampler into the surrounding soil, or

$$N_{60} = \frac{N\eta_H \eta_B \eta_S \eta_R}{60} \tag{3.6}$$

Table 3.5 Variations of $\eta_H, \eta_B, \eta_S,$ and η_R [Eq. (3.6)]

1. Variation of η_H

Country	Hammer type	Hammer release	η_H (%)
Japan	Donut	Free fall	78
	Donut	Rope and pulley	67
United States	Safety	Rope and pulley	60
	Donut	Rope and pulley	45
Argentina	Donut	Rope and pulley	45
China	Donut	Free fall	60
	Donut	Rope and pulley	50

2. Variation of η_B

Diameter		η_B
mm	in.	
60–120	2.4–4.7	1
150	6	1.05
200	8	1.15

3. Variation of η_S

Variable	η_S
Standard sampler	1.0
With liner for dense sand and clay	0.8
With liner for loose sand	0.9

4. Variation of η_R

Rod length		η_R
m	ft	
>10	>30	1.0
6–10	20–30	0.95
4–6	12–20	0.85
0–4	0–12	0.75

where

N_{60} = standard penetration number, corrected for field conditions
$\bar{N}$ = measured penetration number
η_H = hammer efficiency (%)
η_B = correction for borehole diameter
η_S = sampler correction
η_R = correction for rod length

Variations of $\eta_H, \eta_B, \eta_S,$ and η_R, based on recommendations by Seed et al. (1985) and Skempton (1986), are summarized in Table 3.5.

Correlations for N_{60} in Cohesive Soil

Besides compelling the geotechnical engineer to obtain soil samples, standard penetration tests provide several useful correlations. For example, the consistency of clay soils can be estimated from the standard penetration number, N_{60}. In order to achieve that, Szechy and Vargi (1978) calculated the *consistency index* (CI) as

$$CI = \frac{LL - w}{LL - PL} \tag{3.7}$$

where

w = natural moisture content (%)
LL = liquid limit
PL = plastic limit

Table 3.6 Approximate Correlation between CI, N_{60}, and q_u

Standard penetration number, N_{60}	Consistency	CI	Unconfined compression strength, q_u	
			(kN/m²)	(lb/ft²)
<2	Very soft	<0.5	<25	500
2–8	Soft to medium	0.5–0.75	25–80	500–1700
8–15	Stiff	0.75–1.0	80–150	1700–3100
15–30	Very stiff	1.0–1.5	150–400	3100–8400
>30	Hard	>1.5	>400	8400

The approximate correlation between CI, N_{60}, and the unconfined compression strength (q_u) is given in Table 3.6.

Hara, et al. (1971) also suggested the following correlation between the undrained shear strength of clay (cu) and N_{60}.

$$\frac{c_u}{p_a} = 0.29 N_{60}^{0.72} \tag{3.8}$$

where p_a = atmospheric pressure (≈ 100 kN/m²; ≈ 2000 lb/in²).

The overconsolidation ratio, OCR, of a natural clay deposit can also be correlated with the standard penetration number. On the basis of the regression analysis of 110 data points, Mayne and Kemper (1988) obtained the relationship

$$OCR = 0.193 \left(\frac{N_{60}}{\sigma_o'} \right)^{0.689} \tag{3.9}$$

where σ_o' = effective vertical stress in MN/m².

It is important to point out that any correlation between c_u, OCR, and N_{60} is only approximate.

Using the field test results of Mayne and Kemper (1988) and others (112 data points), Kulhawy and Mayne (1990) suggested the approximate correlation

$$OCR = 0.58 \frac{N_{60} p_a}{\sigma_o'} \tag{3.10}$$

Kulhawy and Mayne (1990) have also provided an approximate correlation for the preconsolidation pressure (σ_c') of clay as

$$\sigma_c' = 0.47 N_{60} p_a \tag{3.11}$$

Correction for N_{60} in Granular Soil

In granular soils, the value of N_{60} is affected by the effective overburden pressure, σ_o'. For that reason, the value of N_{60} obtained from field exploration under different effective overburden pressures should be changed to correspond to a standard value of σ_o'. That is,

$$(N_1)_{60} = C_N N_{60} \tag{3.12}$$

where

$(N_1)_{60}$ = value of N_{60} corrected to a standard value of $\sigma_a' = p_a [\approx 100 \text{ kN/m}^2 \ (2000 \text{ lb/ft}^2)]$
$\quad C_N$ = correction factor
$\quad N_{60}$ = value of N obtained from field exploration [Eq. (3.6)]

In the past, a number of empirical relations were proposed for C_N. Some of the relationships are given next. The most commonly cited relationships are those of Liao and Whitman (1986) and Skempton (1986).

In the following relationships for C_N, note that σ_o' is the effective overburden pressure and p_a = atmospheric pressure ($\approx 100 \text{ kN/m}^2$, or $\approx 2000 \text{ lb/ft}^2$)
Liao and Whitman's relationship (1986):

$$C_N = \left[\frac{1}{\left(\dfrac{\sigma_o'}{p_a} \right)} \right]^{0.5} \tag{3.13}$$

Skempton's relationship (1986):

$$C_N = \frac{2}{1 + \left(\dfrac{\sigma_o'}{p_a} \right)} \quad \text{(for normally consolidated fine sand)} \tag{3.14}$$

$$C_N = \frac{3}{2 + \left(\dfrac{\sigma_o'}{p_a} \right)} \quad \text{(for normally consolidated coarse sand)} \tag{3.15}$$

$$C_N = \frac{1.7}{0.7 + \left(\dfrac{\sigma_o'}{p_a} \right)} \quad \text{(for overconsolidated sand)} \tag{3.16}$$

Seed et al.'s relationship (1975):

$$C_N = 1 - 1.25 \log\left(\frac{\sigma_o'}{p_a} \right) \tag{3.17}$$

Peck et al.'s relationship (1974):

$$C_N = 0.77 \log\left[\frac{20}{\left(\dfrac{\sigma_o'}{p_a} \right)} \right] \left(\text{for } \frac{\sigma_o'}{p_a} \geq 0.25 \right) \tag{3.18}$$

Table 3.7 Variation of C_N

$\dfrac{\sigma'_o}{p_a}$	Eq. (3.13)	Eq. (3.14)	Eq. (3.15)	Eq. (3.16)	Eq. (3.17)	Eq. (3.18)	Eqs. (3.19) and (3.20)
0.25	2.00	1.60	1.33	1.78	1.75	1.47	2.00
0.50	1.41	1.33	1.20	1.17	1.38	1.23	1.33
0.75	1.15	1.14	1.09	1.17	1.15	1.10	1.00
1.00	1.00	1.00	1.00	1.00	1.00	1.00	0.94
1.50	0.82	0.80	0.86	0.77	0.78	0.87	0.84
2.00	0.71	0.67	0.75	0.63	0.62	0.77	0.76
3.00	0.58	0.50	0.60	0.46	0.40	0.63	0.65
4.00	0.50	0.40	0.60	0.36	0.25	0.54	0.55

The C_N column header spans Eq. (3.13) through Eqs. (3.19) and (3.20).

Bazaraa (1967):

$$C_N = \frac{4}{1 + 4\left(\dfrac{\sigma'_o}{p_a}\right)} \left(\text{for } \frac{\sigma'_o}{p_a} \le 0.75 \right) \tag{3.19}$$

$$C_N = \frac{4}{3.25 + \left(\dfrac{\sigma'_o}{p_a}\right)} \left(\text{for } \frac{\sigma'_o}{p_a} > 0.75 \right) \tag{3.20}$$

Table 3.7 shows the comparison of C_N derived using various relationships cited above. It can be seen that the magnitude of the correction factor estimated by using any one of the relationships is approximately the same, considering the uncertainties involved in conducting the standard penetration tests. Hence, it is recommended that Eq. (3.13) may be used for all calculations.

Example 3.1

Following are the results of a standard penetration test in sand. Determine the corrected standard penetration number, $(N_1)_{60}$, at various depths. Note that the water table was not observed within a depth of 10.5 m below the ground surface. Assume that the average unit weight of sand is 17.3 kN/m^3. Use Eq. (3.13).

Depth, z (m)	N_{60}
1.5	8
3.0	7
4.5	12
6.0	14
7.5	13

Solution

From Eq. (3.13)

$$C_N = \left[\frac{1}{\left(\dfrac{\sigma_0'}{p_a} \right)} \right]^{0.5}$$

$$p_a \approx 100 \text{ kN/m}^2$$

Now the following table can be prepared.

Depth, z (m)	σ_0' (kN/m^2)	C_N	N_{60}	$(N_1)_{60}$
1.5	25.95	1.96	8	≈ 16
3.0	51.90	1.39	7	≈ 10
4.5	77.85	1.13	12	≈ 14
6.0	103.80	0.98	14	≈ 14
7.5	129.75	0.87	13	≈ 11

Correlation between N_{60} and Relative Density of Granular Soil

Kulhawy and Mayne (1990) modified an empirical relationship for relative density that was given by Marcuson and Bieganousky (1977), which can be expressed as

$$D_r(\%) = 12.2 + 0.75 \left[222N_{60} + 2311 - 711\text{OCR} - 779\left(\frac{\sigma_0'}{p_a} \right) - 50C_u^2 \right]^{0.5} \qquad (3.21)$$

where

D_r = relative density
σ_0' = effective overburden pressure
C_u = uniformity coefficient of sand
$\text{OCR} = \dfrac{\text{preconsolidation pressure}, \sigma_c'}{\text{effective overburden pressure}, \sigma_0'}$
p_a = atmospheric pressure

Meyerhof (1957) developed a correlation between D_r and N_{60} as

$$N_{60} = \left[17 + 24\left(\frac{\sigma_0'}{p_a} \right) \right] D_r^2$$

or

$$D_r = \left\{ \frac{N_{60}}{\left[17 + 24\left(\dfrac{\sigma_0'}{p_a} \right) \right]} \right\}^{0.5} \qquad (3.22)$$

Equation (3.22) provides a reasonable estimate only for clean, medium fine sand.

Cubrinovski and Ishihara (1999) also proposed a correlation between N_{60} and the relative density of sand (D_r) that can be expressed as

$$D_r(\%) = \left[\frac{N_{60}\left(0.23 + \dfrac{0.06}{D_{50}}\right)^{1.7}}{9} \left(\frac{1}{\dfrac{\sigma_o'}{p_a}}\right) \right]^{0.5} (100) \tag{3.23}$$

where

p_a = atmospheric pressure (≈ 100 kN/m^2, or ≈ 2000 lb/ft^2)
D_{50} = sieve size through which 50% of the soil will pass (mm)

Kulhawy and Mayne (1990) correlated the corrected standard penetration number and the relative density of sand in the form

$$D_r(\%) = \left[\frac{(N_1)_{60}}{C_p C_A C_{OCR}} \right]^{0.5} (100) \tag{3.24}$$

where

C_P = grain-size correlations factor = $60 + 25 \log D_{50}$ $\qquad$ (3.25)

C_A = correlation factor for aging = $1.2 + 0.05 \log\left(\dfrac{t}{100}\right)$ $\qquad$ (3.26)

C_{OCR} = correlation factor for overconsolidation = $OCR^{0.18}$ $\qquad$ (3.27)
D_{50} = diameter through which 50% soil will pass through (mm)
t = age of soil since deposition (years)
OCR = overconsolidation ratio

Skempton (1986) suggested that, for sands with a relative density greater than 35%,

$$\frac{(N_1)_{60}}{D_r^2} \approx 60 \tag{3.28}$$

where $(N_1)_{60}$ should be multiplied by 0.92 for coarse sands and 1.08 for fine sands.

Correlation between Angle of Friction and Standard Penetration Number

The peak friction angle, ϕ', of granular soil has also been correlated with N_{60} or $(N_1)_{60}$ by several investigators. Some of these correlations are as follows:

1. Peck, Hanson, and Thornburn (1974) give a correlation between N_{60} and ϕ' in a graphical form, which can be approximated as (Wolff, 1989)

$$\phi'(\text{deg}) = 27.1 + 0.3N_{60} - 0.00054[N_{60}]^2 \tag{3.29}$$

2. Schmertmann (1975) provided the correlation between N_{60}, σ'_o, and ϕ'. Mathematically, the correlation can be approximated as (Kulhawy and Mayne, 1990)

$$\phi' = \tan^{-1}\left[\frac{N_{60}}{12.2 + 20.3\left(\dfrac{\sigma'_o}{p_a}\right)}\right]^{0.34} \tag{3.30}$$

where

N_{60} = field standard penetration number
σ'_o = effective overburden pressure
p_a = atmospheric pressure in the same unit as σ'_o
ϕ' = soil friction angle

3. Hatanaka and Uchida (1996) provided a simple correlation between ϕ' and $(N_1)_{60}$ that can be expressed as

$$\phi' = \sqrt{20(N_1)_{60}} + 20 \tag{3.31}$$

The following qualifications should be noted when standard penetration resistance values are used in the preceding correlations to estimate soil parameters:

1. The equations are approximate.
2. Because the soil is not homogeneous, the values of N_{60} obtained from a given borehole vary widely.
3. In soil deposits that contain large boulders and gravel, standard penetration numbers may be erratic and unreliable.

Although approximate, with correct interpretation the standard penetration test provides a good evaluation of soil properties. The primary sources of error in standard penetration tests are inadequate cleaning of the borehole, careless measurement of the blow count, eccentric hammer strikes on the drill rod, and inadequate maintenance of water head in the borehole. Figure 3.17 shows approximate borderline values for D_r, N_{60}, $(N_1)_{60}$, ϕ' and $\dfrac{(N_1)_{60}}{D_r^2}$.

Correlation between Modulus of Elasticity and Standard Penetration Number

The modulus of elasticity of granular soils (E_s) is an important parameter in estimating the elastic settlement of foundations. A first-order estimation for E_s was given by Kulhawy and Mayne (1990) as

$$\frac{E_s}{p_a} = \alpha N_{60} \tag{3.32}$$

where

p_a = atmospheric pressure (same unit as E_s)

$$\alpha = \begin{cases} 5 \text{ for sands with fines} \\ 10 \text{ for clean normally consolidated sand} \\ 15 \text{ for clean overconsolidated sand} \end{cases}$$

	*Very loose	Loose	Medium dense		Dense	Very dense	
#D_r (%)	0	15	35		65	85	100
*N_{60}		4	10		30	50	
##$(N_1)_{60}$		3	8		25	42	
**ϕ'(deg)		28	30		36	41	
##$(N_1)_{60}/D_r^2$			65		59	58	

*Terzaghi & Peck (1948); #Gibb & Holtz (1957); ##Skempton (1986); **Peck et al. (1974)

Figure 3.17 Approximate borderline values for D_r, N_{60}, $(N_1)_{60}$, and $\dfrac{(N_1)_{60}}{D_r^2}$ (After Sivakugan and Das, 2010. With permission from J. Ross Publishing Co. Fort Lauderdale, FL)

Example 3.2

Refer to Example 3.1. Using Eq. (3.30), estimate the average soil friction angle, ϕ'. From $z = 0$ to $z = 7.5$ m.

Solution

From Eq. (3.30)

$$\phi' = \tan^{-1}\left[\frac{N_{60}}{12.2 + 20.3\left(\dfrac{\sigma_a'}{p_a}\right)}\right]^{0.34}$$

$$p_a = 100 \text{ kN/m}^2$$

Now the following table can be prepared.

Depth, z (m)	σ_0' (kN/m²)	N_{60}	ϕ' (deg) [Eq. (3.30)]
1.5	25.95	8	37.5
3.0	51.9	7	33.8
4.5	77.85	12	36.9
6.0	103.8	14	36.7
7.5	129.75	13	34.6

Average $\phi' \approx 36°$ ∎

3.16 Sampling with a Scraper Bucket

When the soil deposits are sand mixed with pebbles, obtaining samples by split spoon with a spring core catcher may not be possible because the pebbles may prevent the springs from closing. In such cases, a scraper bucket may be used to obtain disturbed

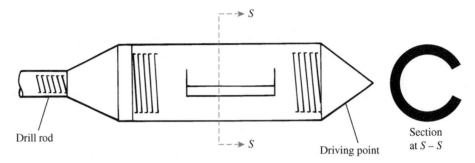

Figure 3.18 Scraper bucket

representative samples (Figure 3.18). The scraper bucket has a driving point and can be attached to a drilling rod. The sampler is driven down into the soil and rotated, and the scrapings from the side fall into the bucket.

3.17 Sampling with a Thin-Walled Tube

Thin-walled tubes are sometimes referred to as *Shelby tubes*. They are made of seamless steel and are frequently used to obtain undisturbed clayey soils. The most common thin-walled tube samplers have outside diameters of 50.8 mm (2 in.) and 76.2 mm (3 in.). The bottom end of the tube is sharpened. The tubes can be attached to drill rods (Figure 3.19). The drill rod with the sampler attached is lowered to the bottom of the borehole, and the sampler is pushed into the soil. The soil sample inside the tube is then pulled out. The two ends are sealed, and the sampler is sent to the laboratory for testing. Figure 3.20 shows the sequence of sampling with a thin-walled tube in the field.

Samples obtained in this manner may be used for consolidation or shear tests. A thin-walled tube with a 50.8-mm (2-in.) outside diameter has an inside diameter of about 47.63 mm ($1\frac{7}{8}$ in.). The area ratio is

$$A_R(\%) = \frac{D_o^2 - D_i^2}{D_i^2}(100) = \frac{(50.8)^2 - (47.63)^2}{(47.63)^2}(100) = 13.75\%$$

Increasing the diameters of samples increases the cost of obtaining them.

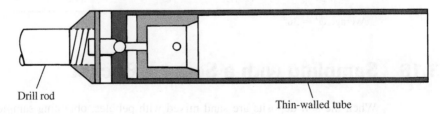

Figure 3.19 Thin-walled tube

(a)

(b)

Figure 3.20 Sampling with a thin-walled tube: (a) tube being attached to drill rod; (b) tube sampler pushed into soil (*Courtesy of Khaled Sobhan, Florida Atlantic University, Boca Raton, Florida*)

(c)

Figure 3.20 (*continued*) (c) recovery of soil sample (*Courtesy of Khaled Sobhan, Florida Atlantic University, Boca Raton, Florida*)

3.18 Sampling with a Piston Sampler

When undisturbed soil samples are very soft or larger than 76.2 mm (3 in.) in diameter, they tend to fall out of the sampler. Piston samplers are particularly useful under such conditions. There are several types of piston sampler; however, the sampler proposed by Osterberg (1952) is the most useful (see Figures 3.21a and 3.21b). It consists of a thin-walled tube with a piston. Initially, the piston closes the end of the tube. The sampler is lowered to the bottom of the borehole (Figure 3.21a), and the tube is pushed into the soil hydraulically, past the piston. Then the pressure is released through a hole in the piston rod (Figure 3.21b). To a large extent, the presence of the piston prevents distortion in the sample by not letting the soil squeeze into the sampling tube very fast and by not admitting excess soil. Consequently, samples obtained in this manner are less disturbed than those obtained by Shelby tubes.

3.19 Observation of Water Tables

The presence of a water table near a foundation significantly affects the foundation's load-bearing capacity and settlement, among other things. The water level will change seasonally. In many cases, establishing the highest and lowest possible levels of water during the life of a project may become necessary.

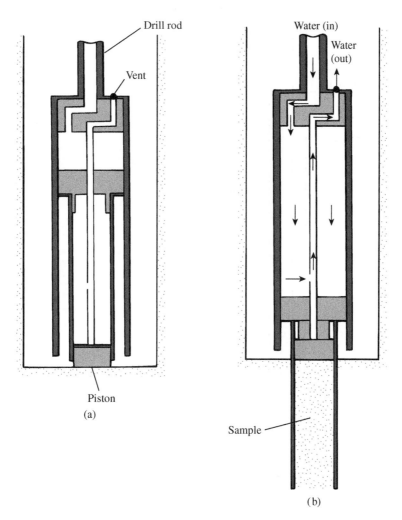

Figure 3.21 Piston sampler: (a) sampler at the bottom of borehole; (b) tube pushed into the soil hydraulically

If water is encountered in a borehole during a field exploration, that fact should be recorded. In soils with high hydraulic conductivity, the level of water in a borehole will stabilize about 24 hours after completion of the boring. The depth of the water table can then be recorded by lowering a chain or tape into the borehole.

In highly impermeable layers, the water level in a borehole may not stabilize for several weeks. In such cases, if accurate water-level measurements are required, a *piezometer* can be used. A piezometer basically consists of a porous stone or a perforated pipe with a plastic standpipe attached to it. Figure 3.22 shows the general placement of a piezometer in a borehole. This procedure will allow periodic checking until the water level stabilizes.

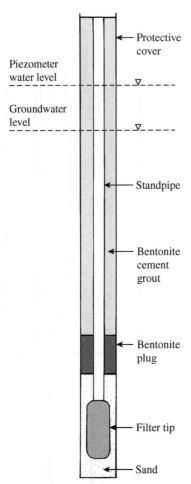

Protective cover

Piezometer water level

Groundwater level

Standpipe

Bentonite cement grout

Bentonite plug

Filter tip

Sand

Figure 3.22 Casagrande-type piezometer (*Courtesy of N. Sivakugan, James Cook University, Australia.*)

3.20   Vane Shear Test

The *vane shear test* (ASTM D-2573) may be used during the drilling operation to determine the *in situ* undrained shear strength (c_u) of clay soils—particularly soft clays. The vane shear apparatus consists of four blades on the end of a rod, as shown in Figure 3.23. The height, H, of the vane is twice the diameter, D. The vane can be either rectangular or tapered (see Figure 3.23). The dimensions of vanes used in the field are given in Table 3.8. The vanes of the apparatus are pushed into the soil at the bottom of a borehole without disturbing the soil appreciably. Torque is applied at the top of the rod to rotate the vanes at a standard rate of 0.1°/sec. This rotation will induce failure in a soil of cylindrical shape surrounding the vanes. The maximum torque, T, applied to cause failure is measured. Note that

$$T = f(c_u, H, \text{ and } D) \tag{3.33}$$

or

$$c_u = \frac{T}{K} \tag{3.34}$$

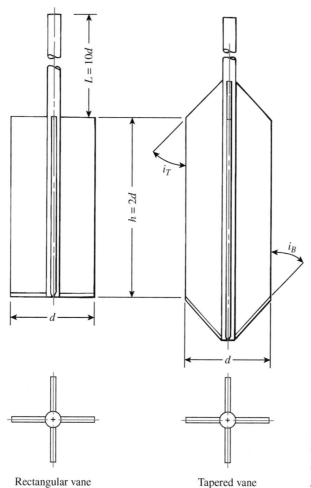

Rectangular vane   Tapered vane

Figure 3.23 Geometry of field vane (After ASTM, 2014) (Based on *Annual Book of ASTM Standards, Vol. 04.08.*)

Table 3.8 ASTM Recommended Dimensions of Field Vanes[a] (Based on *Annual Book of ASTM Standards, Vol. 04.08.*)

Casing size	Diameter, d mm (in.)	Height, h mm (in.)	Thickness of blade mm (in.)	Diameter of rod mm (in.)
AX	38.1 ($1\frac{1}{2}$)	76.2 (3)	1.6 ($\frac{1}{16}$)	12.7 ($\frac{1}{2}$)
BX	50.8 (2)	101.6 (4)	1.6 ($\frac{1}{16}$)	12.7 ($\frac{1}{2}$)
NX	63.5 ($2\frac{1}{2}$)	127.0 (5)	3.2 ($\frac{1}{8}$)	12.7 ($\frac{1}{2}$)
101.6 mm (4 in.)[b]	92.1 ($3\frac{5}{8}$)	184.1 ($7\frac{1}{4}$)	3.2 ($\frac{1}{8}$)	12.7 ($\frac{1}{2}$)

[a]The selection of a vane size is directly related to the consistency of the soil being tested; that is, the softer the soil, the larger the vane diameter should be.
[b]Inside diameter.

According to ASTM (2014), for rectangular vanes,

$$K = \frac{\pi d^2}{2}\left(h + \frac{d}{3}\right) \tag{3.35}$$

If $h/d = 2$,

$$K = \frac{7\pi d^3}{6} \tag{3.36}$$

Thus,

$$c_u = \frac{6T}{7\pi d^3} \tag{3.37}$$

For tapered vanes,

$$K = \frac{\pi d^2}{12}\left(\frac{d}{\cos i_T} + \frac{d}{\cos i_B} + 6h\right) \tag{3.38}$$

The angles i_T and i_B are defined in Figure 3.23.

Field vane shear tests are moderately rapid and economical and are used extensively in field soil-exploration programs. The test gives good results in soft and medium-stiff clays and gives excellent results in determining the properties of sensitive clays.

Sources of significant error in the field vane shear test are poor calibration of torque measurement and damaged vanes. Other errors may be introduced if the rate of rotation of the vane is not properly controlled.

For actual design purposes, the undrained shear strength values obtained from field vane shear tests $[c_{u(\text{VST})}]$ are too high, and it is recommended that they be corrected according to the equation

$$c_{u(\text{corrected})} = \lambda c_{u(\text{VST})} \tag{3.39}$$

where λ = correction factor.

Several correlations have been given previously for the correction factor λ. The most commonly used correlation for λ is that given by Bjerrum (1972), which can be expressed as

$$\lambda = 1.7 - 0.54 \log[\text{PI}(\%)] \tag{3.40a}$$

Morris and Williams (1994) provided the following correlations:

$$\lambda = 1.18e^{-0.08(\text{PI})} + 0.57 \ (\text{for PI} > 5) \tag{3.40b}$$

$$\lambda = 7.01e^{-0.08(\text{LL})} + 0.57 \ (\text{where LL is in } \%) \tag{3.40c}$$

The field vane shear strength can be correlated with the preconsolidation pressure and the overconsolidation ratio of the clay. Using 343 data points, Mayne and

Mitchell (1988) derived the following empirical relationship for estimating the preconsolidation pressure of a natural clay deposit:

$$\sigma_c' = 7.04[c_{u(\text{field})}]^{0.83}$$

(3.41)

Here,

σ_c' = preconsolidation pressure (kN/m^2)

$c_{u(\text{field})}$ = field vane shear strength (kN/m^2)

The overconsolidation ratio, OCR, also can be correlated to $c_{u(\text{field})}$ according to the equation

$$\text{OCR} = \beta\frac{c_{u(\text{field})}}{\sigma_o'}$$

(3.42)

where σ_o' = effective overburden pressure.

The magnitudes of β developed by various investigators are given below.

• Mayne and Mitchell (1988):

$$\beta = 22[\text{PI}(\%)]^{-0.48}$$

(3.43)

• Hansbo (1957):

$$\beta = \frac{222}{w(\%)}$$

(3.44)

• Larsson (1980):

$$\beta = \frac{1}{0.08 + 0.0055(\text{PI})}$$

(3.45)

Example 3.3

Refer to Figure 3.23. Vane shear tests (tapered vane) were conducted in the clay layer. The vane dimensions were 63.5 mm (d) × 127 m (h), and $i_T = i_B = 45°$. For a test at a certain depth in the clay, the torque required to cause failure was 20 N · m. For the clay, liquid limit was 50 and plastic limit was 18. Estimate the undrained cohesion of the clay for use in the design by using each equation:

a. Bjerrum's λ relationship (Eq. 3.40a)
b. Morris and Williams' λ and PI relationship (Eq. 3.40b)

      c. Morris and Williams' λ and LL relationship (Eq. 3.40c)
      d. Estimate the preconsolidation pressure of clay, σ_c'.

Solution

Part a

Given: $h/d = 127/63.5 = 2$

From Eq. (3.38),

$$
\begin{aligned}
K &= \frac{\pi d^2}{12}\left(\frac{d}{\cos i_T} + \frac{d}{\cos i_B} + 6h\right) \\
&= \frac{\pi (0.0635)^2}{12}\left[\frac{0.0635}{\cos 45} + \frac{0.0635}{\cos 45} + 6(0.127)\right] \\
&= (0.001056)(0.0898 + 0.0898 + 0.762) \\
&= 0.000994
\end{aligned}
$$

From Eq. (3.34),

$$
\begin{aligned}
c_{u(\text{VST})} &= \frac{T}{K} = \frac{20}{0.000994} \\
&= 20{,}121 \text{ N/m}^2 \approx 20.12 \text{ kN/m}^2
\end{aligned}
$$

From Eqs. (3.40a) and (3.39),

$$
\begin{aligned}
c_{u(\text{corrected})} &= [1.7 - 0.54 \log{(\text{PI\%})}]c_{u(\text{VST})} \\
&= [1.7 - 0.54 \log(50 - 18)](20.12) \\
&= \mathbf{17.85 \text{ kN/m}^2}
\end{aligned}
$$

Part b

From Eqs. (3.40b) and (3.39),

$$
\begin{aligned}
c_{u(\text{corrected})} &= [1.18e^{-0.08(\text{PI})} + 0.57]c_{u(\text{VST})} \\
&= [1.18e^{-0.08(50 - 18)} + 0.57](20.12) \\
&= \mathbf{13.3 \text{ kN/m}^2}
\end{aligned}
$$

Part c

From Eqs. (3.40c) and (3.39),

$$
\begin{aligned}
c_{u(\text{corrected})} &= [7.01e^{-0.08(\text{LL})} + 0.57]c_{u(\text{VST})} \\
&= [7.01e^{-0.08(50)} + 0.57](20.12) \\
&= \mathbf{14.05 \text{ kN/m}^2}
\end{aligned}
$$

Part d

From Eq. (3.41)

$$
\sigma_c' = 7.04[c_{u(\text{VST})}]^{0.83} = 7.04(20.12)^{0.83} = \mathbf{85 \text{ kN/m}^2}
$$

3.21 Cone Penetration Test

The cone penetration test (CPT), originally known as the Dutch cone penetration test, is a versatile sounding method that can be used to determine the materials in a soil profile and estimate their engineering properties. The test is also called the *static penetration test,* and no boreholes are necessary to perform it. In the original version, a 60° cone with a base area of 10 cm^2 (1.55 in.^{2}) was pushed into the ground at a steady rate of about 20 mm/sec ($\approx$0.8 in./sec), and the resistance to penetration (called the point resistance) was measured.

The cone penetrometers in use at present measure (a) the *cone resistance* (q_c) to penetration developed by the cone, which is equal to the vertical force applied to the cone, divided by its horizontally projected area; and (b) the *frictional resistance* (f_c), which is the resistance measured by a sleeve located above the cone with the local soil surrounding it. The frictional resistance is equal to the vertical force applied to the sleeve, divided by its surface area—actually, the sum of friction and adhesion.

Generally, two types of penetrometers are used to measure q_c and f_c:

1. *Mechanical friction-cone penetrometer* (Figure 3.24). The tip of this penetrometer is connected to an inner set of rods. The tip is first advanced about 40 mm, giving the cone resistance. With further thrusting, the tip engages the friction sleeve. As the inner

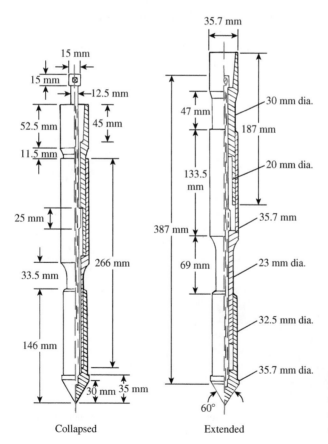

Figure 3.24 Mechanical friction-cone penetrometer (After ASTM, 2001) (Based on *Annual Book of ASTM Standards, Vol. 04.08.*)

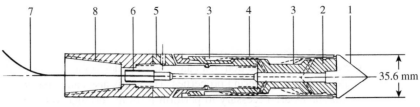

1 Conical point (10 cm^2)
2 Load cell
3 Strain gauges
4 Friction sleeve (150 cm^2)
5 Adjustment ring
6 Waterproof bushing
7 Cable
8 Connection with rods

Figure 3.25 Electric friction-cone penetrometer (After ASTM, 2001) (Based on *Annual Book of ASTM Standards, Vol. 04.08.*)

rod advances, the rod force is equal to the sum of the vertical force on the cone and sleeve. Subtracting the force on the cone gives the side resistance.

2. *Electric friction-cone penetrometer* (Figure 3.25). The tip of this penetrometer is attached to a string of steel rods. The tip is pushed into the ground at the rate of 20 mm/sec. Wires from the transducers are threaded through the center of the rods and continuously measure the cone and side resistances. Figure 3.26 shows a photograph of an electric friction-cone penetrometer.

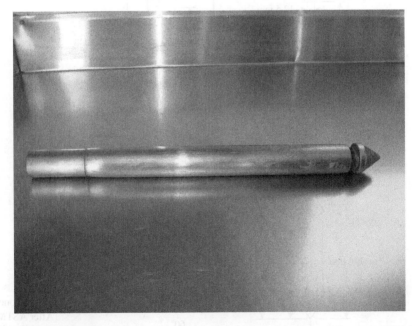

Figure 3.26 Photograph of an electric friction-cone penetrometer (*Courtesy of Sanjeev Kumar, Southern Illinois University, Carbondale, Illinois*)

Figure 3.27 shows the sequence of a cone penetration test in the field. A truck-mounted CPT rig is shown in Figure 3.27a. A hydraulic ram located inside the truck pushes the cone into the ground. Figure 3.27b shows the cone penetrometer in the truck being put in the proper location. Figure 3.27c shows the progress of the CPT. Figure 3.28 shows the results of penetrometer test in a soil profile with friction measurement by an electric friction-cone penetrometer.

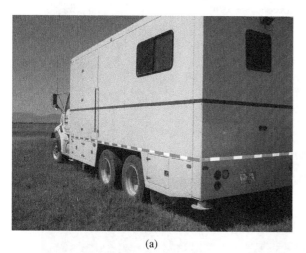

(a)

(b)

Figure 3.27 Cone penetration test in field: (a) mounted CPT rig; (b) cone penetrometer being set in proper location (*Courtesy of Sanjeev Kumar, Southern Illinois University, Carbondale, Illinois*)

(c)

Figure 3.27 (continued) (c) test in progress (*Courtesy of Sanjeev Kumar, Southern Illinois University, Carbondale, Illinois*)

Several correlations that are useful in estimating the properties of soils encountered during an exploration program have been developed for the point resistance (q_c) and the friction ratio (F_r) obtained from the cone penetration tests. The friction ratio is defined as

$$F_r = \frac{\text{frictional resistance}}{\text{cone resistance}} = \frac{f_c}{q_c} \tag{3.46}$$

In a more recent study on several soils in Greece, Anagnostopoulos et al. (2003) expressed F_r as

$$F_r(\%) = 1.45 - 1.36 \log D_{50} \text{ (electric cone)} \tag{3.47}$$

and

$$F_r(\%) = 0.7811 - 1.611 \log D_{50} \text{ (mechanical cone)} \tag{3.48}$$

where D_{50} = size through which 50% of soil will pass through (mm).

The D_{50} for soils based on which Eqs. (3.47) and (3.48) have been developed ranged from 0.001 mm to about 10 mm.

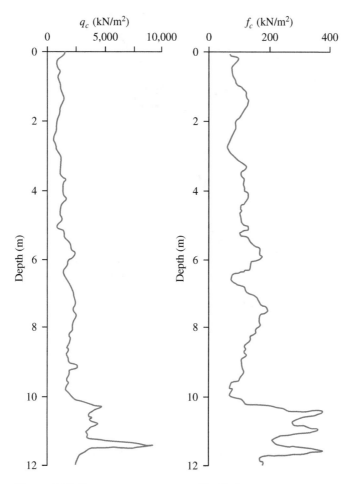

Figure 3.28 Cone penetrometer test with friction measurement

As in the case of standard penetration tests, several correlations have been developed between q_c and other soil properties. Some of these correlations are presented next.

Correlation between Relative Density (D_r) and q_c for Sand

Lancellotta (1983) and Jamiolkowski et al. (1985) showed that the relative density of *normally consolidated sand, D_r,* and q_c can be correlated according to the formula (Figure 3.29).

$$D_r(\%) = A + B \log_{10}\left(\frac{q_c}{\sqrt{\sigma_o'}}\right)$$

(3.49)

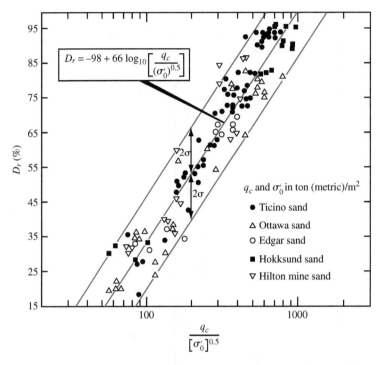

Figure 3.29 Relationship between D_r and q_c (Based on Lancellotta, 1983, and Jamiolski et al., 1985)

The preceding relationship can be rewritten as (Kulhawy and Mayne, 1990)

$$D_r(\%) = 68 \left[\log \left(\frac{q_c}{\sqrt{p_a \cdot \sigma_0'}} \right) - 1 \right] \tag{3.50}$$

where

p_a = atmospheric pressure ($\approx 100 \text{ kN/m}^2$)

σ_o' = vertical effective stress

Kulhawy and Mayne (1990) proposed the following relationship to correlate D_r, q_c, and the vertical effective stress σ_o':

$$D_r = \sqrt{ \left[\frac{1}{305 Q_c \text{OCR}^{1.8}} \right] \left[\frac{\dfrac{q_c}{p_a}}{\left(\dfrac{\sigma_o'}{p_a} \right)^{0.5}} \right] } \tag{3.51}$$

In this equation,

OCR = overconsolidation ratio
p_a = atmospheric pressure
Q_c = compressibility factor

The recommended values of Q_c are as follows:

Highly compressible sand = 0.91
Moderately compressible sand = 1.0
Low compressible sand = 1.09

Correlation between q_c and Drained Friction Angle (ϕ') for Sand

On the basis of experimental results, Robertson and Campanella (1983) suggested the variation of D_r, σ'_o, and ϕ' for normally consolidated quartz sand. This relationship can be expressed as (Kulhawy and Mayne, 1990)

$$\phi' = \tan^{-1}\left[0.1 + 0.38 \log\left(\frac{q_c}{\sigma'_o}\right)\right] \tag{3.52}$$

Based on the cone penetration tests on the soils in the Venice Lagoon (Italy), Ricceri et al. (2002) proposed a similar relationship for soil with classifications of ML and SP-SM as

$$\phi' = \tan^{-1}\left[0.38 + 0.27 \log\left(\frac{q_c}{\sigma'_o}\right)\right] \tag{3.53}$$

In a more recent study, Lee et al. (2004) developed a correlation between ϕ', q_c, and the horizontal effective stress (σ'_h) in the form

$$\phi' = 15.575\left(\frac{q_c}{\sigma'_h}\right)^{0.1714} \tag{3.54}$$

Correlation between q_c and N_{60}

For granular soils, several correlations have been proposed to correlate q_c and N_{60} (N_{60} = standard penetration resistance) against the mean grain size (D_{50} in mm). These correlations are of the form,

$$\frac{\left(\frac{q_c}{p_a}\right)}{N_{60}} = cD_{50}^a \tag{3.55}$$

Table 3.9 shows the values of c and a as developed from various studies.

Correlations of Soil Types

Robertson and Campanella (1986) provided the correlations shown in Figure 3.30 between q_c and the friction ratio [Eq. (3.46)] to identify various types of soil encountered in the field.

Table 3.9 Values of c and a [Eq. (3.55)]

Investigator		c	a
Burland and Burbidge (1985)	Upper limit	15.49	0.33
	Lower limit	4.9	0.32
Robertson and Campanella (1983)	Upper limit	10	0.26
	Lower limit	5.75	0.31
Kulhawy and Mayne (1990)		5.44	0.26
Anagnostopoulos et al. (2003)		7.64	0.26

Correlations for Undrained Shear Strength (c_u), Preconsolidation Pressure (σ'_c), and Overconsolidation Ratio (OCR) for Clays

The undrained shear strength, c_u, can be expressed as

$$c_u = \frac{q_c - \sigma_o}{N_K} \qquad (3.56)$$

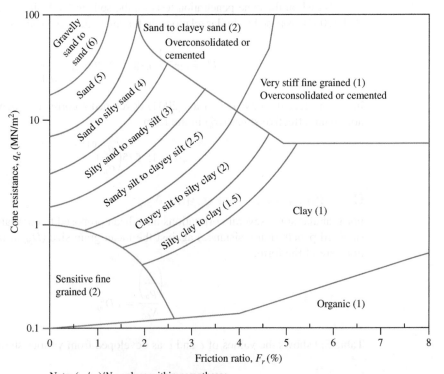

Note: $(q_c/p_a)/N_{60}$ values within parentheses

Figure 3.30 Robertson and Campanella's correlation (1986) between q_c, F_r, and the type of soil (Based on Robertson and Campanella, 1986)

where

σ_o = total vertical stress
N_K = bearing capacity factor

The bearing capacity factor, N_K, may vary from 11 to 19 for normally consolidated clays and may approach 25 for overconsolidated clay. According to Mayne and Kemper (1988)

$$N_K = 15 \text{ (for electric cone)}$$

and

$$N_K = 20 \text{ (for mechanical cone)}$$

Based on tests in Greece, Anagnostopoulos et al. (2003) determined

$$N_K = 17.2 \text{ (for electric cone)}$$

and

$$N_K = 18.9 \text{ (for mechanical cone)}$$

These field tests also showed that

$$c_u = \frac{f_c}{1.26} \text{ (for mechanical cones)} \tag{3.57}$$

and

$$c_u = f_c \text{ (for electrical cones)} \tag{3.58}$$

Mayne and Kemper (1988) provided correlations for preconsolidation pressure (σ_c') and overconsolidation ratio (OCR) as

$$\underset{\underset{\text{MN/m}^2}{\uparrow}}{\sigma_c'} = 0.243 (\underset{\underset{\text{MN/m}^2}{\uparrow}}{q_c})^{0.96} \tag{3.59}$$

and

$$OCR = 0.37 \left(\frac{q_c - \sigma_o}{\sigma_o'} \right)^{1.01} \tag{3.60}$$

where σ_o and σ_o' = total and effective stress, respectively.

Example 3.4

At a depth of 12.5 m in a *moderately compressible sand deposit*, a cone penetration test showed $q_c = 20 \text{ MN/m}^2$. For the sand given: $\gamma = 16 \text{ kN/m}^3$ and OCR = 2. Estimate the relative density of the sand. Use Eq. (3.51).

Solution
Vertical effective stress $\sigma_o' = (12.5)(16) = 200 \text{ kN/m}^2$.
Q_c (moderately compressible sand) ≈ 1.

From Eq. (3.51),

$$D_r = \sqrt{\frac{1}{305(OCR)^{1.8}}\left[\frac{\left(\dfrac{q_c}{p_a}\right)}{\left(\dfrac{\sigma'_o}{p_a}\right)^{0.5}}\right]}$$

$$= \sqrt{\frac{1}{(305)(2)^{1.8}}\left[\frac{\left(\dfrac{20{,}000 \text{ kN/m}^2}{100 \text{ kN/m}^2}\right)}{\left(\dfrac{200 \text{ kN/m}^2}{100 \text{ kN/m}^2}\right)^{0.5}}\right]}$$

$$= \sqrt{(0.00094)(141.41)} = 0.365$$

Hence,

$$D_r = \mathbf{36.5\%}$$

$\blacksquare$

3.22   Pressuremeter Test (PMT)

The pressuremeter test is an *in situ* test conducted in a borehole. It was originally developed by Menard (1956) to measure the strength and deformability of soil. It has also been adopted by ASTM as Test Designation 4719. The Menard-type PMT consists essentially of a probe with three cells. The top and bottom ones are *guard cells* and the middle one is the *measuring cell,* as shown schematically in Figure 3.31a. The test is conducted in a prebored hole with a diameter that is between 1.03 and 1.2 times the nominal diameter of the probe. The probe that is most commonly used has a diameter of 58 mm and a length of 420 mm. The probe cells can be expanded by either liquid or gas. The guard cells are expanded to reduce the end-condition effect on the measuring cell, which has a volume (V_o) of 535 cm^3. Following are the dimensions for the probe diameter and the diameter of the borehole, as recommended by ASTM:

Probe diameter (mm)	Borehole diameter	
	Nominal (mm)	Maximum (mm)
44	45	53
58	60	70
74	76	89

In order to conduct a test, the measuring cell volume, V_o, is measured and the probe is inserted into the borehole. Pressure is applied in increments and the new volume of the cell is measured. The process is continued until the soil fails or until the pressure limit of the device is reached. The soil is considered to have failed when the total volume of the expanded cavity (V) is about twice the volume of the original cavity. After the completion of the test, the probe is deflated and advanced for testing at another depth.

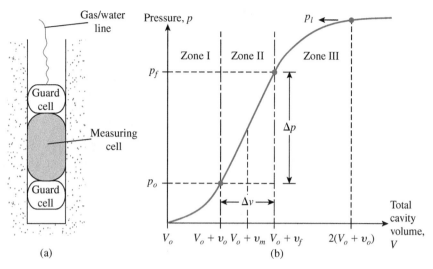

Figure 3.31 (a) Pressuremeter; (b) plot of pressure versus total cavity volume

The results of the pressuremeter test are expressed in the graphical form of pressure versus volume, as shown in Figure 3.31b. In the figure, Zone I represents the reloading portion during which the soil around the borehole is pushed back into the initial state (i.e., the state it was in before drilling). The pressure p_o represents the *in situ* total horizontal stress. Zone II represents a pseudoelastic zone in which the cell volume versus cell pressure is practically linear. The pressure p_f represents the creep, or yield, pressure. The zone marked III is the plastic zone. The pressure p_l represents the limit pressure. Figure 3.32 shows some photographs for a pressuremeter test in the field.

The pressuremeter modulus, E_p, of the soil is determined with the use of the theory of expansion of an infinitely thick cylinder. Thus,

$$E_p = 2(1 + \mu_s)(V_o + v_m)\left(\frac{\Delta p}{\Delta v}\right) \tag{3.61}$$

where

$$v_m = \frac{v_o + v_f}{2}$$
$$\Delta p = p_f - p_o$$
$$\Delta v = v_f - v_o$$
μ_s = Poisson's ratio (which may be assumed to be 0.33)

The limit pressure p_l is usually obtained by extrapolation and not by direct measurement.

In order to overcome the difficulty of preparing the borehole to the proper size, self-boring pressuremeters (SBPMTs) have also been developed. The details concerning SBPMTs can be found in the work of Baguelin et al. (1978).

Figure 3.32 Pressuremeter test in the field: (a) the pressuremeter probe; (b) drilling the bore hole by wet rotary method; (c) pressuremeter control unit with probe in the background; (d) getting ready to insert the pressuremeter probe into the bore hole (*Courtesy of Jean-Louis Briaud, Texas A&M University, College Station, Texas*)

Correlations between various soil parameters and the results obtained from the pressuremeter tests have been developed by various investigators. Kulhawy and Mayne (1990) proposed that, for clays,

$$\sigma_c' = 0.45 p_l \tag{3.62}$$

where σ_c' = preconsolidation pressure.

On the basis of the cavity expansion theory, Baguelin et al. (1978) proposed that

$$c_u = \frac{(p_l - p_o)}{N_p} \tag{3.63}$$

where

c_u = undrained shear strength of a clay

$N_p = 1 + \ln\left(\dfrac{E_p}{3c_u}\right)$

Typical values of N_p vary between 5 and 12, with an average of about 8.5. Ohya et al. (1982) (see also Kulhawy and Mayne, 1990) correlated E_p with field standard penetration numbers (N_{60}) for sand and clay as follows:

$$\text{Clay: } E_p(\text{kN/m}^2) = 1930 \, N_{60}^{0.63} \tag{3.64}$$

$$\text{Sand: } E_p(\text{kN/m}^2) = 908 \, N_{60}^{0.66} \tag{3.65}$$

3.23 Dilatometer Test

The use of the flat-plate dilatometer test (DMT) is relatively recent (Marchetti, 1980; Schmertmann, 1986). The equipment essentially consists of a flat plate measuring 220 mm (length) × 95 mm (width) × 14 mm (thickness)(8.66 in. × 3.74 in. × 0.55 in.). A thin, flat, circular, expandable steel membrane having a diameter of 60 mm (2.36 in.) is located flush at the center on one side of the plate (Figure 3.33a). Figure 3.34 shows two flat-plate dilatometers with other instruments for conducting a test in the field. The dilatometer probe is inserted into the ground with a cone penetrometer testing rig (Figure 3.33b). Gas and electric lines extend from the surface control box, through the penetrometer rod, and into the blade. At the required depth, high-pressure nitrogen gas is used to inflate the membrane. Two pressure readings are taken:

1. The pressure *A* required to "lift off" the membrane.
2. The pressure *B* at which the membrane expands 1.1 mm (0.4 in.) into the surrounding soil.

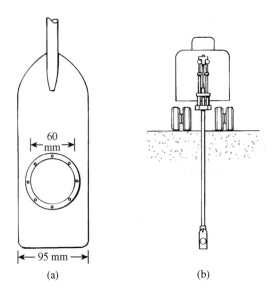

60 mm

95 mm

(a)

(b)

Figure 3.33 (a) Schematic diagram of a flat-plate dilatometer; (b) dilatometer probe inserted into ground

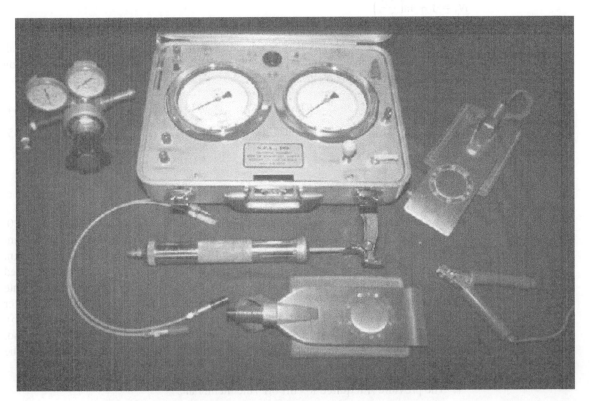

Figure 3.34 Dilatometer and other equipment (*Courtesy of N. Sivakugan, James Cook University, Australia*)

The A and B readings are corrected as follows (Schmertmann, 1986):

$$\text{Contact stress, } p_o = 1.05(A + \Delta A - Z_m) - 0.05(B - \Delta B - Z_m) \tag{3.66}$$

$$\text{Expansion stress, } p_1 = B - Z_m - \Delta B \tag{3.67}$$

where

ΔA = vacuum pressure required to keep the membrane in contact with its seating

ΔB = air pressure required inside the membrane to deflect it outward to a center expansion of 1.1 mm

Z_m = gauge pressure deviation from zero when vented to atmospheric pressure

The test is normally conducted at depths 200 to 300 mm apart. The result of a given test is used to determine three parameters:

1. Material index, $I_D = \dfrac{p_1 - p_o}{p_o - u_o}$

2. Horizontal stress index, $K_D = \dfrac{p_o - u_o}{\sigma_o'}$

3. Dilatometer modulus, $E_D(\text{kN/m}^2) = 34.7(p_1 \text{ kN/m}^2 - p_o \text{ kN/m}^2)$

where

u_o = pore water pressure

σ_o' = *in situ* vertical effective stress

Figure 3.35 shows the results of a dilatometer test conducted in Bangkok soft clay and reported by Shibuya and Hanh (2001). Based on his initial tests, Marchetti (1980) provided the following correlations.

$$K_o = \left(\frac{K_D}{1.5}\right)^{0.47} - 0.6 \tag{3.68}$$

$$\text{OCR} = (0.5K_D)^{1.56} \tag{3.69}$$

$$\frac{c_u}{\sigma_o'} = 0.22 \qquad \text{(for normally consolidated clay)} \tag{3.70}$$

$$\left(\frac{c_u}{\sigma_o'}\right)_{\text{OC}} = \left(\frac{c_u}{\sigma_o'}\right)_{\text{NC}} (0.5K_D)^{1.25} \tag{3.71}$$

$$E_s = (1 - \mu_s^2)E_D \tag{3.72}$$

where

K_o = coefficient of at-rest earth pressure

OCR = overconsolidation ratio

OC = overconsolidated soil

NC = normally consolidated soil

E_s = modulus of elasticity

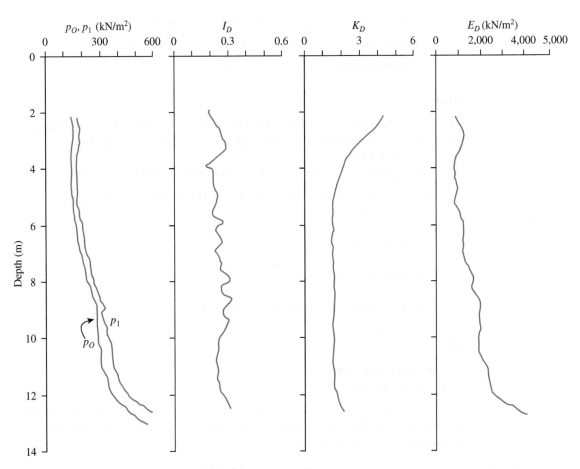

Figure 3.35 A dilatometer test result conducted on soft Bangkok clay (Based on Lancellotta, 1983, and Jamiolski et al., 1985)

Other relevant correlations using the results of dilatometer tests are as follows:

- For undrained cohesion in clay (Kamei and Iwasaki, 1995):

$$c_u = 0.35\ \sigma_0'\ (0.47K_D)^{1.14} \tag{3.73}$$

- For soil friction angle (ML and SP-SM soils) (Ricceri et al., 2002):

$$\phi' = 31 + \frac{K_D}{0.236 + 0.066K_D} \tag{3.74a}$$

$$\phi'_{ult} = 28 + 14.6\ \log K_D - 2.1(\log K_D)^2 \tag{3.74b}$$

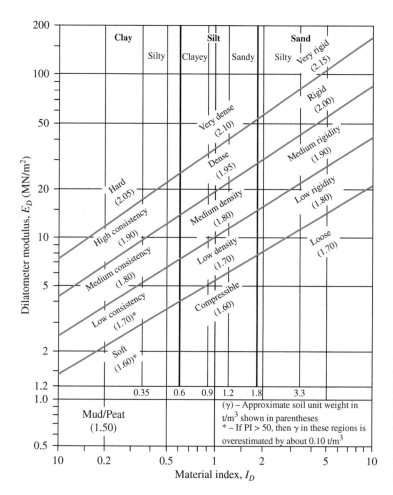

Figure 3.36 Chart for determination of soil description and unit weight (After Schmertmann, 1986) (*Note:* 1 t/m³ = 9.81 kN/m³) (Based on Schmertmann, J.H. (1986). "Suggested method for performing that flat dilatometer test," *Geotechnical Testing Journal*, ASTM, Vol. 9, No. 2, pp. 93-101, Fig. 2.)

Schmertmann (1986) also provided a correlation between the material index (I_D) and the dilatometer modulus (E_D) for a determination of the nature of the soil and its unit weight (γ). This relationship is shown in Figure 3.36.

3.24 Iowa Borehole Shear Test

The Iowa borehole shear test is a simple device to determine the shear strength parameters of soil at a given depth during subsoil exploration. The shear device consists of two grooved plates that are pushed into the borehole (Figure 3.37). A controlled normal force (N) can be applied to each of the grooved plates. Shear failure in soil close to the plates is induced by applying a vertical force S, after allowing the soil to consolidate under the

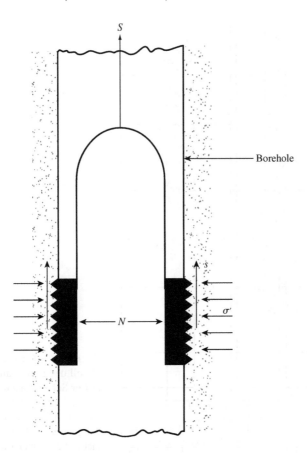

Figure 3.37 Iowa borehole shear test

normal stress (5 minutes in sand and 10 to 20 minutes in clay). So, the effective normal stress (σ') on the wall of the borehole can be given as

$$\sigma' = \frac{N}{A} \tag{3.75}$$

where A = area of each plate in contact with the soil

Similarly, the shear stress at failure (s) is

$$s = \frac{S}{2A} \tag{3.76}$$

The test could be repeated with a number of increasing normal forces (N) without removing the shearing device. The results can be plotted in graphic form (Figure 3.38) to obtain the shear strength parameters (that is, cohesion c' and angle of friction ϕ') of the soil. The shear strength parameters obtained in this manner are likely to represent those of a consolidated drained test.

Figure 3.39 shows the photograph of a shear head and a hand pump.

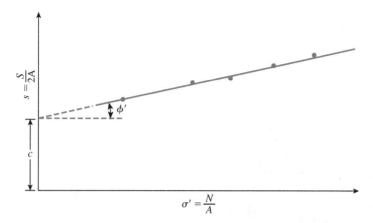

Figure 3.38 Variation of *s* with σ' from Iowa borehole shear test

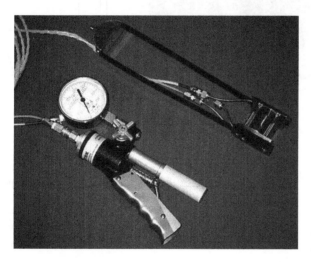

Figure 3.39 Photograph of shear head and a hand pump (*Courtesy of R. L. Handy, Iowa State University, Ames, Iowa*)

3.25 *K_0* Stepped-Blade Test

In the 1970s, the K_0 stepped-blade test for measuring lateral *in situ* stress (and hence K_0 as the at-rest earth pressure coefficient) was developed by Dr. Richard L. Handy at Iowa State University. Figure 3.40a shows a K_0 stepped-blade test in progress. The long blade consists of four steps, 100 mm apart, ranging from 3 mm thin to 7.5 mm thick from its bottom to its top (Figure 3.40b). Even the thickest step is thinner than the dilatometer; therefore, the soil disturbance is relatively less. Each step carries a pneumatic pressure cell flush with the flat surface that comes in contact with the soil when pushed into it.

The test is conducted in a borehole where the first blade is pushed into the soil at the bottom of the hole and the stress in the bottom step, σ_1, is measured. The second blade is pushed into the soil and the stress in the bottom two steps (σ_1 and σ_2) is measured. This is repeated until all of the steps are in the soil, giving 14 ($= 1 + 2 + 3 + 4 + 4$) stress measurements. The fifth step has the same thickness as the fourth but with no pressure

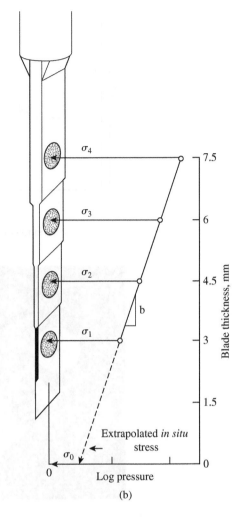

(a)

Figure 3.40 K_0 stepped-blade test: (a) Test in
progress in the field; (b) Schematic diagram of the
blade (*Courtesy of R. L. Handy, Iowa State University,
Ames, Iowa*)

cell (see Figure 3.40b). As shown in Figure 3.40b, the logarithm of stress is plotted against
the blade thickness. The stress corresponding to zero blade thickness, σ_0, is extrapolated
from the figure and is taken as the *total in situ* horizontal stress from which K_0 can be
computed once the pore water pressure is known from the groundwater table depth. The
pressure should increase with blade thickness. Any data that do not show an increase in
stress with an increase in step thickness must be discarded, and only the remaining data
should be used in estimating the *in situ* horizontal stress.

3.26 Coring of Rocks

When a rock layer is encountered during a drilling operation, rock coring may be necessary. To core rocks, a *core barrel* is attached to a drilling rod. A *coring bit* is attached

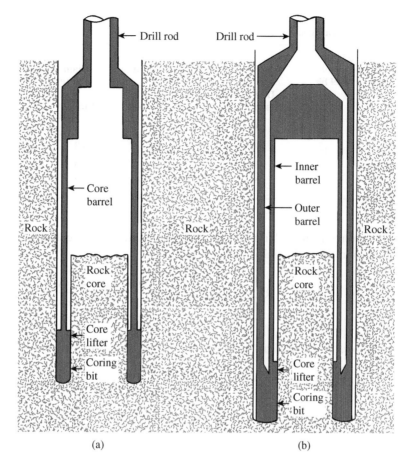

Figure 3.41 Rock coring: (a) single-tube core barrel; (b) double-tube core barrel

to the bottom of the barrel (Fig. 3.41). The cutting elements may be diamond, tungsten, carbide, and so on. Table 3.10 summarizes the various types of core barrel and their sizes, as well as the compatible drill rods commonly used for exploring foundations. The coring is advanced by rotary drilling. Water is circulated through the drilling rod during coring, and the cutting is washed out.

Two types of core barrel are available: the *single-tube core barrel* (Figure 3.41a) and the *double-tube core barrel* (Figure 3.41b). Rock cores obtained by single-tube core barrels can be highly disturbed and fractured because of torsion. Rock cores smaller than the BX size tend to fracture during the coring process. Figure 3.42 shows the photograph of a diamond coring bit. Figure 3.43 shows the end and side views of a diamond coring bit attached to a double-tube core barrel.

When the core samples are recovered, the depth of recovery should be properly recorded for further evaluation in the laboratory. Based on the length of the rock core

Table 3.10 Standard Size and Designation of Casing, Core Barrel, and Compatible Drill Rod

Casing and core barrel designation	Outside diameter of core barrel bit		Drill rod designation	Outside diameter of drill rod		Diameter of borehole		Diameter of core sample	
	(mm)	(in.)		(mm)	(in.)	(mm)	(in.)	(mm)	(in.)
EX	36.51	$1\frac{7}{16}$	E	33.34	$1\frac{5}{16}$	38.1	$1\frac{1}{2}$	22.23	$\frac{7}{8}$
AX	47.63	$1\frac{7}{8}$	A	41.28	$1\frac{5}{8}$	50.8	2	28.58	$1\frac{1}{8}$
BX	58.74	$2\frac{5}{16}$	B	47.63	$1\frac{7}{8}$	63.5	$2\frac{1}{2}$	41.28	$1\frac{5}{8}$
NX	74.61	$2\frac{15}{16}$	N	60.33	$2\frac{3}{8}$	76.2	3	53.98	$2\frac{1}{8}$

recovered from each run, the following quantities may be calculated for a general evaluation of the rock quality encountered:

$$\text{Recovery ratio} = \frac{\text{length of core recovered}}{\text{theoretical length of rock cored}} \tag{3.77}$$

Rock quality designation (RQD)

$$= \frac{\Sigma \text{ length of recovered pieces equal to or larger than 101.6 mm (4 in.)}}{\text{theoretical length of rock cored}} \tag{3.78}$$

Figure 3.42 Diamond coring bit (*Courtesy of Braja M. Das, Henderson, Nevada*)

(a)

(b)

Figure 3.43 Diamond coring bit attached to a double-tube core barrel: (a) end view; (b) side view (*Courtesy of Professional Service Industries, Inc. (PSI), Waukesha, Wisconsin*)

Table 3.11 Relation between *in situ* Rock Quality and RQD

RQD	Rock quality
0–0.25	Very poor
0.25–0.5	Poor
0.5–0.75	Fair
0.75–0.9	Good
0.9–1	Excellent

A recovery ratio of unity indicates the presence of intact rock; for highly fractured rocks, the recovery ratio may be 0.5 or smaller. Table 3.11 presents the general relationship (Deere, 1963) between the RQD and the *in situ* rock quality.

3.27 Preparation of Boring Logs

The detailed information gathered from each borehole is presented in a graphical form called the *boring log*. As a borehole is advanced downward, the driller generally should record the following information in a standard log:

1. Name and address of the drilling company
2. Driller's name
3. Job description and number
4. Number, type, and location of boring
5. Date of boring
6. Subsurface stratification, which can be obtained by visual observation of the soil brought out by auger, split-spoon sampler, and thin-walled Shelby tube sampler
7. Elevation of water table and date observed, use of casing and mud losses, and so on
8. Standard penetration resistance and the depth of SPT
9. Number, type, and depth of soil sample collected
10. In case of rock coring, type of core barrel used and, for each run, the actual length of coring, length of core recovery, and RQD

This information should never be left to memory, because doing so often results in erroneous boring logs.

After completion of the necessary laboratory tests, the geotechnical engineer prepares a finished log that includes notes from the driller's field log and the results of tests conducted in the laboratory. Figure 3.44 shows a typical boring log. These logs have to be attached to the final soil-exploration report submitted to the client. The figure also lists the classifications of the soils in the left-hand column, along with the description of each soil (based on the Unified Soil Classification System).

3.28 Geophysical Exploration

Several types of geophysical exploration techniques permit a rapid evaluation of subsoil characteristics. These methods also allow rapid coverage of large areas and are less

Boring Log

Name of the Project Two-story apartment building

Location Johnson & Olive St. Date of Boring March 2, 2005

Boring No. _3_ Type of _Hollow-stem auger_ Ground _____60.8 m_____
  Boring   Elevation

Soil description	Depth (m)	Soil sample type and number	N_{60}	w_n (%)	Comments
Light brown clay (fill)	1				
Silty sand (SM)	2	SS-1	9	8.2	
	3	SS-2	12	17.6	LL = 38 PI = 11
°G.W.T. ▼ 3.5 m	4				
Light gray clayey silt (ML)	5	ST-1		20.4	LL = 36 $q_u = 112$ kN/m^2
	6	SS-3	11	20.6	
Sand with some gravel (SP)	7				
End of boring @ 8 m	8	SS-4	27	9	

N_{60} = standard penetration number
w_n = natural moisture content
LL = liquid limit; PI = plasticity index
q_u = unconfined compression strength
SS = split-spoon sample; ST = Shelby tube sample

Groundwater table
observed after one
week of drilling

Figure 3.44 A typical boring log

expensive than conventional exploration by drilling. However, in many cases, definitive interpretation of the results is difficult. For that reason, such techniques should be used for preliminary work only. Here, we discuss three types of geophysical exploration technique: the seismic refraction survey, cross-hole seismic survey, and resistivity survey.

Seismic Refraction Survey

Seismic refraction surveys are useful in obtaining preliminary information about the thickness of the layering of various soils and the depth to rock or hard soil at a site. Refraction surveys are conducted by impacting the surface, such as at point *A* in Figure 3.45a, and observing the first arrival of the disturbance (stress waves) at several other points (e.g., *B*, *C*, *D*, . . .). The impact can be created by a hammer blow or by a small explosive charge. The first arrival of disturbance waves at various points can be recorded by geophones.

The impact on the ground surface creates two types of *stress wave: P waves* (or *plane waves*) and *S waves* (or *shear waves*). *P* waves travel faster than *S* waves; hence, the

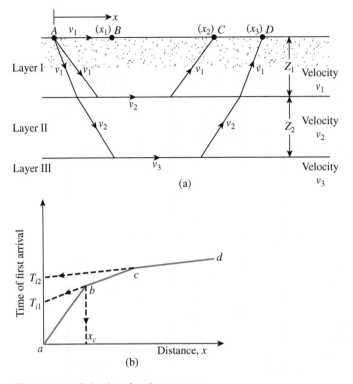

Figure 3.45 Seismic refraction survey

first arrival of disturbance waves will be related to the velocities of the P waves in various layers. The velocity of P waves in a medium is

$$v = \sqrt{\frac{E_s}{\left(\dfrac{\gamma}{g}\right)} \frac{(1 - \mu_s)}{(1 - 2\mu_s)(1 + \mu_s)}} \qquad (3.79)$$

where

E_s = modulus of elasticity of the medium
γ = unit weight of the medium
g = acceleration due to gravity
μ_s = Poisson's ratio

To determine the velocity v of P waves in various layers and the thicknesses of those layers, we use the following procedure:

Step 1. Obtain the times of first arrival, $t_1, t_2, t_3, \ldots$, at various distances $x_1, x_2, x_3, \ldots$ from the point of impact.

Step 2. Plot a graph of time t against distance x. The graph will look like the one shown in Figure 3.45b.

Step 3. Determine the slopes of the lines *ab, bc, cd,* ... :

$$\text{Slope of } ab = \frac{1}{v_1}$$

$$\text{Slope of } bc = \frac{1}{v_2}$$

$$\text{Slope of } cd = \frac{1}{v_3}$$

Here, $v_1, v_2, v_3, \ldots$ are the *P*-wave velocities in layers I, II, III, ..., respectively (Figure 3.45a).

Step 4. Determine the thickness of the top layer:

$$Z_1 = \frac{1}{2}\sqrt{\frac{v_2 - v_1}{v_2 + v_1}}\, x_c \tag{3.80}$$

The value of x_c can be obtained from the plot, as shown in Figure 3.45b.

Step 5. Determine the thickness of the second layer:

$$Z_2 = \frac{1}{2}\left[T_{i2} - 2Z_1\frac{\sqrt{v_3^2 - v_1^2}}{v_3 v_1} \right]\frac{v_3 v_2}{\sqrt{v_3^2 - v_2^2}} \tag{3.81}$$

Here, T_{i2} is the time intercept of the line *cd* in Figure 3.45b, extended backwards.

(For detailed derivatives of these equations and other related information, see Dobrin, 1960, and Das, 1992).

The velocities of *P* waves in various layers indicate the types of soil or rock that are present below the ground surface. The range of the *P*-wave velocity that is generally encountered in different types of soil and rock at shallow depths is given in Table 3.12.

Table 3.12 Range of *P*-Wave Velocity in Various Soils and Rocks

	P-wave velocity	
Type of soil or rock	**m/sec**	**ft/sec**
Soil		
Sand, dry silt, and fine-grained topsoil	200–1000	650–3300
Alluvium	500–2000	1650–6600
Compacted clays, clayey gravel, and dense clayey sand	1000–2500	3300–8200
Loess	250–750	800–2450
Rock		
Slate and shale	2500–5000	8200–16,400
Sandstone	1500–5000	4900–16,400
Granite	4000–6000	13,100–19,700
Sound limestone	5000–10,000	16,400–32,800

In analyzing the results of a refraction survey, two limitations need to be kept in mind:

1. The basic equations for the survey—that is, Eqs. (3.80) and (3.81)—are based on the assumption that the *P*-wave velocity $v_1 < v_2 < v_3 < \ldots$.
2. When a soil is saturated below the water table, the *P*-wave velocity may be deceptive. *P* waves can travel with a velocity of about 1500 m/sec (5000 ft/sec) through water. For dry, loose soils, the velocity may be well below 1500 m/sec. However, in a saturated condition, the waves will travel through water that is present in the void spaces with a velocity of about 1500 m/sec. If the presence of groundwater has not been detected, the *P*-wave velocity may be erroneously interpreted to indicate a stronger material (e.g., sandstone) than is actually present *in situ*. In general, geophysical interpretations should always be verified by the results obtained from borings.

Example 3.5

The results of a refraction survey at a site are given in the following table:

Distance of geophone from the source of disturbance (m)	Time of first arrival (sec × 10³)
2.5	11.2
5	23.3
7.5	33.5
10	42.4
15	50.9
20	57.2
25	64.4
30	68.6
35	71.1
40	72.1
50	75.5

Determine the *P*-wave velocities and the thickness of the material encountered.

Solution
Velocity
In Figure 3.46, the times of first arrival of the *P* waves are plotted against the distance of the geophone from the source of disturbance. The plot has three straight-line segments. The velocity of the top three layers can now be calculated as:

$$\text{Slope of segment } 0a = \frac{1}{v_1} = \frac{\text{time}}{\text{distance}} = \frac{23 \times 10^{-3}}{5.25}$$

or

$$v_1 = \frac{5.25 \times 10^3}{23} = 228 \text{ m/sec (top layer)}$$

$$\text{Slope of segment } ab = \frac{1}{v_2} = \frac{13.5 \times 10^{-3}}{11}$$

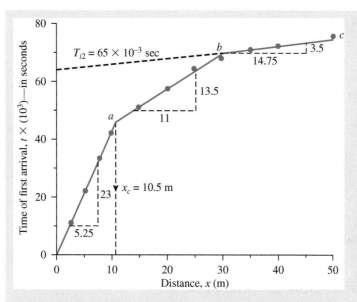

Figure 3.46 Plot of first arrival time of *P* wave versus distance of geophone from source of disturbance

or

$$v_2 = \frac{11 \times 10^3}{13.5} = \textbf{814.8 m/sec (middle layer)}$$

$$\text{Slope of segment } bc = \frac{1}{v_3} = \frac{3.5 \times 10^{-3}}{14.75}$$

or

$$v_3 = \textbf{4214 m/sec (third layer)}$$

Comparing the velocities obtained here with those given in Table 3.12 indicates that the third layer is a *rock layer*.

Thickness of Layers
From Figure 3.46, $x_c = 10.5$ m, so

$$Z_1 = \frac{1}{2}\sqrt{\frac{v_2 - v_1}{v_2 + v_1}}x_c$$

Thus,

$$Z_1 = \frac{1}{2}\sqrt{\frac{814.8 - 228}{814.8 + 228}} \times 10.5 = \textbf{3.94 m}$$

Again, from Eq. (3.81)

$$Z_2 = \frac{1}{2}\left[T_{i2} - \frac{2Z_1 \sqrt{v_3^2 - v_1^2}}{(v_3 v_1)} \right] \frac{(v_3)(v_2)}{\sqrt{v_3^2 - v_2^2}}$$

The value of T_{i2} (from Figure 3.46) is 65×10^{-3} sec. Hence,

$$Z_2 = \frac{1}{2}\left[65 \times 10^{-3} - \frac{2(3.94)\sqrt{(4214)^2 - (228)^2}}{(4214)(228)} \right] \frac{(4214)(814.8)}{\sqrt{(4214)^2 - (814.8)^2}}$$

$$= \frac{1}{2}(0.065 - 0.0345)830.47 = \textbf{12.66 m}$$

Thus, the rock layer lies at a depth of $Z_1 + Z_2 = 3.94 + 12.66 = $ **16.60 m from the surface of the ground.** ∎

Cross-Hole Seismic Survey

The velocity of shear waves created as the result of an impact to a given layer of soil can be effectively determined by the *cross-hole seismic survey* (Stokoe and Woods, 1972). The principle of this technique is illustrated in Figure 3.47, which shows two holes drilled into the ground a distance L apart. A vertical impulse is created at the bottom of one borehole by means of an impulse rod. The shear waves thus generated are recorded by a vertically sensitive transducer. The velocity of shear waves can be calculated as

$$v_s = \frac{L}{t} \tag{3.82}$$

where t = travel time of the waves.

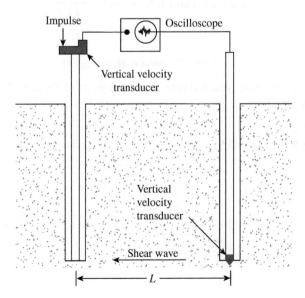

Figure 3.47 Cross-hole method of seismic survey

The shear modulus G_s of the soil at the depth at which the test is taken can be determined from the relation

$$v_s = \sqrt{\frac{G_s}{(\gamma/g)}}$$

or

$$G_s = \frac{v_s^2 \gamma}{g} \qquad (3.83)$$

where

v_s = velocity of shear waves
γ = unit weight of soil
g = acceleration due to gravity

The shear modulus is useful in the design of foundations to support vibrating machinery and the like.

Resistivity Survey

Another geophysical method for subsoil exploration is the *electrical resistivity survey.* The electrical resistivity of any conducting material having a length L and an area of cross section A can be defined as

$$\rho = \frac{RA}{L} \qquad (3.84)$$

where R = electrical resistance.

The unit of resistivity is *ohm-centimeter* or *ohm-meter.* The resistivity of various soils depends primarily on their moisture content and also on the concentration of dissolved ions in them. Saturated clays have a very low resistivity; dry soils and rocks have a high resistivity. The range of resistivity generally encountered in various soils and rocks is given in Table 3.13.

The most common procedure for measuring the electrical resistivity of a soil profile makes use of four electrodes driven into the ground and spaced equally along a straight line. The procedure is generally referred to as the *Wenner method* (Figure 3.48a). The two outside electrodes are used to send an electrical current I (usually a dc current with

Table 3.13 Representative Values of Resistivity

Material	Resistivity (ohm · m)
Sand	500–1500
Clays, saturated silt	0–100
Clayey sand	200–500
Gravel	1500–4000
Weathered rock	1500–2500
Sound rock	>5000

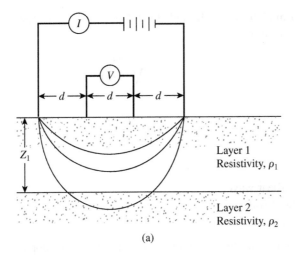

(a)

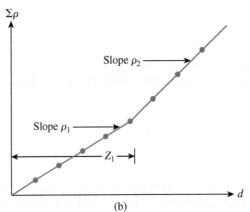

(b)

Figure 3.48 Electrical resistivity survey: (a) Wenner method; (b) empirical method for determining resistivity and thickness of each layer

nonpolarizing potential electrodes) into the ground. The current is typically in the range of 50 to 100 milliamperes. The voltage drop, V, is measured between the two inside electrodes. If the soil profile is homogeneous, its electrical resistivity is

$$\rho = \frac{2\pi d V}{I} \tag{3.85}$$

In most cases, the soil profile may consist of various layers with different resistivities, and Eq. (3.85) will yield the *apparent resistivity*. To obtain the *actual resistivity* of various layers and their thicknesses, one may use an empirical method that involves conducting tests at various electrode spacings (i.e., d is changed). The sum of the apparent resistivities, $\Sigma\rho$, is plotted against the spacing d, as shown in Figure 3.48b. The plot thus obtained has relatively straight segments, the slopes of which give the resistivity of individual layers. The thicknesses of various layers can be estimated as shown in Figure 3.48b.

The resistivity survey is particularly useful in locating gravel deposits within a fine-grained soil.

3.29 Subsoil Exploration Report

At the end of all soil exploration programs, the soil and rock specimens collected in the field are subject to visual observation and appropriate laboratory testing. (The basic soil tests were described in Chapter 2.) After all the required information has been compiled, a soil exploration report is prepared for use by the design office and for reference during future construction work. Although the details and sequence of information in such reports may vary to some degree, depending on the structure under consideration and the person compiling the report, each report should include the following items:

1. A description of the scope of the investigation
2. A description of the proposed structure for which the subsoil exploration has been conducted
3. A description of the location of the site, including any structures nearby, drainage conditions, the nature of vegetation on the site and surrounding it, and any other features unique to the site
4. A description of the geological setting of the site
5. Details of the field exploration—that is, number of borings, depths of borings, types of borings involved, and so on
6. A general description of the subsoil conditions, as determined from soil specimens and from related laboratory tests, standard penetration resistance and cone penetration resistance, and so on
7. A description of the water-table conditions
8. Recommendations regarding the foundation, including the type of foundation recommended, the allowable bearing pressure, and any special construction procedure that may be needed; alternative foundation design procedures should also be discussed in this portion of the report
9. Conclusions and limitations of the investigations

The following graphical presentations should be attached to the report:

1. A site location map
2. A plan view of the location of the borings with respect to the proposed structures and those nearby
3. Boring logs
4. Laboratory test results
5. Other special graphical presentations

The exploration reports should be well planned and documented, as they will help in answering questions and solving foundation problems that may arise later during design and construction.

Problems

3.1 For a Shelby tube, given: outside diameter = 3 in. and inside diameter 2.874 in. What is the area ratio of the tube?

3.2 A soil profile is shown in Figure P3.2 along with the standard penetration numbers in the clay layer. Use Eqs. (3.8) and (3.9) to determine the variation of c_u and OCR with depth. What is the average value of c_u and OCR?

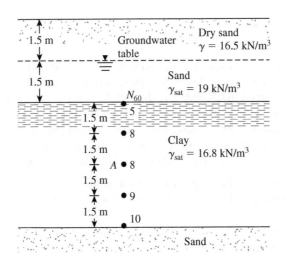

Figure P3.2

3.3 Refer to Figure P3.2. Use Eqs. (3.10) and (3.11) to determine the variation of OCR and preconsolidation pressure σ'_c.

3.4 Following is the variation of the field standard penetration number (N_{60}) in a sand deposit:

Depth (m)	N_{60}
1.5	6
3	8
4.5	9
6	8
7.9	13
9	14

The groundwater table is located at a depth of 6 m. Given: the dry unit weight of sand from 0 to a depth of 6 m is 18 kN/m³, and the saturated unit weight of sand for depth 6 to 12 m is 20.2 kN/m³. Use the relationship given in Eq. (3.13) to calculate the corrected penetration numbers.

3.5 Redo Problem 3.4 using Eq. (3.14).

3.6 For the soil profile described in Problem 3.4, estimate an average peak soil friction angle. Use Eq. (3.31).

3.7 Repeat Problem 3.6 using Eq. (3.30).

3.8 Repeat Problem 3.6 using Eq. (3.29).

3.9 Refer to Problem 3.4. Using Eq. (3.22), determine the average relative density of sand.

3.10 Refer to Problem 3.4. Using Eq. (3.28), determine the average relative density of the sand. Assume it is a fine sand. Use Eq. (3.13) to obtain $(N_1)_{60}$.

3.11 The following table gives the variation of the field standard penetration number (N_{60}) in a sand deposit:

Depth (m)	N_{60}
1.5	5
3.0	11
4.5	14
6.0	18
7.5	16
9.0	21

The groundwater table is located at a depth of 12 m. The dry unit weight of sand from 0 to a depth of 12 m is 17.6 kN/m³. Assume that the mean grain size (D_{50}) of the sand deposit to be about 0.8 mm. Estimate the variation of the relative density with depth for sand. Use Eq. (3.23).

3.12 Following are the standard penetration numbers determined from a sandy soil in the field:

Depth (ft)	Unit weight of soil (lb/ft²)	N_{60}
10	106	7
15	106	9
20	106	11
25	118	16
30	118	18
35	118	20
40	118	22

Using Eq. (3.30), determine the variation of the peak soil friction angle, ϕ'. Estimate an average value of ϕ' for the design of a shallow foundation. (*Note:* For depth greater than 20 ft, the unit weight of soil is 118 lb/ft³.)

3.13 Refer to Problem 3.12. Assume that the sand is clean and normally consolidated. Estimate the average value of the modulus of elasticity between depths of 20 ft and 30 ft.

3.14 Following are the details for a soil deposit in sand:

Depth (m)	Effective overburden pressure (kN/m²)	Field standard penetration number, N_{60}
3.0	55	9
4.5	82	11
6.0	98	12

Assume the uniformity coefficient (C_u) of the sand to be 2.8 and an overconsolidation ratio (OCR) of 2. Estimate the average relative density of the sand between the depth of 3 to 6 m. Use Eq. (3.21).

3.15 Refer to Figure P3.2. Vane shear tests were conducted in the clay layer. The vane (tapered) dimensions were 63.5 mm $(d) \times$ 127 mm (h), $i_B = i_T = 45°$ (see Figure 3.23). For the test at A, the torque required to cause failure was 51 N $\cdot$ m. For the clay, given: liquid limit $= 46$ and plastic limit $= 21$. Estimate the undrained cohesion of the clay for use in the design by using Bjerrum's λ relationship [Eq. (3.40a)].

3.16 Refer to Problem 3.15. Estimate the overconsolidation ratio of the clay. Use Eqs. (3.42) and (3.43).

3.17 a. A vane shear test was conducted in a saturated clay. The height and diameter of the rectangular vane were 4 in. and 2 in., respectively. During the test, the maximum torque applied was 230 lb-in. Determine the undrained shear strength of the clay.

   b. The clay soil described in part (a) has a liquid limit of 58 and a plastic limit of 29. What would be the corrected undrained shear strength of the clay for design purposes? Use Bjerrum's relationship for λ [Eq. (3.40a)].

3.18 Refer to Problem 3.17. Determine the overconsolidation ratio for the clay. Use Eqs. (3.42) and (3.45). Use $\sigma'_0 = 1340$ lb/ft^2.

3.19 In a deposit of normally consolidated dry sand, a cone penetration test was conducted. Following are the results:

Depth (m)	Point resistance of cone, q_c (MN/m^2)
1.5	2.06
3.0	4.23
4.5	6.01
6.0	8.18
7.5	9.97
9.0	12.42

Assuming the dry unit weight of sand to be 16 kN/m^3, estimate the average peak friction angle, ϕ', of the sand. Use Eq. (3.53).

3.20 Refer to Problem 3.19. Using Eq. (3.51), determine the variation of the relative density with depth. Use $Q_c = 1$.

3.21 Refer to Problem 3.19. Use Eq. (3.55) and Kulhawy and Mayne factors for a and c to predict the variation of N_{60} with depth. Given:mean grain size $D_{50} = 0.2$ mm.

3.22 In the soil profile shown in Figure P3.22, if the cone penetration resistance (q_c) at A (as determined by an electric friction-cone penetrometer) is 0.8 MN/m.^{2}, estimate

   a. The undrained cohesion, c_u

   b. The overconsolidation ratio, OCR

3.23 In a pressuremeter test in a soft saturated clay, the measuring cell volume $V_o = 535$ cm^3, $p_o = 42.4$ kN/m^2, $p_f = 326.5$ kN/m^2, $v_o = 46$ cm^3, and $v_f = 180$ cm^3. Assuming Poisson's ratio (μ_s) to be 0.5 and using Figure 3.31, calculate the pressuremeter modulus (E_p).

3.24 A dilatometer test was conducted in a clay deposit. The groundwater table was located at a depth of 3 m below the surface. At a depth of 8 m below the surface, the contact

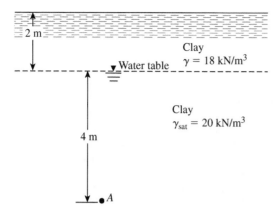

Figure P3.22

pressure (p_o) was 280 kN/m² and the expansion stress (p_1) was 350 kN/m². Determine the following:

a. Coefficient of at-rest earth pressure, K_o
b. Overconsolidation ratio, OCR
c. Modulus of elasticity, E_s
Assume σ_o' at a depth of 8 m to be 95 kN/m² and $\mu_s = 0.35$.

3.25 A dilatometer test was conducted in a sand deposit at a depth of 6 m. The ground-water table was located at a depth of 2 m below the ground surface. Given, for the sand: $\gamma_d = 14.5$ kN/m³ and $\gamma_{sat} = 19.8$ kN/m³. The contact stress during the test was 260 kN/m². Estimate the soil friction angle, ϕ'.

3.26 The P-wave velocity in a soil is 105 m/sec. Assuming Poisson's ratio to be 0.32, calculate the modulus of elasticity of the soil. Assume that the unit weight of soil is 18 kN/m³.

3.27 The results of a refraction survey (Figure 3.45a) at a site are given in the following table. Determine the thickness and the P-wave velocity of the materials encountered.

Distance from the source of disturbance (m)	Time of first arrival of P-waves (sec $\times$ 10³)
2.5	5.08
5.0	10.16
7.5	15.24
10.0	17.01
15.0	20.02
20.0	24.2
25.0	27.1
30.0	28.0
40.0	31.1
50.0	33.9

References

AMERICAN SOCIETY FOR TESTING AND MATERIALS (2001). *Annual Book of ASTM Standards,* Vol. 04.08, West Conshohocken, PA.

AMERICAN SOCIETY FOR TESTING AND MATERIALS (2014). *Annual book of ASTM Standards,* Vol. 04.08, West Conshohocken, PA.

AMERICAN SOCIETY OF CIVIL ENGINEERS (1972). "Subsurface Investigation for Design and Construction of Foundations of Buildings," *Journal of the Soil Mechanics and Foundations Division,* American Society of Civil Engineers, Vol. 98, No. SM5, pp. 481–490.

ANAGNOSTOPOULOS, A., KOUKIS, G., SABATAKAKIS, N., and TSIAMBAOS, G. (2003). "Empirical Correlations of Soil Parameters Based on Cone Penetration Tests (CPT) for Greek Soils," *Geotechnical and Geological Engineering,* Vol. 21, No. 4, pp. 377–387.

BAGUELIN, F., JÉZÉQUEL, J. F., and SHIELDS, D. H. (1978). *The Pressuremeter and Foundation Engineering,* Trans Tech Publications, Clausthal, Germany.

BALDI, G., BELLOTTI, R., GHIONNA, V., and JAMIOLKOWSKI, M. (1982). "Design Parameters for Sands from CPT," *Proceedings, Second European Symposium on Penetration Testing,* Amsterdam, Vol. 2, pp. 425–438.

BAZARAA, A. (1967). *Use of the Standard Penetration Test for Estimating Settlements of Shallow Foundations on Sand,* Ph.D. Dissertation, Civil Engineering Department, University of Illinois, Champaign-Urbana, Illinois.

BJERRUM, L. (1972). "Embankments on Soft Ground," *Proceedings of the Specialty Conference,* American Society of Civil Engineers, Vol. 2, pp. 1–54.

BURLAND, J. B. AND BURBIDGE, M. C. (1985). "Settlement of Foundations on Sand and Gravel," *Proceedings, Institute of Civil Engineers,* Part I, Vol. 7, pp. 1325–1381.

CRUDEN, D. M. and VARNES, D. J. (1996). "Landslide Types and Processes," *Special Report 247,* Transportation Research Board, pp. 36–75.

CUBRINOVSKI, M. and ISHIHARA, K. (1999). "Empirical Correlations between SPT *N*-Values and Relative Density for Sandy Soils," *Soils and Foundations,* Vol. 39, No. 5, pp. 61–92.

DAS, B. M. (1992). *Principles of Soil Dynamics,* PWS Publishing Company, Boston.

DEERE, D. U. (1963). "Technical Description of Rock Cores for Engineering Purposes," *Felsmechanik und Ingenieurgeologie,* Vol. 1, No. 1, pp. 16–22.

DOBRIN, M. B. (1960). *Introduction to Geophysical Prospecting,* McGraw-Hill, New York.

HANSBO, S. (1957). *A New Approach to the Determination of the Shear Strength of Clay by the Fall Cone Test,* Swedish Geotechnical Institute, Report No. 114.

HARA, A., OHATA, T., and NIWA, M. (1971). "Shear Modulus and Shear Strength of Cohesive Soils," *Soils and Foundations,* Vol. 14, No. 3, pp. 1–12.

HATANAKA, M. and UCHIDA, A. (1996). "Empirical Correlation between Penetration Resistance and Internal Friction Angle of Sandy Soils," *Soils and Foundations,* Vol. 36, No. 4, pp. 1–10.

JAMIOLKOWSKI, M., LADD, C. C., GERMAINE, J. T., and LANCELLOTTA, R. (1985). "New Developments in Field and Laboratory Testing of Soils," *Proceedings, 11th International Conference on Soil Mechanics and Foundation Engineering,* Vol. 1, pp. 57–153.

KAMEI, T. and IWASAKI, K. (1995). "Evaluation of Undrained Shear Strength of Cohesive Soils using a Flat Dilatometer," *Soils and Foundations,* Vol. 35, No. 2, pp. 111–116.

KOLB, C. R. and SHOCKLEY, W. G. (1959). "Mississippi Valley Geology: Its Engineering Significance" *Proceedings,* American Society of Civil Engineers, Vol. 124, pp. 633–656.

KULHAWY, F. H. and MAYNE, P. W. (1990). *Manual on Estimating Soil Properties for Foundation Design,* Electric Power Research Institute, Palo Alto, California.

LANCELLOTTA, R. (1983). *Analisi di Affidabilità in Ingegneria Geotecnica,* Atti Istituto Scienza Construzioni, No. 625, Politecnico di Torino.

LARSSON, R. (1980). "Undrained Shear Strength in Stability Calculation of Embankments and Foundations on Clay," *Canadian Geotechnical Journal,* Vol. 17, pp. 591–602.

LEE, J., SALGADO, R. and CARRARO, A. H. (2004). "Stiffness Degradation and Shear Strength of Silty Sand," *Canadian Geotechnical Journal*, Vol. 41, No. 5, pp. 831–843.

LIAO, S. S. C. AND WHITMAN, R. V. (1986). "Overburden Correction Factors for SPT in Sand," *Journal of Geotechnical Engineering,* American Society of Civil Engineers, Vol. 112, No. 3, pp. 373–377.

MARCHETTI, S. (1980). "*In Situ* Test by Flat Dilatometer," *Journal of Geotechnical Engineering Division,* ASCE, Vol. 106, GT3, pp. 299–321.

MARCUSON, W. F. III, AND BIEGANOUSKY, W. A. (1977). "SPT and Relative Density in Coarse Sands," *Journal of Geotechnical Engineering Division,* American Society of Civil Engineers, Vol. 103, No. 11, pp. 1295–1309.

MAYNE, P. W. AND KEMPER, J. B. (1988). "Profiling OCR in Stiff Clays by CPT and SPT," *Geotechnical Testing Journal,* ASTM, Vol. 11, No. 2, pp. 139–147.

MAYNE, P. W. AND MITCHELL, J. K. (1988). "Profiling of Overconsolidation Ratio in Clays by Field Vane," *Canadian Geotechnical Journal,* Vol. 25, No. 1, pp. 150–158.

MENARD, L. (1956). *An Apparatus for Measuring the Strength of Soils in Place,* master's thesis, University of Illinois, Urbana, Illinois.

MEYERHOF, G. G. (1957). "Discussion on Research on Determining the Density of Sands by Spoon Penetration Testing," *Proceedings, Fourth International Conference on Soil Mechanics and Foundation Engineering,* Vol. 3, p. 110.

MORRIS, P. M. and WILLIAMS, D. T. (1994). "Effective Stress Vane Shear Strength Correction Factor Correlations," *Canadian Geotechnical Journal,* Vol. 31, No. 3, pp. 335–342.

OHYA, S., IMAI, T., AND MATSUBARA, M. (1982). "Relationships between *N* Value by SPT and LLT Pressuremeter Results," *Proceedings, 2nd European Symposium on Penetration Testing,* Vol. 1, Amsterdam, pp. 125–130.

OSTERBERG, J. O. (1952). "New Piston-Type Soil Sampler," *Engineering News-Record,* April 24.

PECK, R. B., HANSON, W. E., and THORNBURN, T. H. (1974). *Foundation Engineering,* 2d ed., Wiley, New York.

RICCERI, G., SIMONINI, P., and COLA, S. (2002). "Applicability of Piezocone and Dilatometer to Characterize the Soils of the Venice Lagoon" *Geotechnical and Geological Engineering,* Vol. 20, No. 2, pp. 89–121.

ROBERTSON, P. K. and CAMPANELLA, R. G. (1983). "Interpretation of Cone Penetration Tests. Part I: Sand," *Canadian Geotechnical Journal,* Vol. 20, No. 4, pp. 718–733.

SCHMERTMANN, J. H. (1975). "Measurement of *In Situ* Shear Strength," *Proceedings, Specialty Conference on* In Situ *Measurement of Soil Properties,* ASCE, Vol. 2, pp. 57–138.

SCHMERTMANN, J. H. (1986). "Suggested Method for Performing the Flat Dilatometer Test," *Geotechnical Testing Journal,* ASTM, Vol. 9, No. 2, pp. 93–101.

SEED, H. B., ARANGO, I., AND CHAN, C. K. (1975). *Evaluation of Soil Liquefaction Potential during Earthquakes,* Report No. EERC 75-28, Earthquake Engineering Research Center, University of California, Berkeley.

SEED, H. B., TOKIMATSU, K., HARDER, L. F., and CHUNG, R. M. (1985). "Influence of SPT Procedures in Soil Liquefaction Resistance Evaluations," *Journal of Geotechnical Engineering,* ASCE, Vol. 111, No. 12, pp. 1425–1445.

SHIBUYA, S. and HANH, L. T. (2001). "Estimating Undrained Shear Strength of Soft Clay Ground Improved by Pre-Loading with PVD—Case History in Bangkok," *Soils and Foundations,* Vol. 41, No. 4, pp. 95–101.

SIVAKUGAN, N. and DAS, B. (2010). *Geotechnical Engineering—A Practical Problem Solving Approach,* J. Ross Publishing Co., Fort Lauderdale, FL.

SKEMPTON, A. W. (1986). "Standard Penetration Test Procedures and the Effect in Sands of Overburden Pressure, Relative Density, Particle Size, Aging and Overconsolidation," *Geotechnique,* Vol. 36, No. 3, pp. 425–447.

SOWERS, G. B. and SOWERS, G. F. (1970). *Introductory Soil Mechanics and Foundations,* 3d ed., Macmillan, New York.

STOKOE, K. H. and WOODS, R. D. (1972). "*In Situ* Shear Wave Velocity by Cross-Hole Method," *Journal of Soil Mechanics and Foundations Division,* American Society of Civil Engineers, Vol. 98, No. SM5, pp. 443–460.

SZECHY, K. and VARGA, L. (1978). *Foundation Engineering—Soil Exploration and Spread Foundation,* Akademiai Kiado, Budapest, Hungary.

WOLFF, T. F. (1989). "Pile Capacity Prediction Using Parameter Functions," in *Predicted and Observed Axial Behavior of Piles, Results of a Pile Prediction Symposium,* sponsored by the Geotechnical Engineering Division, ASCE, Evanston, IL, June, 1989, ASCE Geotechnical Special Publication No. 23, pp. 96–106.

CHAPTER 2

Soil Compaction

6.1 Introduction

In the construction of highway embankments, earth dams, and many other engineering structures, loose soils must be compacted to increase their unit weights. Compaction increases the strength characteristics of soils, which increase the bearing capacity of foundations constructed over them. Compaction also decreases the amount of undesirable settlement of structures and increases the stability of slopes of embankments. Smooth-wheel rollers, sheepsfoot rollers, rubber-tired rollers, and vibratory rollers are generally used in the field for soil compaction. Vibratory rollers are used mostly for the densification of granular soils. Vibroflot devices are also used for compacting granular soil deposits to a considerable depth. Compaction of soil in this manner is known as *vibroflotation*. This chapter discusses in some detail the principles of soil compaction in the laboratory and in the field.

This chapter includes elaboration of the following:

- Laboratory compaction test methods
- Factors affecting compaction in general
- Empirical relationships related to compaction
- Structure and properties of compacted cohesive soils
- Field compaction
- Tests for quality control of field compaction
- Special compaction techniques in the field

6.2 Compaction—General Principles

Compaction, in general, is the densification of soil by removal of air, which requires mechanical energy. The degree of compaction of a soil is measured in terms of its dry unit weight. When water is added to the soil during compaction, it acts as a softening agent on the soil particles. The soil particles slip over each other and move into a densely packed position. The dry unit weight after compaction first increases as the moisture content

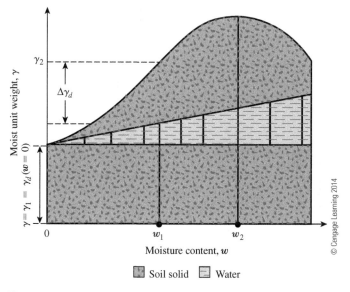

Figure 6.1 Principles of compaction

increases. (See Figure 6.1.) Note that at a moisture content $w = 0$, the moist unit weight (γ) is equal to the dry unit weight (γ_d), or

$$\gamma = \gamma_{d(w=0)} = \gamma_1$$

When the moisture content is gradually increased and the same compactive effort is used for compaction, the weight of the soil solids in a unit volume gradually increases. For example, at $w = w_1$,

$$\gamma = \gamma_2$$

However, the dry unit weight at this moisture content is given by

$$\gamma_{d(w=w_1)} = \gamma_{d(w=0)} + \Delta\gamma_d$$

Beyond a certain moisture content $w = w_2$ (Figure 6.1), any increase in the moisture content tends to reduce the dry unit weight. This phenomenon occurs because the water takes up the spaces that would have been occupied by the solid particles. The moisture content at which the maximum dry unit weight is attained is generally referred to as the *optimum moisture content.*

The laboratory test generally used to obtain the maximum dry unit weight of compaction and the optimum moisture content is called the *Proctor compaction test* (Proctor, 1933). The procedure for conducting this type of test is described in the following section.

6.3 Standard Proctor Test

In the Proctor test, the soil is compacted in a mold that has a volume of 944 cm³ ($\frac{1}{30}$ ft³). The diameter of the mold is 101.6 mm (4 in.). During the laboratory test, the mold is attached to a baseplate at the bottom and to an extension at the top (Figure 6.2a). The soil is mixed with varying amounts of water and then compacted in three equal layers by a hammer (Figure 6.2b) that delivers 25 blows to each layer. The hammer has a mass of

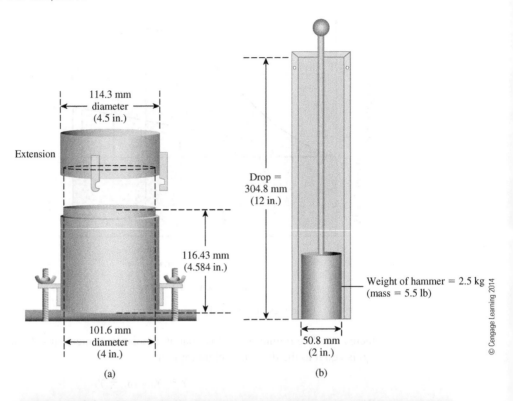

Figure 6.2 Standard Proctor test equipment: (a) mold; (b) hammer; (c) photograph of laboratory equipment used for test (*Courtesy of Braja M. Das, Henderson, Nevada*)

2.5 kg (6.5 lb) and has a drop of 30.5 mm (12 in.). Figure 6.2c is a photograph of the laboratory equipment required for conducting a standard Proctor test.

For each test, the moist unit weight of compaction, γ, can be calculated as

$$\gamma = \frac{W}{V_m} \tag{6.1}$$

where W = weight of the compacted soil in the mold

V_m = volume of the mold $\left[944 \text{ cm}^3 \left(\frac{1}{30} \text{ ft}^3 \right) \right]$

For each test, the moisture content of the compacted soil is determined in the laboratory. With the known moisture content, the dry unit weight can be calculated as

$$\gamma_d = \frac{\gamma}{1 + \dfrac{w\,(\%)}{100}} \tag{6.2}$$

where $w\,(\%)$ = percentage of moisture content.

The values of γ_d determined from Eq. (6.2) can be plotted against the corresponding moisture contents to obtain the maximum dry unit weight and the optimum moisture content for the soil. Figure 6.3 shows such a plot for a silty-clay soil.

The procedure for the standard Proctor test is elaborated in ASTM Test Designation D-698 (ASTM, 2010) and AASHTO Test Designation T-99 (AASHTO, 1982).

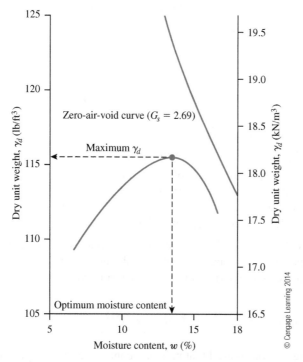

Figure 6.3 Standard Proctor compaction test results for a silty clay

For a given *moisture content w* and *degree of saturation S,* the dry unit weight of compaction can be calculated as follows. From Chapter 3 [Eq. (3.17)], for any soil,

$$\gamma_d = \frac{G_s \gamma_w}{1 + e}$$

where G_s = specific gravity of soil solids
γ_w = unit weight of water
e = void ratio

and, from Eq. (3.19),

$$Se = G_s w$$

or

$$e = \frac{G_s w}{S}$$

Thus,

$$\gamma_d = \frac{G_s \gamma_w}{1 + \dfrac{G_s w}{S}} \tag{6.3}$$

For a given moisture content, the theoretical maximum dry unit weight is obtained when no air is in the void spaces—that is, when the degree of saturation equals 100%. Hence, the maximum dry unit weight at a given moisture content with zero air voids can be obtained by substituting $S = 1$ into Eq. (6.3), or

$$\gamma_{zav} = \frac{G_s \gamma_w}{1 + w G_s} = \frac{\gamma_w}{w + \dfrac{1}{G_s}} \tag{6.4}$$

where γ_{zav} = zero-air-void unit weight.

To obtain the variation of γ_{zav} with moisture content, use the following procedure:

1. Determine the specific gravity of soil solids.
2. Know the unit weight of water (γ_w).
3. Assume several values of w, such as 5%, 10%, 15%, and so on.
4. Use Eq. (6.4) to calculate γ_{zav} for various values of w.

Figure 6.3 also shows the variation of γ_{zav} with moisture content and its relative location with respect to the compaction curve. Under no circumstances should any part of the compaction curve lie to the right of the zero-air-void curve.

6.4 Factors Affecting Compaction

The preceding section showed that moisture content has a strong influence on the degree of compaction achieved by a given soil. Besides moisture content, other important factors that affect compaction are soil type and compaction effort (energy per unit volume).

The importance of each of these two factors is described in more detail in the following two sections.

Effect of Soil Type

The soil type—that is, grain-size distribution, shape of the soil grains, specific gravity of soil solids, and amount and type of clay minerals present—has a great influence on the maximum dry unit weight and optimum moisture content. Figure 6.4 shows typical compaction curves obtained from four soils. The laboratory tests were conducted in accordance with ASTM Test Designation D-698.

Note also that the bell-shaped compaction curve shown in Figure 6.3 is typical of most clayey soils. Figure 6.4 shows that for sands, the dry unit weight has a general tendency first to decrease as moisture content increases and then to increase to a maximum value with further increase of moisture. The initial decrease of dry unit weight with increase of moisture content can be attributed to the capillary tension effect. At lower moisture contents, the capillary tension in the pore water inhibits the tendency of the soil particles to move around and be compacted densely.

Lee and Suedkamp (1972) studied compaction curves for 35 soil samples. They observed that four types of compaction curves can be found. These curves are shown in Figure 6.5. The following table is a summary of the type of compaction curves encountered in various soils with reference to Figure 6.5.

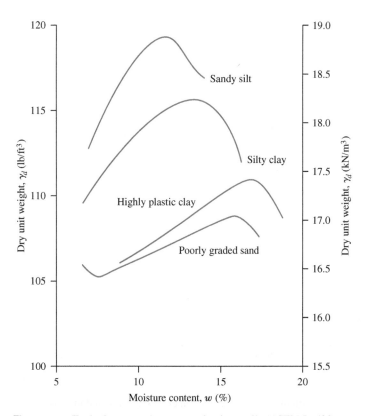

Figure 6.4 Typical compaction curves for four soils (*ASTM D-698*)

Type of compaction curve (Figure 6.5)	Description of curve	Liquid limit
A	Bell shaped	Between 30 to 70
B	1-1/2 peak	Less than 30
C	Double peak	Less than 30 and those greater than 70
D	Odd shaped	Greater than 70

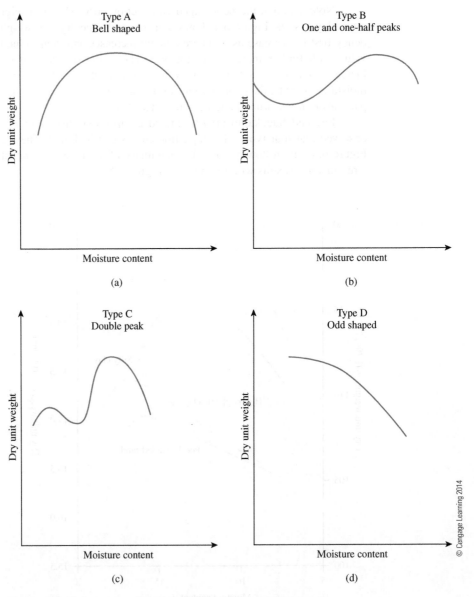

Figure 6.5 Various types of compaction curves encountered in soils

Effect of Compaction Effort

The compaction energy per unit volume used for the standard Proctor test described in Section 6.3 can be given as

$$E = \frac{\left(\begin{array}{c}\text{Number}\\\text{of blows}\\\text{per layer}\end{array}\right) \times \left(\begin{array}{c}\text{Number}\\\text{of}\\\text{layers}\end{array}\right) \times \left(\begin{array}{c}\text{Weight}\\\text{of}\\\text{hammer}\end{array}\right) \times \left(\begin{array}{c}\text{Height of}\\\text{drop of}\\\text{hammer}\end{array}\right)}{\text{Volume of mold}} \quad (6.5)$$

or, in SI units,

$$E = \frac{(25)(3)\left(\dfrac{2.5 \times 9.81}{1000}\,\text{kN}\right)(0.305\ \text{m})}{944 \times 10^{-6}\ \text{m}^3} = 594\ \text{kN-m/m}^3 \approx 600\ \text{kN-m/m}^3$$

In English units,

$$E = \frac{(25)(3)(5.5)(1)}{\left(\dfrac{1}{30}\right)} = 12{,}375\ \text{ft-lb/ft}^3 \approx 12{,}400\ \text{ft-lb/ft}^3$$

If the compaction effort per unit volume of soil is changed, the moisture–unit weight curve also changes. This fact can be demonstrated with the aid of Figure 6.6, which shows

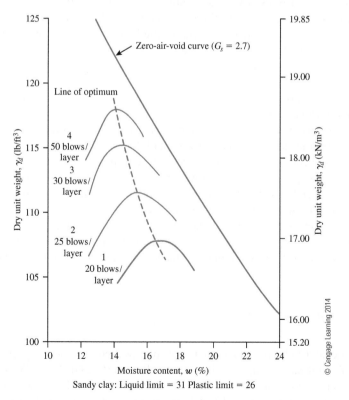

Figure 6.6 Effect of compaction energy on the compaction of a sandy clay

four compaction curves for a sandy clay. The standard Proctor mold and hammer were used to obtain these compaction curves. The number of layers of soil used for compaction was three for all cases. However, the number of hammer blows per each layer varied from 20 to 50, which varied the energy per unit volume.

From the preceding observation and Figure 6.6, we can see that

1. As the compaction effort is increased, the maximum dry unit weight of compaction is also increased.
2. As the compaction effort is increased, the optimum moisture content is decreased to some extent.

The preceding statements are true for all soils. Note, however, that the degree of compaction is not directly proportional to the compaction effort.

6.5 Modified Proctor Test

With the development of heavy rollers and their use in field compaction, the standard Proctor test was modified to better represent field conditions. This revised version sometimes is referred to as the *modified Proctor test* (ASTM Test Designation D-1557 and AASHTO Test Designation T-180). For conducting the modified Proctor test, the same mold is used with a volume of 944 cm^3 $\left(\frac{1}{30}\text{ ft}^3\right)$, as in the case of the standard Proctor test. However, the soil is compacted in five layers by a hammer that has a mass of 4.54 kg (10 lb). The drop of the hammer is 457 mm (18 in.). The number of hammer blows for each layer is kept at 25 as in the case of the standard Proctor test. Figure 6.7 shows a comparison between the hammers used in standard and modified Proctor tests.

The compaction energy for this type of compaction test can be calculated as 2700 kN-m/m^3 (56,000 ft-lb/lb^3).

Because it increases the compactive effort, the modified Proctor test results in an increase in the maximum dry unit weight of the soil. The increase in the maximum dry unit weight is accompanied by a decrease in the optimum moisture content.

In the preceding discussions, the specifications given for Proctor tests adopted by ASTM and AASHTO regarding the volume of the mold and the number of blows are generally those adopted for fine-grained soils that pass through the U.S. No. 4 sieve. However, under each test designation, there are three suggested methods that reflect the mold size, the number of blows per layer, and the maximum particle size in a soil aggregate used for testing. A summary of the test methods is given in Table 6.1.

6.6 Empirical Relationships

Omar et al. (2003) presented the results of modified Proctor compaction tests on 311 soil samples. Of these samples, 45 were gravelly soil (GP, GP-GM, GW, GW-GM, and GM), 264 were sandy soil (SP, SP-SM, SW-SM, SW, SC-SM, SC, and SM), and two were clay with low plasticity (CL). All compaction tests were conducted using ASTM 1557

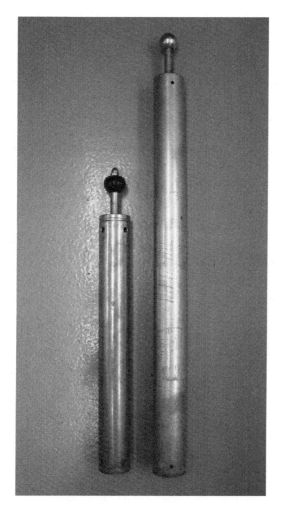

Figure 6.7 Comparison between standard Proctor hammer (left) and modified Proctor hammer (right) (*Courtesy of Braja M. Das, Henderson, Nevada*)

method C to avoid over-size correction. Based on the tests, the following correlations were developed.

$$\rho_{d(\max)} \ (\text{kg/m}^3) = [4{,}804{,}574G_s - 195.55(LL)^2 + 156{,}971 \ (\text{R\#4})^{0.5}$$
$$- \ 9{,}527{,}830]^{0.5} \tag{6.6}$$

$$\ln(w_{\text{opt}}) = 1.195 \times 10^{-4} \ (LL)^2 - 1.964G_s - 6.617 \times 10^{-5} \ (\text{R\#4})$$
$$+ \ 7.651 \tag{6.7}$$

where $\rho_{d(\max)}$ = maximum dry density (kg/m^3)
$\quad\quad\ w_{\text{opt}}$ = optimum moisture content(%)
$\quad\quad\ G_s$ = specific gravity of soil solids
$\quad\quad\ LL$ = liquid limit, in percent
$\quad\quad$ R#4 = percent retained on No. 4 sieve

Table 6.1 Summary of Standard and Modified Proctor Compaction Test Specifications (*ASTM D-698 and D-1557*)

	Description	Method A	Method B	Method C
Physical data for the tests	Material	Passing No. 4 sieve	Passing 9.5 mm ($\frac{3}{8}$ in.) sieve	Passing 19 mm ($\frac{3}{4}$ in.) sieve
	Use	Used if 20% or less by weight of material is retained on No. 4 (4.75 mm) sieve	Used if more than 20% by weight of material is retained on No. 4 (4.75 mm) sieve and 20% or less by weight of material is retained on 9.5 mm ($\frac{3}{8}$ in.) sieve	Used if more than 20% by weight of material is retained on 9.5 mm ($\frac{3}{8}$ in.) sieve and less than 30% by weight of material is retained on 19 mm ($\frac{3}{4}$ in.) sieve
	Mold volume	944 cm^3 ($\frac{1}{30}$ ft^3)	944 cm^3 ($\frac{1}{30}$ ft^3)	2124 cm^3 ($\frac{1}{13.33}$ ft^3)
	Mold diameter	101.6 mm (4 in.)	101.6 mm (4 in.)	152.4 mm (6 in.)
	Mold height	116.4 mm (4.584 in.)	116.4 mm (4.584 in.)	116.4 mm (4.584 in.)
Standard Proctor test	Weight of hammer	24.4 N (5.5 lb)	24.4 N (5.5 lb)	24.4 N (5.5 lb)
	Height of drop	305 mm (12 in.)	305 mm (12 in.)	305 mm (12 in.)
	Number of soil layers	3	3	3
	Number of blows/layer	25	25	56
Modified Proctor test	Weight of hammer	44.5 N (10 lb)	44.5 N (10 lb)	44.5 N (10 lb)
	Height of drop	457 mm (18 in.)	457 mm (18 in.)	457 mm (18 in.)
	Number of soil layers	5	5	5
	Number of blows/layer	25	25	56

For granular soils with less than 12% fines (i.e., finer than No. 200 sieve), relative density may be a better indicator for end product compaction specification in the field. Based on laboratory compaction tests on 55 clean sands (less than 5% finer than No. 200 sieve), Patra et al. (2010) provided the following relationships

$$D_r = AD_{50}^{-B} \tag{6.8}$$

$$A = 0.216 \ln E - 0.850 \tag{6.9}$$

$$B = -0.03 \ln E + 0.306 \tag{6.10}$$

where D_r = maximum relative density of compaction achieved with compaction energy E (kN-m/m^3)

D_{50} = median grain size (mm)

Gurtug and Sridharan (2004) proposed correlations for optimum moisture content and maximum dry unit weight with the plastic limit (*PL*) of cohesive soils. These correlations can be expressed as:

$$w_{opt}(\%) = [1.95 - 0.38(\log E)](PL) \tag{6.11}$$

$$\gamma_{d(max)}(kN/m^3) = 22.68e^{-0.0183w_{opt}(\%)} \tag{6.12}$$

where *PL* = plastic limit (%)

E = compaction energy (kN-m/m^3)

For modified Proctor test, $E = 2700$ kN/m^3. Hence,

$$w_{opt}(\%) \approx 0.65 \, (PL)$$

and

$$\gamma_{d(max)}(kN/m^3) = 22.68e^{-0.012(PL)}$$

Osman et al. (2008) analyzed a number of laboratory compaction test results on fine-grained (cohesive) soil, including those provided by Gurtug and Sridharan (2004). Based on this study, the following correlations were developed:

$$w_{opt}(\%) = (1.99 - 0.165 \ln E)(PI) \tag{6.13}$$

and

$$\gamma_{d(max)}(kN/m^3) = L - Mw_{opt} \tag{6.14}$$

where

$$L = 14.34 + 1.195 \ln E \tag{6.15}$$

$$M = -0.19 + 0.073 \ln E \tag{6.16}$$

w_{opt} = optimum water content (%)

PI = plasticity index (%)

$\gamma_{d(max)}$ = maximum dry unit weight (kN/m^3)

E = compaction energy (kN-m/m^3)

Matteo et al. (2009) analyzed the results of 71 fine-grained soils and provided the following correlations for optimum water content (w_{opt}) and maximum dry unit weight [$\gamma_{d(max)}$] for modified Proctor tests ($E = 2700$ kN-m/m^3):

$$w_{opt}(\%) = -0.86(LL) + 3.04\left(\frac{LL}{G_s}\right) + 2.2 \tag{6.17}$$

and

$$\gamma_{d(max)}(kN/m^3) = 40.316(w_{opt}^{-0.295})(PI^{0.032}) - 2.4 \tag{6.18}$$

where *LL* = liquid limit (%)

PI = plasticity index (%)

G_s = specific gravity of soil solids

Example 6.1

The laboratory test results of a standard Proctor test are given in the following table.

Volume of mold (ft³)	Weight of moist soil in mold (lb)	Moisture content, w (%)
$\frac{1}{30}$	3.78	10
$\frac{1}{30}$	4.01	12
$\frac{1}{30}$	4.14	14
$\frac{1}{30}$	4.12	16
$\frac{1}{30}$	4.01	18
$\frac{1}{30}$	3.90	20

© Cengage Learning 2014

a. Determine the maximum dry unit weight of compaction and the optimum moisture content.
b. Calculate and plot γ_d versus the moisture content for degree of saturation, $S = 80$, 90, and 100% (i.e., γ_{zav}). Given: $G_s = 2.7$.

Solution
Part a

The following table can be prepared.

Volume of mold V_m (ft³)	Weight of soil, W (lb)	Moist unit weight, γ (lb/ft³)[a]	Moisture content, w (%)	Dry unit weight, γ_d (lb/ft³)[b]
$\frac{1}{30}$	3.78	113.4	10	103.1
$\frac{1}{30}$	4.01	120.3	12	107.4
$\frac{1}{30}$	4.14	124.2	14	108.9
$\frac{1}{30}$	4.12	123.6	16	106.6
$\frac{1}{30}$	4.01	120.3	18	101.9
$\frac{1}{30}$	3.90	117.0	20	97.5

© Cengage Learning 2014

$$^a\gamma = \frac{W}{V_m}$$

$$^b\gamma_d = \frac{\gamma}{1 + \dfrac{w\%}{100}}$$

The plot of γ_d versus w is shown at the bottom of Figure 6.8. From the plot, we see that the maximum dry unit weight $\gamma_{d(max)} = $ **109 lb/ft³** and the optimum moisture content is **14.4%**.

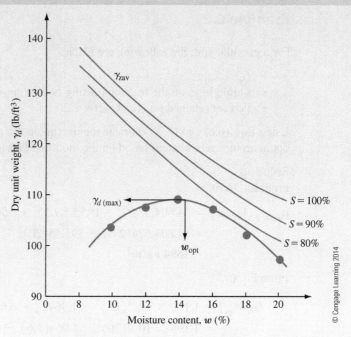

Figure 6.8 Moisture content–unit weight curves

Part b

From Eq. (6.3),

$$\gamma_d = \frac{G_s \gamma_w}{1 + \dfrac{G_s w}{S}}$$

The following table can be prepared.

		γ_d (lb/ft³)		
G_s	w (%)	$S = 80\%$	$S = 90\%$	$S = 100\%$
2.7	8	132.7	135.9	138.6
2.7	10	126.0	129.6	132.7
2.7	12	119.9	123.9	127.3
2.7	14	114.4	118.6	122.3
2.7	16	109.4	113.8	117.7
2.7	18	104.8	109.4	113.4
2.7	20	100.6	105.3	109.4

The plot of γ_d versus w for the various degrees of saturation is also shown in Figure 6.8.

Example 6.2

For a granular soil, the following are given:

- $G_s = 2.6$
- Liquid limit on the fraction passing No. 40 sieve $= 20$
- Percent retained on No. 4 sieve $= 20$

Using Eqs. (6.6) and (6.7), estimate the maximum dry density of compaction and the optimum moisture content based on the modified Proctor test.

Solution

From Eq. (6.6),

$$
\begin{aligned}
\rho_{d(\max)} \ (\text{kg/m}^3) &= [4{,}804{,}574G_s - 195.55(LL)^2 + 156{,}971(R\#4)^{0.5} - 9{,}527{,}830]^{0.5} \\
&= [4{,}804{,}574(2.6) - 195.55(20)^2 + 156{,}971(20)^{0.5} - 9{,}527{,}830]^{0.5} \\
&= \mathbf{1894 \ kg/m^3}
\end{aligned}
$$

From Eq. (6.7),

$$
\begin{aligned}
\ln(w_{\text{opt}}) &= 1.195 \times 10^{-4}(LL)^2 - 1.964G_s - 6.617 \times 10^{-5}(R\#4) + 7{,}651 \\
&= 1.195 \times 10^{-4}(20)^2 - 1.964(2.6) - 6.617 \times 10^{-5}(20) + 7{,}651 \\
&= 2.591 \\
w_{\text{opt}} &= \mathbf{13.35\%}
\end{aligned}
$$

Example 6.3

For a sand with 4% finer than No. 200 sieve, estimate the maximum relative density of compaction that may be obtained from a modified Proctor test. Given $D_{50} = 1.4$ mm.

Solution

For modified Proctor test, $E = 2696$ kN-m/m^3.

From Eq. (6.9),

$$
A = 0.216 \ln E - 0.850 = (0.216)(\ln 2696) - 0.850 = 0.856
$$

From Eq. (6.10),

$$
B = -0.03 \ln E + 0.306 = -(0.03)(\ln 2696) + 0.306 = 0.069
$$

From Eq. (6.8),

$$
D_r = AD_{50}^{-B} = (0.856)(1.4)^{-0.069} = 0.836 = \mathbf{83.6\%}
$$

Example 6.4

For a silty clay soil given $LL = 43$ and $PL = 18$. Estimate the maximum dry unit weight of compaction that can be achieved by conducting a modified Proctor test. Use Eq. (6.14).

Solution
For modified Proctor test, $E = 2696$ kN-m/m^3.

From Eqs. (6.15) and (6.16),

$$L = 14.34 + 1.195 \ln E = 14.34 + 1.195 \ln (2696) = 23.78$$
$$M = -0.19 + 0.073 \ln E = -0.19 + 0.073 \ln (2696) = 0.387$$

From Eq. (6.13),

$$w_{opt}(\%) = (1.99 - 0.165 \ln E)(PI)$$
$$= [1.99 - 0.165 \ln (2696)](43 - 18)$$
$$= 17.16\%$$

From Eq. (6.14),

$$\gamma_{d(max)} = L - Mw_{opt} = 23.78 - (0.387)(17.16) = \mathbf{17.14\ kN/m^3}$$

6.7 Structure of Compacted Clay Soil

Lambe (1958a) studied the effect of compaction on the structure of clay soils, and the results of his study are illustrated in Figure 6.9. If clay is compacted with a moisture content on the dry side of the optimum, as represented by point *A,* it will possess a flocculent structure. This type of structure results because, at low moisture content, the diffuse double layers of ions surrounding the clay particles cannot be fully developed; hence, the interparticle repulsion is reduced. This reduced repulsion results in a more random particle orientation and a lower dry unit weight. When the moisture content of compaction is increased, as shown by point *B,* the diffuse double layers around the particles expand, which increases the repulsion between the clay particles and gives a lower degree of flocculation and a higher dry unit weight. A continued increase in moisture content from *B* to *C* expands the double layers more. This expansion results in a continued increase of repulsion between the particles and thus a still greater degree of particle orientation and a more or less dispersed structure. However, the dry unit weight decreases because the added water dilutes the concentration of soil solids per unit volume.

At a given moisture content, higher compactive effort yields a more parallel orientation to the clay particles, which gives a more dispersed structure. The particles are closer and the soil has a higher unit weight of compaction. This phenomenon can be seen by comparing point *A* with point *E* in Figure 6.9.

Figure 6.10 shows the variation in the degree of particle orientation with molding water content for compacted Boston blue clay. Works of Seed and Chan (1959) have shown similar results for compacted kaolin clay.

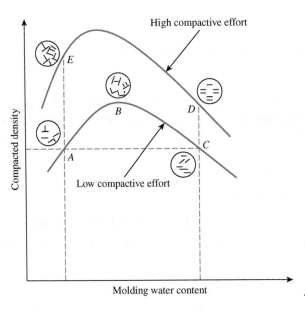

Figure 6.9 Effect of compaction on structure of clay soils (*Redrawn after Lambe, 1958a. With permission from ASCE.*)

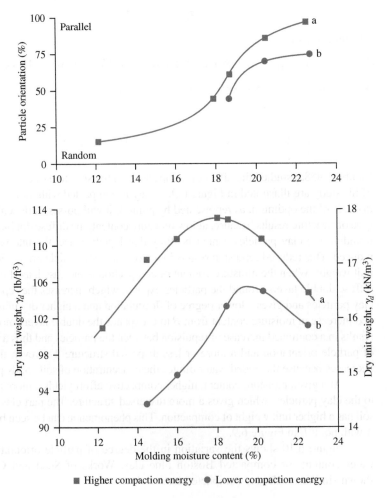

Figure 6.10 Orientation against moisture content for Boston blue clay (*After Lambe, 1958a. With permission from ASCE.*)

6.8 Effect of Compaction on Cohesive Soil Properties

Compaction induces variations in the structure of cohesive soils. Results of these structural variations include changes in hydraulic conductivity, compressibility, and strength. Figure 6.11 shows the results of permeability tests (Chapter 7) on Jamaica sandy clay. The samples used for the tests were compacted at various moisture contents

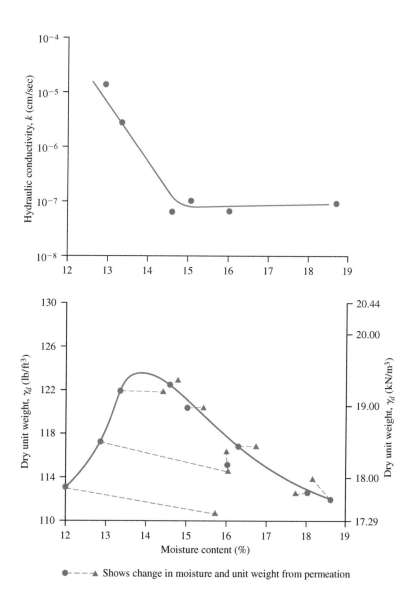

●– – –▲ Shows change in moisture and unit weight from permeation

Figure 6.11 Effect of compaction on hydraulic conductivity of clayey soil (*Redrawn after Lambe, 1958b. With permission from ASCE.*)

by the same compactive effort. The hydraulic conductivity, which is a measure of how easily water flows through soil, decreases with the increase of moisture content. It reaches a minimum value at approximately the optimum moisture content. Beyond the optimum moisture content, the hydraulic conductivity increases slightly. The high value of the hydraulic conductivity on the dry side of the optimum moisture content is due to the random orientation of clay particles that results in larger pore spaces.

One-dimensional compressibility characteristics (Chapter 11) of clay soils compacted on the dry side of the optimum and compacted on the wet side of the optimum are shown in Figure 6.12. Under lower pressure, a soil that is compacted on the wet side of the optimum is more compressible than a soil that is compacted on the dry side of the optimum. This is shown in Figure 6.12a. Under high pressure, the trend is exactly the opposite, and this is shown in Figure 6.12b. For samples compacted on

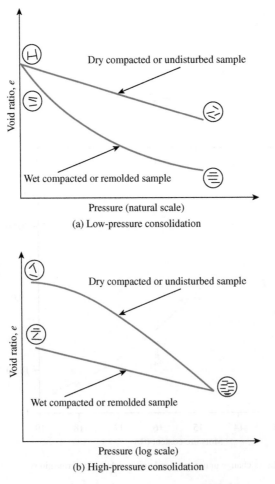

(a) Low-pressure consolidation

(b) High-pressure consolidation

Figure 6.12 Effect of compaction on one-dimensional compressibility of clayey soil (*Redrawn after Lambe, 1958b. With permission from ASCE.*)

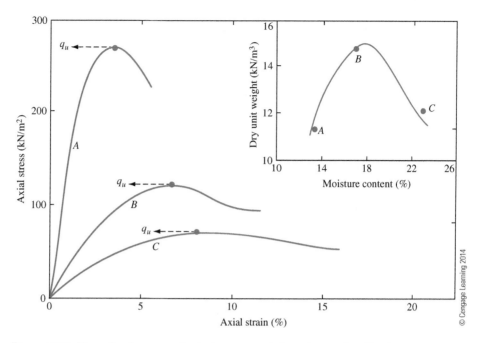

Figure 6.13 Unconfined compression test on compacted specimens of a silty clay

the dry side of the optimum, the pressure tends to orient the particles normal to its direction of application. The space between the clay particles is also reduced at the same time. However, for samples compacted on the wet side of the optimum, pressure merely reduces the space between the clay particles. At very high pressure, it is possible to have identical structures for samples compacted on the dry and wet sides of optimum.

The strength of compacted clayey soils (Chapter 12) generally decreases with the molding moisture content. This is shown in Figure 6.13, which is the result of several unconfined compression-strength tests on compacted specimens of a silty clay soil. The test specimens were prepared by kneading compaction. The insert in Figure 6.13 shows the relationship between dry unit weight and moisture content for the soil. Note that specimens A, B, and C have been compacted, respectively, on the dry side of the optimum moisture content, near optimum moisture content, and on the wet side of the optimum moisture content. The unconfined compression strength, q_u, is greatly reduced for the specimen compacted on the wet side of the optimum moisture content.

Some expansive clays in the field do not stay compacted, but expand upon entry of water and shrink with loss of moisture. This shrinkage and swelling of soil can cause serious distress to the foundations of structures. The nature of variation of expansion and shrinkage of expansive clay is shown in Figure 6.14. Laboratory observations such as this will help soils engineers to adopt a moisture content for compaction to minimize swelling and shrinkage.

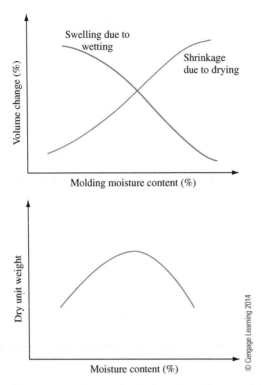

Figure 6.14 Nature of variation of swelling and shrinkage of expansive clay

6.9 Field Compaction

Compaction Equipment

Most of the compaction in the field is done with rollers. The four most common types of rollers are

1. Smooth-wheel rollers (or smooth-drum rollers)
2. Pneumatic rubber-tired rollers
3. Sheepsfoot rollers
4. Vibratory rollers

Smooth-wheel rollers (Figure 6.15) are suitable for proof rolling subgrades and for finishing operation of fills with sandy and clayey soils. These rollers provide 100% coverage under the wheels, with ground contact pressures as high as 310 to 380 kN/m² (45 to 55 lb/in²). They are not suitable for producing high unit weights of compaction when used on thicker layers.

Pneumatic rubber-tired rollers (Figure 6.16) are better in many respects than the smooth-wheel rollers. The former are heavily loaded with several rows of tires. These tires are closely spaced—four to six in a row. The contact pressure under the tires can range from 600 to 700 kN/m² (85 to 100 lb/in²), and they produce about 70 to 80% coverage. Pneumatic rollers can be used for sandy and clayey soil compaction. Compaction is achieved by a combination of pressure and kneading action.

Figure 6.15 Smooth-wheel roller (*Ingram Compaction LLC*)

Figure 6.16 Pneumatic rubber-tired roller (*Ingram Compaction LLC*)

Figure 6.17 Sheepsfoot roller (*SuperStock/Alamy*)

Sheepsfoot rollers (Figure 6.17) are drums with a large number of projections. The area of each projection may range from 25 to 85 cm^2 ($\approx$4 to 13 in^2). These rollers are most effective in compacting clayey soils. The contact pressure under the projections can range from 1400 to 7000 kN/m^2 (200 to 1000 lb/in^2). During compaction in the field, the initial passes compact the lower portion of a lift. Compaction at the top and middle of a lift is done at a later stage.

Vibratory rollers are extremely efficient in compacting granular soils. Vibrators can be attached to smooth-wheel, pneumatic rubber-tired, or sheepsfoot rollers to provide vibratory effects to the soil. Figure 6.18 demonstrates the principles of vibratory rollers. The vibration is produced by rotating off-center weights.

Handheld vibrating plates can be used for effective compaction of granular soils over a limited area. Vibrating plates are also gang-mounted on machines. These plates can be used in less restricted areas.

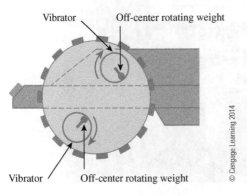

Figure 6.18 Principles of vibratory rollers

Field Compaction and Factors Affecting Field Compaction

For field compaction, soil is spread in layers and a predetermined amount of water is sprayed (Figure 6.19) on each layer (lift) of soil, after which compaction is initiated by a desired roller.

In addition to soil type and moisture content, other factors must be considered to achieve the desired unit weight of compaction in the field. These factors include the thickness of lift, the intensity of pressure applied by the compacting equipment, and the area over which the pressure is applied. These factors are important because the pressure applied at the surface decreases with depth, which results in a decrease in the degree of soil compaction. During compaction, the dry unit weight of soil also is affected by the number of roller passes. Figure 6.20 shows the growth curves for a silty clay soil. The dry unit weight of a soil at a given moisture content increases to a certain point with the number of roller passes. Beyond this point, it remains approximately constant. In most cases, about 10 to 15 roller passes yield the maximum dry unit weight economically attainable.

Figure 6.21a shows the variation in the unit weight of compaction with depth for a poorly graded dune sand for which compaction was achieved by a vibratory drum roller. Vibration was produced by mounting an eccentric weight on a single rotating shaft within the drum cylinder. The weight of the roller used for this compaction was 55.6 kN (12.5 kip), and the drum diameter was 1.19 m (47 in.). The lifts were kept at 2.44 m (8 ft). Note that, at any given depth, the dry unit weight of compaction increases with the number of roller passes. However, the rate of increase in unit weight gradually decreases after about 15 passes. Another fact to note from Figure 6.21a is the variation of dry unit weight with depth for any given number of roller passes. The dry unit weight and hence

Figure 6.19 Spraying of water on each lift of soil before compaction in the field (*Courtesy of N. Sivakugan, James Cook University, Australia*)

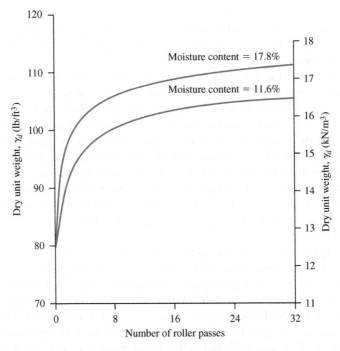

Silty clay: Liquid limit = 43   Plasticity index= 19

Figure 6.20 Growth curves for a silty clay—relationship between dry unit weight and number of passes of 84.5 kN (19 kip) three-wheel roller when the soil is compacted in 229 mm (9 in) loose layers at different moisture contents (*From Full-Scale Field Tests on 3-Wheel Power Rollers. In Highway Research Bulletin 272, Highway Research Board, National Research Council, Washington, D.C., 1960, Figure 15, p. 23. Reproduced with permission of the Transportation Research Board.*)

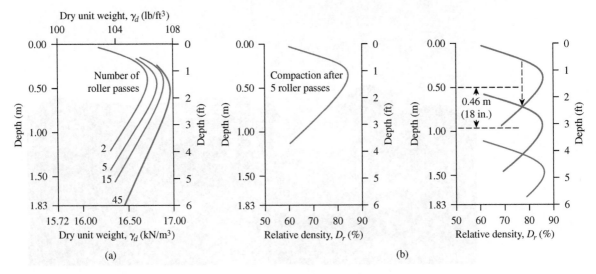

Figure 6.21 (a) Vibratory compaction of a sand—variation of dry unit weight with number of roller passes; thickness of lift = 2.44 m (8 ft); (b) estimation of compaction lift thickness for minimum required relative density of 75% with five roller passes (*After D'Appolonia, Whitman, and D'Appolonia, 1969. With permission from ASCE.*)

the relative density, D_r, reach maximum values at a depth of about 0.5 m (1.5 ft) and gradually decrease at lesser depths. This decrease occurs because of the lack of confining pressure toward the surface. Once the relationship between depth and relative density (or dry unit weight) for a given soil with a given number of roller passes is determined, estimating the approximate thickness of each lift is easy. This procedure is shown in Figure 6.21b (D'Appolonia et al., 1969).

6.10 Specifications for Field Compaction

In most specifications for earthwork, the contractor is instructed to achieve a compacted field dry unit weight of 90 to 95% of the maximum dry unit weight determined in the laboratory by either the standard or modified Proctor test. This is a specification for relative compaction, which can be expressed as

$$R(\%) = \frac{\gamma_{d(\text{field})}}{\gamma_{d(\text{max—lab})}} \times 100 \tag{6.19}$$

where R = relative compaction

For the compaction of granular soils, specifications sometimes are written in terms of the required relative density D_r or the required relative compaction. Relative density should not be confused with relative compaction. From Chapter 3, we can write

$$D_r = \left[\frac{\gamma_{d(\text{field})} - \gamma_{d(\text{min})}}{\gamma_{d(\text{max})} - \gamma_{d(\text{min})}}\right]\left[\frac{\gamma_{d(\text{max})}}{\gamma_{d(\text{field})}}\right] \tag{6.20}$$

Comparing Eqs. (6.19) and (6.20), we see that

$$R = \frac{R_0}{1 - D_r(1 - R_0)} \tag{6.21}$$

where

$$R_0 = \frac{\gamma_{d(\text{min})}}{\gamma_{d(\text{max})}} \tag{6.22}$$

On the basis of observation of 47 soil samples, Lee and Singh (1971) devised a correlation between R and D_r for granular soils:

$$R = 80 + 0.2D_r \tag{6.23}$$

The specification for field compaction based on relative compaction or on relative density is an end product specification. The contractor is expected to achieve a minimum dry unit weight regardless of the field procedure adopted. The most economical compaction condition can be explained with the aid of Figure 6.22. The compaction curves A, B, and C are for the same soil with varying compactive effort. Let curve A represent the conditions of maximum compactive effort that can be obtained from the existing equipment. Let the contractor be required to achieve a minimum dry unit weight of $\gamma_{d(\text{field})} = R\gamma_{d(\text{max})}$. To achieve this, the contractor must ensure that the moisture content w falls

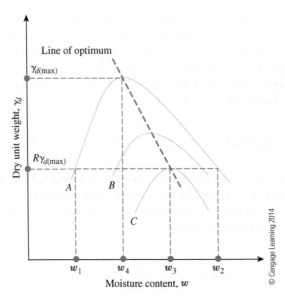

Figure 6.22 Most economical compaction condition

between w_1 and w_2. As can be seen from compaction curve C, the required $\gamma_{d(field)}$ can be achieved with a lower compactive effort at a moisture content $w = w_3$. However, for most practical conditions, a compacted field unit weight of $\gamma_{d(field)} = R\gamma_{d(max)}$ cannot be achieved by the minimum compactive effort. Hence, equipment with slightly more than the minimum compactive effort should be used. The compaction curve B represents this condition. Now we can see from Figure 6.22 that the most economical moisture content is between w_3 and w_4. Note that $w = w_4$ is the optimum moisture content for curve A, which is for the maximum compactive effort.

The concept described in the preceding paragraph, along with Figure 6.22, is attributed historically to Seed (1964) and is elaborated on in more detail in Holtz and Kovacs (1981).

6.11 Determination of Field Unit Weight of Compaction

When the compaction work is progressing in the field, knowing whether the specified unit weight has been achieved is useful. The standard procedures for determining the field unit weight of compaction include

1. Sand cone method
2. Rubber balloon method
3. Nuclear method

Following is a brief description of each of these methods.

Sand Cone Method (ASTM Designation D-1556)
The sand cone device consists of a glass or plastic jar with a metal cone attached at its top (Figure 6.23). The jar is filled with uniform dry Ottawa sand. The combined weight of the jar, the cone, and the sand filling the jar is determined (W_1). In the field, a small hole is

Figure 6.23 Glass jar filled with Ottawa sand with sand cone attached (*Courtesy of Braja M. Das, Henderson, Nevada*)

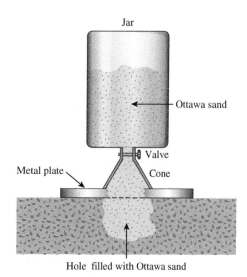

Figure 6.24 Field unit weight determined by sand cone method (*Most economical compaction condition*)

excavated in the area where the soil has been compacted. If the weight of the moist soil excavated from the hole (W_2) is determined and the moisture content of the excavated soil is known, the dry weight of the soil can be obtained as

$$W_3 = \frac{W_2}{1 + \dfrac{w\,(\%)}{100}} \tag{6.24}$$

where w = moisture content.

After excavation of the hole, the cone with the sand-filled jar attached to it is inverted and placed over the hole (Figure 6.24). Sand is allowed to flow out of the jar to fill the hole and the cone. After that, the combined weight of the jar, the cone, and the remaining sand in the jar is determined (W_4), so

$$W_5 = W_1 - W_4 \tag{6.25}$$

where W_5 = weight of sand to fill the hole and cone

The volume of the excavated hole can then be determined as

$$V = \frac{W_5 - W_c}{\gamma_{d(\text{sand})}} \tag{6.26}$$

where W_c = weight of sand to fill the cone only

$\gamma_{d(\text{sand})}$ = dry unit weight of Ottawa sand used

The values of W_c and $\gamma_{d(\text{sand})}$ are determined from the calibration done in the laboratory. The dry unit weight of compaction made in the field then can be determined as follows:

$$\gamma_d = \frac{\text{Dry weight of the soil excavated from the hole}}{\text{Volume of the hole}} = \frac{W_3}{V} \qquad (6.27)$$

Rubber Balloon Method (ASTM Designation D-2167)

The procedure for the rubber balloon method is similar to that for the sand cone method; a test hole is made and the moist weight of soil removed from the hole and its moisture content are determined. However, the volume of the hole is determined by introducing into it a rubber balloon filled with water from a calibrated vessel, from which the volume can be read directly. The dry unit weight of the compacted soil can be determined by using Eq. (6.27). Figure 6.25 shows a calibrated vessel that would be used with a rubber balloon.

Nuclear Method

Nuclear density meters are often used for determining the compacted dry unit weight of soil. The density meters operate either in drilled holes or from the ground surface. It uses a radioactive isotope source. The isotope gives off Gamma rays that radiate back to the meter's detector. Dense soil absorbs more radiation than loose soil. The instrument measures the weight of wet soil per unit volume and the weight of water present in a unit

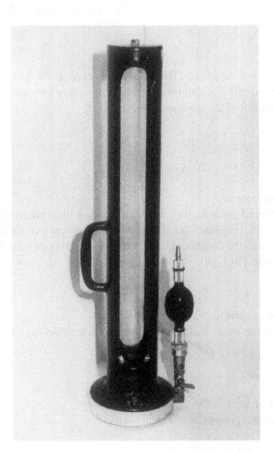

Figure 6.25 Calibrated vessel used with rubber balloon (not shown) (*Courtesy of John Hester, Carterville, Illinois*)

Figure 6.26 Nuclear density meter (*Courtesy of Braja M. Das, Henderson, Nevada*)

volume of soil. The dry unit weight of compacted soil can be determined by subtracting the weight of water from the moist unit weight of soil. Figure 6.26 shows a photograph of a nuclear density meter.

Example 6.5

Laboratory compaction test results for a clayey silt are given in the following table.

Moisture content (%)	Dry unit weight (kN/m³)
6	14.80
8	17.45
9	18.52
11	18.9
12	18.5
14	16.9

© Cengage Learning 2014

Following are the results of a field unit-weight determination test performed on the same soil by means of the sand cone method:

- Calibrated dry density of Ottawa sand = 1570 kg/m³
- Calibrated mass of Ottawa sand to fill the cone = 0.545 kg
- Mass of jar + cone + sand (before use) = 7.59 kg
- Mass of jar + cone + sand (after use) = 4.78 kg
- Mass of moist soil from hole = 3.007 kg
- Moisture content of moist soil = 10.2%

Determine:

a. Dry unit weight of compaction in the field
b. Relative compaction in the field

Solution

Part a

In the field,

Mass of sand used to fill the hole and cone = 7.59 kg − 4.78 kg = 2.81 kg

Mass of sand used to fill the hole = 2.81 kg − 0.545 kg = 2.265 kg

$$\text{Volume of the hole } (V) = \frac{2.265 \text{ kg}}{\text{Dry density of Ottawa sand}}$$

$$= \frac{2.265 \text{ kg}}{1570 \text{ kg/m}^3} = 0.0014426 \text{ m}^3$$

$$\text{Moist density of compacted soil} = \frac{\text{Mass of moist soil}}{\text{Volume of hole}}$$

$$= \frac{3.007}{0.0014426} = 2.084.4 \text{ kg/m}^3$$

$$\text{Moist unit weight of compacted soil} = \frac{(2084.4)(9.81)}{1000} = 20.45 \text{ kN/m}^3$$

Hence,

$$\gamma_d = \frac{\gamma}{1 + \frac{w(\%)}{100}} = \frac{20.45}{1 + \frac{10.2}{100}} = \textbf{18.56 kN/m}^3$$

Part b

The results of the laboratory compaction test are plotted in Figure 6.27. From the plot, we see that $\gamma_{d(\max)} = 19 \text{ kN/m}^3$. Thus, from Eq. (6.19),

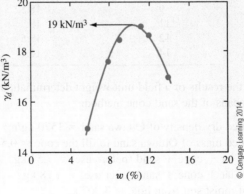

Figure 6.27 Plot of laboratory-compaction test results

© Cengage Learning 2014

$$R = \frac{\gamma_{d(\text{field})}}{\gamma_{d(\max)}} = \frac{18.56}{19.0} = \textbf{97.7\%}$$

Example 6.6

For a given soil, following are the results of compaction tests conducted in the laboratory.

Moisture content (%)	Dry unit weight γ_d (kN/m³)
12	16.34
14	16.93
16	17.24
18	17.20
20	16.75
22	16.23

© Cengage Learning 2014

After compaction of the soil in the field, sand cone tests (control tests) were conducted at five separate locations. Following are the results:

Location	Moisture content (%)	Moist density, ρ (kg/m³)
1	15.2	2055
2	16.4	2060
3	17.2	1971
4	18.8	1980
5	21.1	2104

© Cengage Learning 2014

The specifications require that:

a. γ_d must be at least $0.95\ \gamma_{d(max)}$.
b. Moisture content w should be within $\pm 2\%$ of w_{opt}.

Make necessary calculations to see if the control tests meet the specifications.

Solution
From Eq. (6.4),

$$\gamma_{zav} = \frac{\gamma_w}{w + \dfrac{1}{G_s}}$$

Given: $G_s = 2.72$. Now the following table can be prepared.

w (%)	γ_{zav} (kN/m³)
12	20.12
14	19.33
16	18.59
18	17.91
20	17.28
22	16.70

© Cengage Learning 2014

Figure 6.28 shows the plot of γ_d and γ_{zav}. From the plot, it can be seen that:

$$\gamma_{d(max)} = 17.4 \text{ kN/m}^3$$
$$w_{opt} = 16.8\%$$

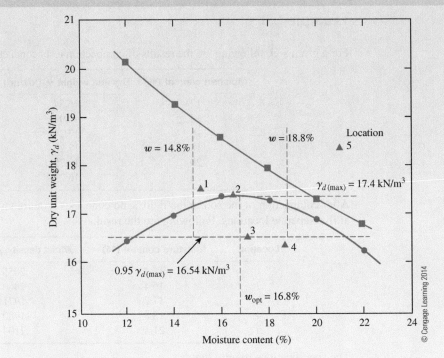

Figure 6.28

Based on the specifications, γ_d must be at least $0.95\gamma_{d(max)} = (0.95)(17.4) = 16.54$ kN/m³ with a moisture content of $16.8\% \pm 2\% = 14.8\%$ to 18.8%. This zone is shown in Figure 6.28.

For the control tests, the following table can be prepared.

Location	w (%)	ρ (kg/m³)	γ_d^* (kN/m³)
1	15.2	2055	17.5
2	16.4	2060	17.36
3	17.1	1971	16.51
4	18.8	1980	16.35
5	21.1	2104	18.41

$$^*\gamma_d(\text{kN/m}^3) = \left[\frac{\rho(\text{kg/m}^3)}{1 + \dfrac{w\,(\%)}{100}} \right] \left(\frac{9.81}{1000} \right)$$

The results of the control tests are also plotted in Figure 6.28. From the plot, it appears that the **tests at locations 1 and 2 meet the specifications.** The test at location 3 is a borderline case. Also note that there is some error for the test in location 5, since it falls above the zero-air-void line.

6.12 Compaction of Organic Soil and Waste Materials

The presence of organic materials in a soil reduces its strength. In many cases, soils with a high organic content are generally discarded as fill material; however, in certain economic circumstances, slightly organic soils are used for compaction. In fact, organic soils are desirable in many circumstances (e.g., for agriculture, decertification, mitigation, and urban planning). The high costs of waste disposal have sparked an interest in the possible use of waste materials (e.g., bottom ash obtained from coal burning, copper slag, paper mill sludge, shredded waste tires mixed with inorganic soil, and so forth) in various landfill operations. Such use of waste materials is one of the major thrusts of present-day environmental geotechnology. Following is a discussion of the compaction characteristics of some of these materials.

Organic Soil

Franklin et al. (1973) conducted several laboratory tests to observe the effect of organic content on the compaction characteristics of soil. In the test program, various natural soils and soil mixtures were tested. Figure 6.29 shows the effect of organic content on the

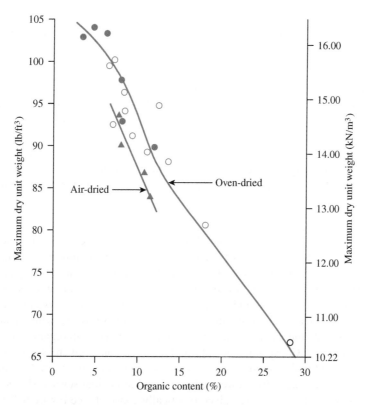

○ Mixture—oven-dried
● Natural sample—oven-dried
▲ Mixture—air-dried

Figure 6.29 Variation of maximum dry unit weight with organic content (*After Franklin, Orozco, and Semrau, 1973. With permission from ASCE.*)

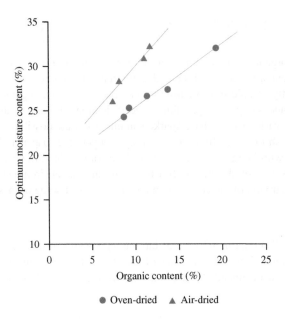

Figure 6.30 Variation of optimum moisture content with organic content (*After Franklin, Orozco, and Semrau, 1973. With permission from ASCE.*)

maximum dry unit weight. When the organic content exceeds 8 to 10%, the maximum dry unit weight of compaction decreases rapidly. Conversely, the optimum moisture content for a given compactive effort increases with an increase in organic content. This trend is shown in Figure 6.30. Likewise, the maximum unconfined compression strength (see Chapter 12) obtained from a compacted soil (with a given compactive effort) decreases with increasing organic content of a soil. From these facts, we can see that soils with organic contents higher than about 10% are undesirable for compaction work.

Soil and Organic Material Mixtures

Lancaster et al. (1996) conducted several modified Proctor tests to determine the effect of organic content on the maximum dry unit weight and optimum moisture content of soil and organic material mixtures. The soils tested consisted of a poorly graded sandy soil (SP-SM) mixed with either shredded redwood bark, shredded rice hulls, or municipal sewage sludge. Figures 6.31 and 6.32 show the variations of maximum dry unit weight of compaction and optimum moisture content, respectively, with organic content. As in Figure 6.29, the maximum dry unit weight decreased with organic content in all cases (see Figure 6.31). Conversely, the optimum moisture content increased with organic content for soil mixed with shredded red-wood or rice hulls (see Figure 6.32), similar to the pattern shown in Figure 6.30. However, for soil and municipal sewage sludge mixtures, the optimum moisture content remained practically constant (see Figure 6.32).

Bottom Ash from Coal Burning and Copper Slag

Laboratory standard Proctor test results for bottom ash from coal-burning power plants and for copper slag are also available in the literature. These waste products have been

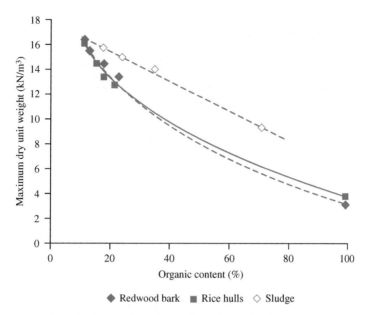

Figure 6.31 Variation of maximum dry unit weight of compaction with organic content—soil and organic material mixture (*Source:* After "The Effect of Organic Content on Soil Compaction," by J. Lancaster, R. Waco, J. Towle, and R. Chaney, 1996. In *Proceedings, Third International Symposium on Environmental Geotechnology,* p. 159. Used with permission of the author)

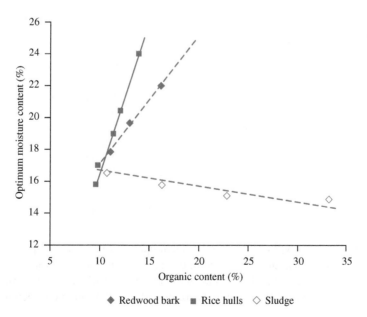

Figure 6.32 Variation of optimum moisture content with organic content—soil and organic material mixtures (*Source:* After "The Effect of Organic Content on Soil Compaction," by J. Lancaster, R. Waco, J. Towle, and R. Chaney, 1996. In *Proceedings, Third International Symposium on Environmental Geotechnology,* p. 159. Used with permission of the author)

Table 6.2 Standard Proctor Test Results of Bottom Ash and Copper Slag

Type	Location	Maximum dry unit weight		Optimum moisture content (%)	Source
		kN/m^3	lb/ft^3		
Bottom ash—	Fort Martin	13.4	85	24.8	Seals, Moulton, and Ruth
bituminous coal	Kammer	16.0	102	13.8	(1972)
(West Virginia)	Kanawha River	11.4	72.6	26.2	
	Mitchell	18.3	116.6	14.6	
	Muskingham	14.3	91.1	22.0	
	Willow Island	14.5	92.4	21.2	
Bottom ash—	Big Stone Power	16.4	104.4	20.5	Das, Selim, and Pfeifle
lignite coal	Plant, South Dakota				(1978)
Copper slag	American Smelter and	19.8	126	18.8	Das, Tarquin, and Jones
	Refinery Company,				(1983)
	El Paso, Texas				

shown to be environmentally safe for use as landfill. A summary of some of these test results is given in Table 6.2.

6.13 Evaluation of Soils as Compaction Material

Table 6.3 provides a general summary of the evaluation of various types of soils as fill material as they relate to roller type, maximum dry unit weight of compaction based on standard Proctor tests, and compaction characteristics. The compressibility and expansion characteristics on compacted soils are as follow (Sowers, 1979):

GW, GP, SW, SP	Practically none
GM, GC, SM, SC	Slight
ML	Slight to medium
MH	High
CL	Medium
CH	Very high

6.14 Special Compaction Techniques

Several special types of compaction techniques have been developed for deep compaction of in-place soils, and these techniques are used in the field for large-scale compaction works. Among these, the popular methods are vibroflotation, dynamic compaction, and blasting. Details of these methods are provided in the following sections.

Vibroflotation

Vibroflotation is a technique for *in situ* densification of thick layers of loose granular soil deposits. It was developed in Germany in the 1930s. The first vibroflotation device was used in the United States about 10 years later. The process involves the use of a *Vibroflot*

Table 6.3 Summary of Evaluation of Fill Materials for Compaction Based on Sowers (1979) and Highway Research Board (1962)

Soil type	Unified classification	Roller(s) for best results	Maximum dry unit weight— standard Proctor compaction		Compaction characteristics
			kN/m³	lb/ft³	
Gravelly	GW	Rubber-tired, steel wheel, vibratory	18.9–20.4	120–130	Good
	GP	Rubber-tired, steel wheel, vibratory	18.1–18.9	115–120	Good
	GM	Rubber-tired, sheepsfoot	18.9–20.4	120–130	Good to fair
	GC	Rubber-tired, sheepsfoot	18.1–19.7	115–125	Good to fair
Sandy	SW	Rubber-tired, vibratory	18.1–19.7	115–125	Good
	SP	Rubber-tired, vibratory	16.5–18.1	105–115	Good
	SM	Rubber-tired, sheepsfoot	17.3–18.9	110–120	Good to fair
	SC	Rubber-tired, sheepsfoot	16.5–18.9	105–120	Good to fair
Silty	ML	Rubber-tired, sheepsfoot	15.7–17.3	100–110	Good to poor
	MH	Rubber-tired, sheepsfoot	13.4–15.7	85–100	Fair to poor
Clayey	CL	Rubber-tired, sheepsfoot	14.1–18.1	90–110	Fair to poor
	CH	Sheepsfoot	13.4–16.5	85–105	Fair to poor

unit (also called the *vibrating unit*), which is about 2.1 m ($\approx$7 ft) long. (As shown in Figure 6.33.) This vibrating unit has an eccentric weight inside it and can develop a centrifugal force, which enables the vibrating unit to vibrate horizontally. There are openings at the bottom and top of the vibrating unit for water jets. The vibrating unit is attached to a follow-up pipe. Figure 6.33 shows the entire assembly of equipment necessary for conducting the field compaction.

The entire vibroflotation compaction process in the field can be divided into four stages (Figure 6.34):

Stage 1: The jet at the bottom of the Vibroflot is turned on and lowered into the ground.

Stage 2: The water jet creates a quick condition in the soil and it allows the vibrating unit to sink into the ground.

Stage 3: Granular material is poured from the top of the hole. The water from the lower jet is transferred to the jet at the top of the vibrating unit. This water carries the granular material down the hole.

Stage 4: The vibrating unit is gradually raised in about 0.3 m ($\approx$1 ft) lifts and held vibrating for about 30 seconds at each lift. This process compacts the soil to the desired unit weight.

The details of various types of Vibroflot units used in the United States are given in Table 6.4. Note that 23 kW (30hp) electric units have been used since the latter part of the 1940s. The 75 kW (100hp) units were introduced in the early 1970s.

The zone of compaction around a single probe varies with the type of Vibroflot used. The cylindrical zone of compaction has a radius of about 2 m ($\approx$6 ft) for a 23 kW (30hp) unit. This radius can extend to about 3 m ($\approx$10 ft) for a 75 kW (100hp) unit.

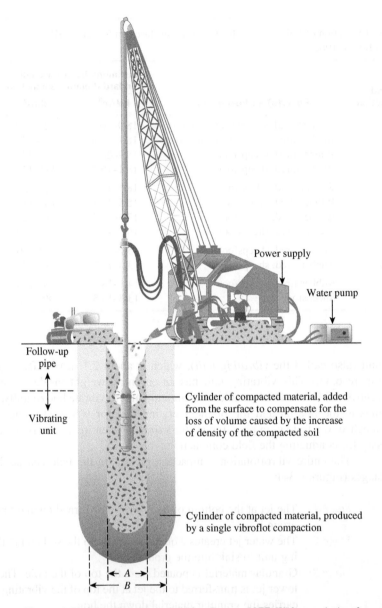

Power supply

Water pump

Follow-up
pipe

Cylinder of compacted material, added
from the surface to compensate for the
loss of volume caused by the increase
of density of the compacted soil

Vibrating
unit

Cylinder of compacted material, produced
by a single vibroflot compaction

← A →

B

Figure 6.33 Vibroflotation unit (*After Brown, 1977. With permission from ASCE.*)

Typical patterns of Vibroflot probe spacings are shown in Figure 6.35. Square and
rectangular patterns generally are used to compact soil for isolated, shallow foundations. Equilateral triangular patterns generally are used to compact large areas. The
capacity for successful densification of *in situ* soil depends on several factors, the most
important of which is the grain-size distribution of the soil and the type of backfill used
to fill the holes during the withdrawal period of the Vibroflot. The range of the grain-
size distribution of *in situ* soil marked Zone 1 in Figure 6.36 is most suitable for compaction by vibroflotation. Soils that contain excessive amounts of fine sand and silt-size

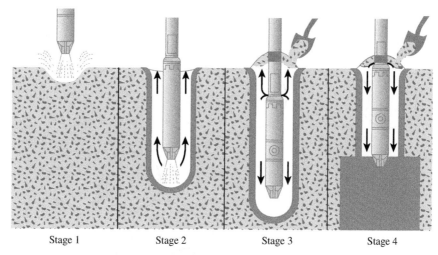

Stage 1   Stage 2   Stage 3   Stage 4

Figure 6.34 Compaction by vibroflotation process (*After Brown, 1977. With permission from ASCE.*)

Table 6.4 Types of Vibroflot Units[*]

Motor type	75 kW electric and hydraulic	23 kW electric
a. Vibrating tip		
Length	2.1 m (7.0 ft)	1.86 m (6.11 ft)
Diameter	406 mm (16 in.)	381 mm (15 in.)
Weight	17.8 kN (4000 lb)	17.8 kN (4000 lb)
Maximum movement when full	12.5 mm (0.49 in.)	7.6 mm (0.3 in.)
Centrifugal force	160 kN (18 ton)	89 kN (10 ton)
b. Eccentric		
Weight	1.2 kN (260 lb)	0.76 kN (170 lb)
Offset	38 mm (1.5 in.)	32 mm (1.25 in.)
Length	610 mm (24 in.)	390 mm (15.25 in.)
Speed	1800 rpm	1800 rpm
c. Pump		
Operating flow rate	0–1.6 m^3/min (0–400 gal/min)	0–0.6 m^3/min (0–150 gal/min)
Pressure	700–1050 kN/m^2 (100–150 lb/in.^{2})	700–1050 kN/m^2 (100–150 lb/in.^{2})
d. Lower follow-up pipe and extensions		
Diameter	305 mm (12 in.)	305 mm (12 in.)
Weight	3.65 kN/m (250 lb/ft)	3.65 kN/m (250 lb/ft)

[*]*After Brown 1977. With permission from ASCE.*

particles are difficult to compact, and considerable effort is needed to reach the proper relative density of compaction. Zone 2 in Figure 6.36 is the approximate lower limit of grain-size distribution for which compaction by vibroflotation is effective. Soil deposits whose grain-size distributions fall in Zone 3 contain appreciable amounts of gravel. For these soils, the rate of probe penetration may be slow and may prove uneconomical in the long run.

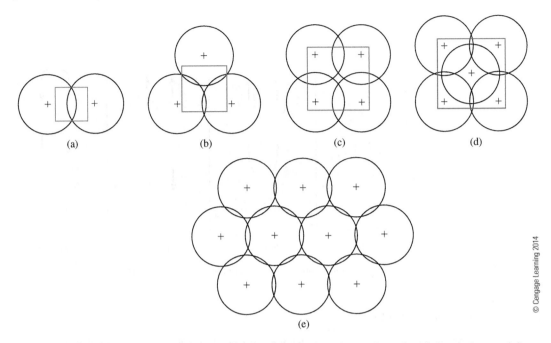

Figure 6.35 Typical patterns of Vibroflot probe spacings for a column foundation (a, b, c, and d) and for compaction over a large area (e)

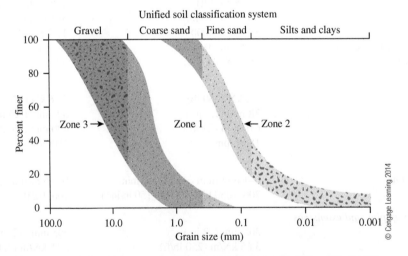

Figure 6.36 Effective range of grain-size distribution of soil for vibroflotation

The grain-size distribution of the backfill material is an important factor that controls the rate of densification. Brown (1977) has defined a quantity called the *suitability number* for rating backfill as

$$S_N = 1.7 \sqrt{\frac{3}{(D_{50})^2} + \frac{1}{(D_{20})^2} + \frac{1}{(D_{10})^2}} \qquad (6.28)$$

where D_{50}, D_{20}, and D_{10} are the diameters (in mm) through which, respectively, 50, 20, and 10% of the material passes.

The smaller the value of S_N, the more desirable the backfill material. Following is a backfill rating system proposed by Brown:

Range of S_N	Rating as backfill
0–10	Excellent
10–20	Good
20–30	Fair
30–50	Poor
>50	Unsuitable

Dynamic Compaction

Dynamic compaction is a technique that has gained popularity in the United States for the densification of granular soil deposits. This process consists primarily of dropping a heavy weight repeatedly on the ground at regular intervals. The weight of the hammer used varies over a range of 80 to 360 kN (18 to 80 kip), and the height of the hammer drop varies between 7.5 and 30.5 m (25 and 100 ft). The stress waves generated by the hammer drops aid in the densification. The degree of compaction achieved at a given site depends on the following three factors:

1. Weight of hammer
2. Height of hammer drop
3. Spacing of locations at which the hammer is dropped

Figure 6.37a shows a dynamic compaction in progress. A site immediately after the completion is shown in Figure 6.37b. Leonards, Cutter, and Holtz (1980) suggested that the significant depth of influence for compaction can be approximated by using the equation

$$D \simeq \left(\tfrac{1}{2}\right)\sqrt{W_H h} \qquad (6.29)$$

where D = significant depth of densification (m)
$\quad W_H$ = dropping weight (metric ton)
$\quad h$ = height of drop (m)

In English units, the preceding equation takes the form

$$D = 0.61\sqrt{W_H h} \qquad (6.30)$$

where the units of D and h are ft, and the unit of W_H is kip.

In 1992, Poran and Rodriguez suggested a rational method for conducting dynamic compaction for granular soils in the field. According to their method, for a hammer of width D having a weight W_H and a drop h, the approximate shape of the densified area will be of the type shown in Figure 6.38 (i.e., a semiprolate spheroid). Note that in this figure $b = DI$ (where DI is the significant depth of densification). Figure 6.39 gives the design chart for a/D and b/D versus $NW_H h/Ab$ (D = width of the hammer if not circular in cross section; A = area of cross section of the hammer; and N = number of required hammer drops). This method uses the following steps.

Step 1: Determine the required significant depth of densification, DI ($= b$).
Step 2: Determine the hammer weight (W_H), height of drop (h), dimensions of the cross section, and thus, the area A and the width D.

(a)

(b)

Figure 6.37 (a) Dynamic compaction in progress; (b) a site after completion of dynamic compaction *(Courtesy of Khaled Sobhan, Florida Atlantic University, Boca Raton, Florida.)*

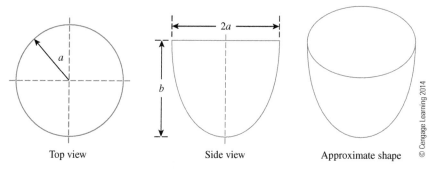

Figure 6.38 Approximate shape of the densified area due to dynamic compaction

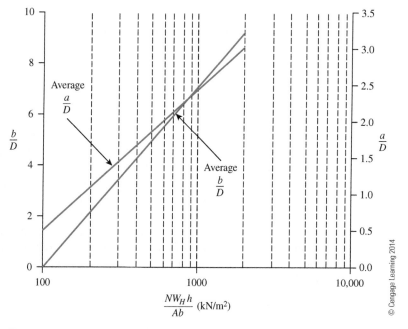

Figure 6.39 Poran and Rodriguez chart for a/D, b/D versus NW_Hh/Ab

Step 3: Determine $DI/D = b/D$.

Step 4: Use Figure 6.39 and determine the magnitude of NW_Hh/Ab for the value of b/D obtained in step 3.

Step 5: Since the magnitudes of W_H, h, A, and b are known (or assumed) from step 2, the number of hammer drops can be estimated from the value of NW_Hh/Ab obtained from step 4.

Step 6: With known values of NW_Hh/Ab, determine a/D and thus a from Figure 6.39.

Step 7: The grid spacing, S_g, for dynamic compaction may now be assumed to be equal to or somewhat less than a. (See Figure 6.40.)

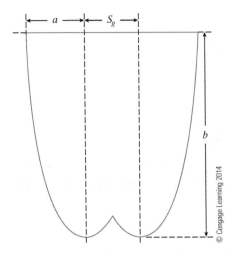

Figure 6.40 Approximate grid spacing for dynamic compaction

Blasting

Blasting is a technique that has been used successfully in many projects (Mitchell, 1970) for the densification of granular soils. The general soil grain sizes suitable for compaction by blasting are the same as those for compaction by vibroflotation. The process involves the detonation of explosive charges, such as 60% dynamite at a certain depth below the ground surface in saturated soil. The lateral spacing of the charges varies from about 3 to 9 m (10 to 30 ft). Three to five successful detonations are usually necessary to achieve the desired compaction. Compaction (up to a relative density of about 80%) up to a depth of about 18 m (60 ft) over a large area can easily be achieved by using this process. Usually, the explosive charges are placed at a depth of about two-thirds of the thickness of the soil layer desired to be compacted. The sphere of influence of compaction by a 60% dynamite charge can be given as follows (Mitchell, 1970):

$$r = \sqrt{\frac{W_{EX}}{C}} \qquad (6.31)$$

where r = sphere of influence
W_{EX} = weight of explosive—60% dynamite
C = 0.0122 when W_{EX} is in kg and r is in m
= 0.0025 when W_{EX} is in lb and r is in ft

Figure 6.41 shows the test results of soil densification by blasting in an area measuring 15 m by 9 m (Mitchell, 1970). For these tests, twenty 2.09 kg (4.6 lb) charges of Gelamite No. 1 (Hercules Powder Company, Wilmington, Delaware) were used.

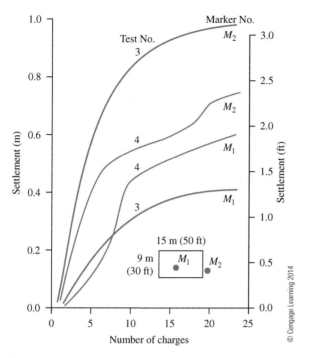

Figure 6.41 Ground settlement as a function of number of explosive charges

Example 6.7

Following are the details for the backfill material used in a vibroflotation project:

- $D_{10} = 0.36$ mm
- $D_{20} = 0.52$ mm
- $D_{50} = 1.42$ mm

Determine the suitability number S_N. What would be its rating as a backfill material?

Solution

From Eq. (6.28),

$$S_N = 1.7\sqrt{\frac{3}{(D_{50})^2} + \frac{1}{(D_{20})^2} + \frac{1}{(D_{10})^2}}$$

$$= 1.7\sqrt{\frac{3}{(1.42)^2} + \frac{1}{(0.52)^2} + \frac{1}{(0.36)^2}}$$

$$= \mathbf{6.1}$$

Rating: **Excellent**

6.15 Summary and General Comments

In this chapter, we have discussed the following:

- Standard and modified Proctor compaction tests are conducted in the laboratory to determine the maximum dry unit weight of compaction [$\gamma_{d(max)}$] and optimum moisture content (w_{opt}) (Sections 6.3 and 6.5).
- $\gamma_{d(max)}$ and w_{opt} are functions of the energy of compaction E.
- Several empirical relations have been presented to estimate $\gamma_{d(max)}$ and w_{opt} for cohesionless and cohesive soils (Section 6.6). Also included in this section is an empirical relationship to estimate the relative density of compaction (D_r) with known median grain size (D_{50}) and energy of compaction (E).
- For a given energy of compaction (E) in a cohesive soil, the hydraulic conductivity and unconfined compression strength, swelling, and shrinkage characteristics are functions of molding moisture content.
- Field compaction is generally carried out by rollers such as smooth-wheel, rubber-tired, sheepsfoot, and vibratory (Section 6.9).
- Control tests to determine the quality of field compaction can be done by using the sand cone method, rubber balloon method, and nuclear method.
- Vibroflotation, dynamic compaction, and blasting are special techniques used for large-scale compaction in the field (Section 6.14).

Laboratory standard and modified Proctor compaction tests described in this chapter are essentially for *impact* or *dynamic* compaction of soil; however, in the laboratory, *static compaction* and *kneading compaction* also can be used. It is important to realize that the compaction of clayey soils achieved by rollers in the field is essentially the kneading type. The relationships of dry unit weight (γ_d) and moisture content (w) obtained by dynamic and kneading compaction are not the same. Proctor compaction test results obtained in the laboratory are used primarily to determine whether the roller compaction in the field is sufficient. The structures of compacted cohesive soil at a similar dry unit weight obtained by dynamic and kneading compaction may be different. This difference, in turn, affects physical properties, such as hydraulic conductivity, compressibility, and strength.

For most fill operations, the final selection of the borrow site depends on such factors as the soil type and the cost of excavation and hauling.

Problems

6.1 Calculate and plot the variation of dry density of a soil in kg/m^3 ($G_s = 2.65$) at $w = 5$, 10, 15, and 20% for degree of saturation, $S = 70$, 80, 90, and 100%.

6.2 Calculate the zero-air-void unit weights (kN/m^3) for a soil with $G_s = 2.68$ at moisture contents of 5, 10, 15, 20, and 25%.

6.3 The results of a standard Proctor test are given in the following table.

    a. Determine the maximum dry unit weight of compaction and the optimum moisture content. Given: mold volume=1/30 ft^3.

    b. Determine the void ratio and the degree of saturation at the optimum moisture content. Given: $G_s = 2.68$.

Trial no.	Weight of moist soil in the mold (lb)	Moisture content (%)
1	3.92	5.0
2	4.12	7.5
3	4.29	10.0
4	4.37	12.5
5	4.45	15
6	4.35	17.5
7	4.20	20.0

© Cengage Learning 2014

6.4 Repeat Problem 6.3 with the following data (use $G_s = 2.7$):

Trial no.	Weight of moist soil in the mold (lb)	Moisture content (%)
1	3.70	9.9
2	3.77	10.6
3	3.91	12.1
4	4.03	13.8
5	4.11	15.1
6	4.14	17.4
7	4.12	19.4
8	4.07	21.2

© Cengage Learning 2014

6.5 The results of a standard Proctor test are given in the following table.
a. Determine the maximum dry density (kg/m^3) of compaction and the optimum moisture content. Given mold volume = 943.3 cm^3.
b. If specification calls for 97% relative compaction in the field, what would be the field dry density and the range of acceptable moisture content?

Trial no.	Mass of moist soil in the mold (kg)	Moisture content (%)
1	1.47	10.0
2	1.83	12.5
3	2.02	15.0
4	1.95	17.5
5	1.73	20.0
6	1.69	22.5

© Cengage Learning 2014

6.6 The *in situ* moist unit weight of a soil is 17.3 kN/m^3 and the moisture content is 16%. The specific gravity of soil solids is 2.72. This soil is to be excavated and transported to a construction site for use in a compacted fill. If the specification calls for the soil to be compacted to a minimum dry unit weight of 18.1 kN/m^3 at the same moisture content of 16%, how many cubic meters of soil from the excavation site are needed to produce 2000 m^3 of compacted fill? How many 20 ton truckloads are needed to transport the excavated soil?

6.7 A proposed embankment fill requires 5000 m^3 of compacted soil. The void ratio of the compacted fill is specified as 0.75. Soil can be transported from one of the four borrow pits as described in the following table. The void ratio, specific gravity of soil solids, and the cost per cubic meter for moving the soil to the proposed construction site are provided in the table.

      a. Determine the volume of each borrow pit soil required to meet the specification of the embankment site

      b. Make necessary calculations to select the borrow pit which would be most cost effective.

Borrow pit	Void ratio	G_s	Cost ($/m³)
I	0.8	2.65	9
II	0.9	2.68	6
III	1.1	2.71	7
IV	0.85	2.74	10

© Cengage Learning 2014

6.8 The maximum and minimum dry unit weights of a sand were determined in the laboratory to be 18.9 and 15.5 kN/m³, respectively. What is the relative compaction in the field if the relative density is 75%?

6.9 The relative compaction of a sand in the field is 93.5%. Given that $\gamma_{d(max)}$ = 16.98 kN/m³ and $\gamma_{d(min)}$ = 14.46 kN/m³, determine the dry unit weight in the field and the relative density of compaction.

6.10 The relative compaction of a sand in the field is 88%. The maximum and minimum dry unit weights of the sand are 118 lb/ft³ and 98 lb/ft³, respectively. Determine:

    a. Dry unit weight in the field

    b. Relative density of compaction

    c. Moist unit weight at a moisture content of 13%

6.11 Following are the results of a field unit weight determination test performed by means of the sand cone method:

- Calibrated dry density of Ottawa sand = 1731 kg/m³
- Mass of Ottawa sand to fill the cone = 0.118 kg
- Mass of jar + cone + sand (before use) = 6.08 kg
- Mass of jar + cone + sand (after use) = 2.86 kg
- Mass of moist soil from hole = 3.34 kg
- Moisture content of moist soil = 12.1%

Determine:

    a. Dry density of compaction in the field.

    b. Relative compaction in the field assuming that the test data in Problem 6.5 represent the same soil as in the field

6.12 The backfill material for a vibroflotation project has the following grain sizes:

- D_{10} = 0.18 mm
- D_{20} = 0.31 mm
- D_{50} = 1.98 mm

Determine the suitability number, S_N, and rate it as a backfill material.

Critical Thinking Problem

6.C.1 Since laboratory or field experiments are generally expensive and time consuming, geotechnical engineers often have to rely on empirical relationships to predict design parameters. Section 6.6 presents such relationships for predicting optimum moisture content and maximum dry unit weight. Let us use some of these equations and compare our results with known experimental data.

The following table presents the results from laboratory compaction tests conducted on a wide range of fine-grained soils using various compactive efforts (E). Based on the soil data given in the table, determine the optimum moisture content and maximum dry unit weight using the empirical relationships presented in Section 6.6.

a. Use the Osman et al. (2008) method [Eqs. (6.13) through (6.16)].

b. Use the Gurtug and Sridharan (2004) method [Eqs. (6.11) and (6.12)].

c. Use the Matteo et al. (2009) method [Eqs. (6.17) and (6.18)].

d. Plot the calculated w_{opt} against the experimental w_{opt}, and the calculated $\gamma_{d(max)}$ with the experimental $\gamma_{d(max)}$. Draw a 45° *line of equality* on each plot.

e. Comment on the predictive capabilities of various methods. What can you say about the inherent nature of empirical models?

Soil	G_s	LL (%)	PL (%)	E (kN-m/m³)	w_{opt} (%)	$\gamma_{d(max)}$ (kN/m³)
1[a]	2.67	17	16	2700[b]	8	20.72
				600[c]	10	19.62
				354[d]	10	19.29
2[a]	2.73	68	21	2700	20	16.00
				600	28	13.80
				354	31	13.02
3	2.68	56	14	2700	15	18.25
				1300[e]	16	17.5
				600	17	16.5
				275[f]	19	15.75
4	2.68	66	27	600	21	15.89
5	2.67	25	21	600	18	16.18
6	2.71	35	22	600	17	16.87
7	2.69	23	18	600	12	18.63
8	2.72	29	19	600	15	17.65

Note:

[a] Tschebotarioff (1951)

[b] Modified Proctor test

[c] Standard Proctor test

[d] Standard Proctor mold and hammer; drop: 305 mm; layers: 3; blows/layer: 15

[e] Modified Proctor mold and hammer; drop: 457 mm; layers: 5; blows/layer: 26

[f] Modified Proctor mold; standard Proctor hammer; drop: 305 mm; layers: 3; blows/layer: 25

© Cengage Learning 2014

References

AMERICAN ASSOCIATION OF STATE HIGHWAY AND TRANSPORTATION OFFICIALS (1982). *AASHTO Materials, Part II,* Washington, D.C.

AMERICAN SOCIETY FOR TESTING AND MATERIALS (2010). *Annual Book of ASTM Standards,* Vol 04.08, West Conshohocken, Pa.

BROWN, E. (1977). "Vibroflotation Compaction of Cohesionless Soils," *Journal of the Geo technical Engineering Division,* ASCE, Vol. 103, No. GT12, 1437–1451.

D'APPOLONIA, D. J., WHITMAN, R. V., and D'APPOLONIA, E. D. (1969). "Sand Compaction with Vibratory Rollers," *Journal of the Soil Mechanics and Foundations Division,* ASCE, Vol. 95, No. SM1, 263–284.

DAS, B. M., SELIM, A. A., and PFEIFLE, T. W. (1978). "Effective Use of Bottom Ash as a Geotechnical Material," *Proceedings,* 5th Annual UMR-DNR Conference and Exposition on Energy, University of Missouri, Rolla, 342–348.

DAS, B. M., TARQUIN, A. J., and JONES, A. D. (1983). "Geotechnical Properties of a Copper Slag," *Transportation Research Record No. 941,* National Research Council, Washington, D.C., 1–4.

FRANKLIN, A. F., OROZCO, L. F., and SEMRAU, R. (1973). "Compaction of Slightly Organic Soils," *Journal of the Soil Mechanics and Foundations Division,* ASCE, Vol. 99, No. SM7, 541–557.

GURTUG, Y., and SRIDHARAN, A. (2004). "Compaction Behaviour and Prediction of Its Characteristics of Fine Grained Soils with Particular Reference to Compaction Energy," *Soils and Foundations,* Vol. 44, No. 5, 27–36.

HIGHWAY RESEARCH BOARD (1962). *Factors influencing Compaction Test Results,* Bulletin 319, Washington, D.C.

HOLTZ, R. D., and KOVACS, W. D. (1981). *An Introduction to Geotechnical Engineering,* Prentice-Hall, Englewood Cliffs, N.J.

JOHNSON, A. W., and SALLBERG, J. R. (1960). "Factors That Influence Field Compaction of Soil," Highway Research Board, *Bulletin No. 272.*

LAMBE, T. W. (1958a). "The Structure of Compacted Clay," *Journal of the Soil Mechanics and Foundations Division,* ASCE, Vol. 84, No. SM2, 1654–1 to 1654–35.

LAMBE, T. W. (1958b). "The Engineering Behavior of Compacted Clay," *Journal of the Soil Mechanics and Foundations Division,* ASCE, Vol. 84, No. SM2, 1655–1 to 1655–35.

LANCASTER, J., WACO, R., TOWLE, J., and CHANEY, R. (1996). "The Effect of Organic Content on Soil Compaction," *Proceedings,* 3rd International Symposium on Environmental Geotechnology, San Diego, 152–161.

LEE, K. W., and SINGH, A. (1971). "Relative Density and Relative Compaction," *Journal of the Soil Mechanics and Foundations Division,* ASCE, Vol. 97, No. SM7, 1049–1052.

LEE, P. Y., and SUEDKAMP, R. J. (1972). "Characteristics of Irregularly Shaped Compaction Curves of Soils," *Highway Research Record No. 381,* National Academy of Sciences, Washington, D.C., 1–9.

LEONARDS, G. A., CUTTER, W. A., and HOLTZ, R. D. (1980). "Dynamic Compaction of Granular Soils," *Journal of the Geotechnical Engineering Division,* ASCE, Vol. 106, No. GT1, 35–44.

MATTEO, L. D., BIGOTTI, F., and RICCO, R. (2009). "Best-Fit Model to Estimate Proctor Properties of Compacted Soil," *Journal of Geotechnical and Geoenvironmental Engineering,* ASCE, Vol. 135, No. 7, 992–996.

MITCHELL, J. K. (1970). "In-Place Treatment of Foundation Soils," *Journal of the Soil Mechanics and Foundations Division,* ASCE, Vol. 96, No. SM1, 73–110.

OMAR, M., ABDALLAH, S., BASMA, A., and BARAKAT, S. (2003). "Compaction Characteristics of Granular Soils in United Arab Emirates," *Geotechnical and Geological Engineering,* Vol. 21, No. 3, 283–295.

OSMAN, S., TOGROL, E., and KAYADELEN, C. (2008). "Estimating Compaction Behavior of Fine-Grained Soils Based on Compaction Energy," *Canadian Geotechnical Journal,* Vol. 45, No. 6, 877–887.

PATRA, C. R., SIVAKUGAN, N., DAS, B. M., and ROUT, S. K. (2010). "Correlation of Relative Density of Clean Sand with Median Grain Size and Compaction Energy," *International Journal of Geotechnical Engineering,* Vol. 4, No. 2, 196–203.

PORAN, C. J., and RODRIGUEZ, J. A. (1992). "Design of Dynamic Compaction," *Canadian Geotechnical Journal,* Vol. 2, No. 5, 796–802.

PROCTOR, R. R. (1933). "Design and Construction of Rolled Earth Dams," *Engineering News Record,* Vol. 3, 245–248, 286–289, 348–351, 372–376.

SEALS, R. K. MOULTON, L. K., and RUTH, E. (1972). "Bottom Ash: An Engineering Material," *Journal of the Soil Mechanics and Foundations Division,* ASCE, Vol. 98, No. SM4, 311–325.

SEED, H. B. (1964). Lecture Notes, CE 271, Seepage and Earth Dam Design, University of California, Berkeley.

SEED, H. B., and CHAN, C. K. (1959). "Structure and Strength Characteristics of Compacted Clays," *Journal of the Soil Mechanics and Foundations Division,* ASCE, Vol. 85, No. SM5, 87–128.

SOWERS, G. F. (1979). *Introductory Soil Mechanics and Foundations: Geotechnical Engineering,* Macmillan, New York.

TSCHEBOTARIOFF, G. P. (1951). *Soil Mechanics, Foundations, and Earth Structures,* McGraw-Hill, New York.

CHAPTER 3

Slope Stability

15.1 Introduction

An exposed ground surface that stands at an angle with the horizontal is called an unrestrained slope. The slope can be natural or man-made. It can fail in various modes. Cruden and Varnes (1996) classified the slope failures into the following five major categories. They are

1. **Fall.** This is the detachment of soil and/or rock fragments that fall down a slope (Figure 15.1). Figure 15.2 shows a fall in which a large amount of soil mass has slid down a slope.
2. **Topple.** This is a forward rotation of soil and/or rock mass about an axis below the center of gravity of mass being displaced (Figure 15.3).
3. **Slide.** This is the downward movement of a soil mass occurring on a surface of rupture (Figure 15.4).
4. **Spread.** This is a form of slide (Figure 15.5) by translation. It occurs by "sudden movement of water-bearing seams of sands or silts overlain by clays or loaded by fills".
5. **Flow.** This is a downward movement of soil mass similar to a viscous fluid (Figure 15.6).

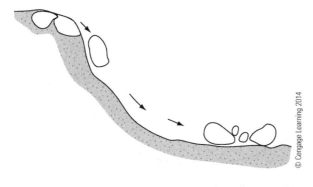

© Cengage Learning 2014

Figure 15.1 "Fall" type of landslide

Figure 15.2 Soil and rock "fall" in a slope (*Courtesy of E.C. Shin, University of Inchon, South Korea*)

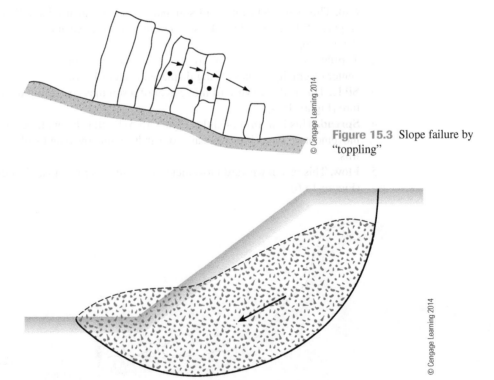

Figure 15.3 Slope failure by "toppling"

Figure 15.4 Slope failure by "sliding"

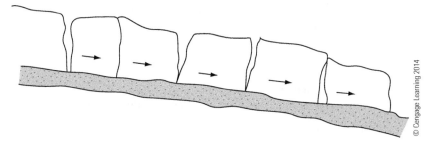

Figure 15.5 Slope failure by lateral "spreading"

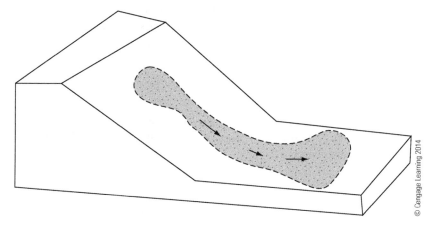

Figure 15.6 Slope failure by "flowing"

This chapter primarily relates to the quantitative analysis that fall under the category of *slide*. We will discuss in detail the following:

- Definition of factor of safety
- Stability of infinite slopes
- Stability of finite slopes with plane and circular failure surfaces
- Analysis of the stability of finite slopes with steady-state seepage and in rapid draw-down conditions

15.2 Factor of Safety

The task of the engineer charged with analyzing slope stability is to determine the factor of safety. Generally, the factor of safety is defined as

$$F_s = \frac{\tau_f}{\tau_d} \tag{15.1}$$

where F_s = factor of safety with respect to strength
τ_f = average shear strength of the soil
τ_d = average shear stress developed along the potential failure surface

The shear strength of a soil consists of two components, cohesion and friction, and may be written as

$$\tau_f = c' + \sigma' \tan \phi' \tag{15.2}$$

where c' = cohesion
ϕ' = angle of friction
σ' = normal stress on the potential failure surface

In a similar manner, we can write

$$\tau_d = c'_d + \sigma' \tan \phi'_d \tag{15.3}$$

where c'_d and ϕ'_d are, respectively, the cohesion and the angle of friction that develop along the potential failure surface. Substituting Eqs. (15.2) and (15.3) into Eq. (15.1), we get

$$F_s = \frac{c' + \sigma' \tan \phi'}{c'_d + \sigma' \tan \phi'_d} \tag{15.4}$$

Now we can introduce some other aspects of the factor of safety—that is, the factor of safety with respect to cohesion, $F_{c'}$, and the factor of safety with respect to friction, $F_{\phi'}$. They are defined as

$$F_{c'} = \frac{c'}{c'_d} \tag{15.5}$$

and

$$F_{\phi'} = \frac{\tan \phi'}{\tan \phi'_d} \tag{15.6}$$

When we compare Eqs. (15.4) through (15.6), we can see that when $F_{c'}$ becomes equal to $F_{\phi'}$, it gives the factor of safety with respect to strength. Or, if

$$\frac{c'}{c'_d} = \frac{\tan \phi'}{\tan \phi'_d}$$

then we can write

$$F_s = F_{c'} = F_{\phi'} \tag{15.7}$$

When F_s is equal to 1, the slope is in a state of impending failure. Generally, a value of 1.5 for the factor of safety with respect to strength is acceptable for the design of a stable slope.

15.3 Stability of Infinite Slopes

In considering the problem of slope stability, let us start with the case of an infinite slope as shown in Figure 15.7. The shear strength of the soil may be given by Eq. (15.2):

$$\tau_f = c' + \sigma' \tan \phi'$$

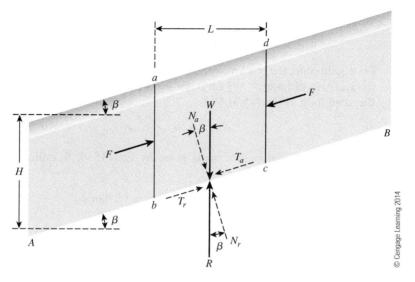

Figure 15.7 Analysis of infinite slope (without seepage)

Assuming that the pore water pressure is zero, we will evaluate the factor of safety against a possible slope failure along a plane AB located at a depth H below the ground surface. The slope failure can occur by the movement of soil above the plane AB from right to left.

Let us consider a slope element $abcd$ that has a unit length perpendicular to the plane of the section shown. The forces, F, that act on the faces ab and cd are equal and opposite and may be ignored. The weight of the soil element is

$$W = (\text{Volume of soil element}) \times (\text{Unit weight of soil}) = \gamma LH \tag{15.8}$$

The weight W can be resolved into two components:

1. Force perpendicular to the plane $AB = N_a = W \cos \beta = \gamma LH \cos \beta$.
2. Force parallel to the plane $AB = T_a = W \sin \beta = \gamma LH \sin \beta$. Note that this is the force that tends to cause the slip along the plane.

Thus, the effective normal stress and the shear stress at the base of the slope element can be given, respectively, as

$$\sigma' = \frac{N_a}{\text{Area of base}} = \frac{\gamma LH \ \cos \beta}{\left(\dfrac{L}{\cos \beta}\right)} = \gamma H \ \cos^2 \beta \tag{15.9}$$

and

$$\tau = \frac{T_a}{\text{Area of base}} = \frac{\gamma LH \ \sin \beta}{\left(\dfrac{L}{\cos \beta}\right)} = \gamma H \ \cos \beta \ \sin \beta \tag{15.10}$$

The reaction to the weight W is an equal and opposite force R. The normal and tangential components of R with respect to the plane AB are

$$N_r = R \ \cos \beta = W \ \cos \beta \tag{15.11}$$

and

$$T_r = R \sin \beta = W \sin \beta \qquad (15.12)$$

For equilibrium, the resistive shear stress that develops at the base of the element is equal to $(T_r)/(\text{Area of base}) = \gamma H \sin \beta \cos \beta$. The resistive shear stress also may be written in the same form as Eq. (15.3):

$$\tau_d = c'_d + \sigma' \tan \phi'_d$$

The value of the normal stress is given by Eq. (15.9). Substitution of Eq. (15.9) into Eq. (15.3) yields

$$\tau_d = c'_d + \gamma H \cos^2 \beta \tan \phi'_d \qquad (15.13)$$

Thus,

$$\gamma H \sin \beta \cos \beta = c'_d + \gamma H \cos^2 \beta \tan \phi'_d$$

or

$$\frac{c'_d}{\gamma H} = \sin \beta \cos \beta - \cos^2 \beta \tan \phi'_d$$

$$= \cos^2 \beta (\tan \beta - \tan \phi'_d) \qquad (15.14)$$

The factor of safety with respect to strength has been defined in Eq. (15.7), from which we get

$$\tan \phi'_d = \frac{\tan \phi'}{F_s} \quad \text{and} \quad c'_d = \frac{c'}{F_s}$$

Substituting the preceding relationships into Eq. (15.14), we obtain

$$F_s = \frac{c'}{\gamma H \cos^2 \beta \tan \beta} + \frac{\tan \phi'}{\tan \beta} \qquad (15.15)$$

For granular soils, $c' = 0$, and the factor of safety, F_s, becomes equal to $(\tan \phi')/(\tan \beta)$. This indicates that in an infinite slope in sand, the value of F_s is independent of the height H and the slope is stable as long as $\beta < \phi'$.

If a soil possesses cohesion and friction, the depth of the plane along which critical equilibrium occurs may be determined by substituting $F_s = 1$ and $H = H_{cr}$ into Eq. (15.15). Thus,

$$H_{cr} = \frac{c'}{\gamma} \frac{1}{\cos^2 \beta (\tan \beta - \tan \phi')} \qquad (15.16)$$

15.4 Infinite Slope with Steady-state Seepage

Figure 15.8a shows an infinite slope. We will assume that there is seepage through the soil and that the groundwater level coincides with the ground surface. The shear strength of the soil is given by

$$\tau_f = c' + \sigma' \tan \phi' \tag{15.17}$$

To determine the factor of safety against failure along the plane AB, consider the slope element $abcd$. The forces that act on the vertical faces ab and cd are equal and opposite. The total weight of the slope element of unit length is

$$W = \gamma_{sat} LH \tag{15.18}$$

where γ_{sat} = saturated unit weight of soil.

The components of W in the directions normal and parallel to plane AB are

$$N_a = W \cos \beta = \gamma_{sat} LH \cos \beta \tag{15.19}$$

and

$$T_a = W \sin \beta = \gamma_{sat} LH \sin \beta \tag{15.20}$$

The reaction to the weight W is equal to R. Thus,

$$N_r = R \cos \beta = W \cos \beta = \gamma_{sat} LH \cos \beta \tag{15.21}$$

and

$$T_r = R \sin \beta = W \sin \beta = \gamma_{sat} LH \sin \beta \tag{15.22}$$

The total normal stress and the shear stress at the base of the element are, respectively,

$$\sigma = \frac{N_r}{\left(\dfrac{L}{\cos \beta}\right)} = \gamma_{sat} H \cos^2 \beta \tag{15.23}$$

and

$$\tau = \frac{T_r}{\left(\dfrac{L}{\cos \beta}\right)} = \gamma_{sat} H \cos \beta \sin \beta \tag{15.24}$$

The resistive shear stress developed at the base of the element also can be given by

$$\tau_d = c'_d + \sigma' \tan \phi_d = c'_d + (\sigma - u) \tan \phi'_d \tag{15.25}$$

where u = pore water pressure. Referring to Figure 15.8b, we see that

$$u = (\text{Height of water in piezometer placed at } f)(\gamma_w) = h\gamma_w$$

and

$$h = \overline{ef} \cos \beta = (H \cos \beta)(\cos \beta) = H \cos^2 \beta$$

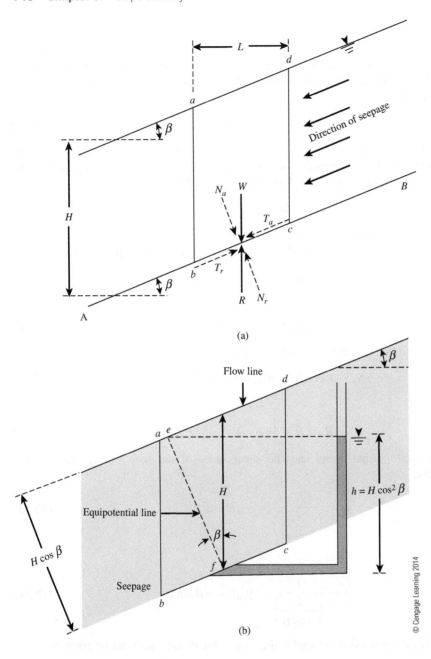

(a)

(b)

Figure 15.8 Analysis of infinite slope (with seepage)

© Cengage Learning 2014

so

$$u = \gamma_w H \cos^2\beta$$

Substituting the values of σ [Eq. (15.23)] and u into Eq. (15.25), we get

$$\tau_d = c'_d + (\gamma_{sat} H \cos^2\beta - \gamma_w H \cos^2\beta) \tan \phi'_d$$
$$= c'_d + \gamma' H \cos^2\beta \tan \phi'_d \qquad\qquad (15.26)$$

Now, setting the right-hand sides of Eqs. (15.24) and (15.26) equal to each other gives

$$\gamma_{sat}H \cos \beta \sin \beta = c'_d + \gamma'H \cos^2\beta \tan \phi'_d$$

or

$$\frac{c'_d}{\gamma_{sat}H} = \cos^2\beta\left(\tan \beta - \frac{\gamma'}{\gamma_{sat}} \tan \phi'_d\right) \tag{15.27}$$

where $\gamma' = \gamma_{sat} - \gamma_w$ = effective unit weight of soil.

The factor of safety with respect to strength can be found by substituting $\tan \phi'_d = (\tan \phi')/F_s$ and $c'_d = c'/F_s$ into Eq. (15.27), or

$$F_s = \frac{c'}{\gamma_{sat}H \cos^2\beta \tan \beta} + \frac{\gamma' \tan \phi'}{\gamma_{sat} \tan \beta} \tag{15.28}$$

Example 15.1

For the infinite slope shown in Figure 15.9 (consider that there is no seepage through the soil), determine:

a. The factor of safety against sliding along the soil–rock interface
b. The height, H, that will give a factor of safety (F_s) of 2 against sliding along the soil–rock interface

Solution
Part a

From Eq. (15.15),

$$F_s = \frac{c'}{\gamma H \cos^2\beta \tan \beta} + \frac{\tan \phi'}{\tan \beta}$$

Given: $c' = 200$ lb/ft^2, $\gamma = 100$ lb/ft^3, $\phi' = 15°$, $\beta = 25°$, and $H = 8$ ft, we have

$$F_s = \frac{200}{(100)(8)(\cos^225)(\tan 25)} + \frac{\tan 15}{\tan 25} = \mathbf{1.23}$$

Part b

From Eq. (15.15),

$$F_s = \frac{c'}{\gamma H \cos^2\beta \tan \beta} + \frac{\tan \phi'}{\tan \beta}$$

$$2 = \frac{200}{(100)(H)(\cos^225)(\tan 25)} + \frac{\tan 15}{\tan 25}$$

$$H = \mathbf{3.66\ ft}$$

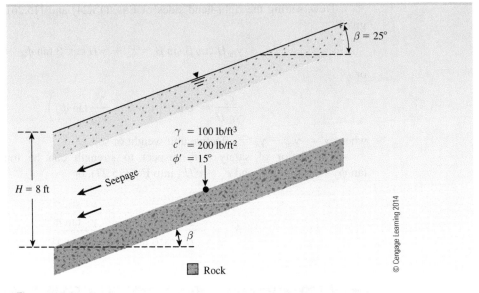

$\beta = 25°$

$\gamma = 100 \text{ lb/ft}^3$
$c' = 200 \text{ lb/ft}^2$
$\phi' = 15°$

$H = 8 \text{ ft}$

Seepage

β

■ Rock

© Cengage Learning 2014

Figure 15.9

Example 15.2

Refer to Figure 15.9. If there is seepage through the soil as shown and the groundwater table coincides with the ground surface, what is the factor of safety, F_s, given $H = 3.66$ ft and $\gamma_{sat} = 118 \text{ lb/ft}^3$?

Solution
From Eq. (15.28),

$$F_s = \frac{c'}{\gamma_{sat} H \cos^2\beta \tan\beta} + \frac{\gamma'}{\gamma_{sat}} \frac{\tan\phi'}{\tan\beta}$$

or

$$F_s = \frac{200}{(118)(3.66)(\cos^2 25)(\tan 25)} + \frac{(118 - 62.4)}{118}\left(\frac{\tan 15}{\tan 25}\right) = \mathbf{1.48}$$

15.5 Finite Slopes—General

When the value of H_{cr} approaches the height of the slope, the slope generally may be considered finite. For simplicity, when analyzing the stability of a finite slope in a homogeneous soil, we need to make an assumption about the general shape of the surface of potential failure. Although considerable evidence suggests that slope failures usually occur on curved failure surfaces, Culmann (1875) approximated the surface of potential failure as a plane. The factor of safety, F_s, calculated by using Culmann's approximation, gives fairly good results for near-vertical slopes only. After extensive investigation of slope failures in the 1920s, a Swedish geotechnical commission recommended that the actual surface of sliding may be approximated to be circularly cylindrical.

Since that time, most conventional stability analyses of slopes have been made by assuming that the curve of potential sliding is an arc of a circle. However, in many circumstances (for example, zoned dams and foundations on weak strata), stability analysis using plane failure of sliding is more appropriate and yields excellent results.

15.6 Analysis of Finite Slopes with Plane Failure Surfaces (Culmann's Method)

Culmann's analysis is based on the assumption that the failure of a slope occurs along a plane when the average shearing stress tending to cause the slip is more than the shear strength of the soil. Also, the most critical plane is the one that has a minimum ratio of the average shearing stress that tends to cause failure to the shear strength of soil.

Figure 15.10 shows a slope of height H. The slope rises at an angle β with the horizontal. AC is a trial failure plane. If we consider a unit length perpendicular to the section of the slope, we find that the weight of the wedge ABC is equal to

$$W = \frac{1}{2}(H)(\overline{BC})(1)(\gamma) = \frac{1}{2}H(H\cot\theta - H\cot\beta)\gamma$$

$$= \frac{1}{2}\gamma H^2\left[\frac{\sin(\beta-\theta)}{\sin\beta\,\sin\theta}\right] \tag{15.29}$$

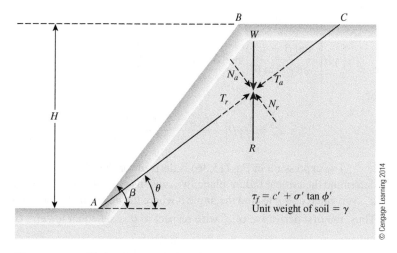

Figure 15.10 Finite slope analysis—Culmann's method

The normal and tangential components of W with respect to the plane AC are as follows.

$$N_a = \text{normal component} = W \cos \theta = \frac{1}{2}\gamma H^2 \left[\frac{\sin(\beta - \theta)}{\sin \beta \, \sin \theta} \right] \cos \theta \qquad (15.30)$$

$$T_a = \text{tangential component} = W \sin \theta = \frac{1}{2}\gamma H^2 \left[\frac{\sin(\beta - \theta)}{\sin \beta \, \sin \theta} \right] \sin \theta \qquad (15.31)$$

The average effective normal stress and the average shear stress on the plane AC are, respectively,

$$\sigma' = \frac{N_a}{(\overline{AC})(1)} = \frac{N_a}{\left(\dfrac{H}{\sin \theta} \right)}$$

$$= \frac{1}{2}\gamma H \left[\frac{\sin(\beta - \theta)}{\sin \beta \, \sin \theta} \right] \cos \theta \, \sin \theta \qquad (15.32)$$

and

$$\tau = \frac{T_a}{(\overline{AC})(1)} = \frac{T_a}{\left(\dfrac{H}{\sin \theta} \right)}$$

$$= \frac{1}{2}\gamma H \left[\frac{\sin(\beta - \theta)}{\sin \beta \, \sin \theta} \right] \sin^2 \theta \qquad (15.33)$$

The average resistive shearing stress developed along the plane AC also may be expressed as

$$\tau_d = c_d' + \sigma' \tan \phi_d'$$

$$= c_d' + \frac{1}{2}\gamma H \left[\frac{\sin(\beta - \theta)}{\sin \beta \, \sin \theta} \right] \cos \theta \, \sin \theta \, \tan \phi_d' \qquad (15.34)$$

Now, from Eqs. (15.33) and (15.34),

$$\frac{1}{2}\gamma H \left[\frac{\sin(\beta - \theta)}{\sin \beta \, \sin \theta} \right] \sin^2 \theta = c_d' + \frac{1}{2}\gamma H \left[\frac{\sin(\beta - \theta)}{\sin \beta \, \sin \theta} \right] \cos \theta \, \sin \theta \, \tan \phi_d' \qquad (15.35)$$

or

$$c_d' = \frac{1}{2}\gamma H \left[\frac{\sin(\beta - \theta)(\sin \theta - \cos \theta \tan \phi_d')}{\sin \beta} \right] \qquad (15.36)$$

The expression in Eq. (15.36) is derived for the trial failure plane AC. In an effort to determine the critical failure plane, we must use the principle of maxima and minima (for a given value of ϕ_d') to find the angle θ where the developed cohesion would be maximum. Thus, the first derivative of c_d' with respect to θ is set equal to zero, or

$$\frac{\partial c_d'}{\partial \theta} = 0 \qquad (15.37)$$

Because γ, H, and β are constants in Eq. (15.36), we have

$$\frac{\partial}{\partial \theta}[\sin(\beta - \theta)(\sin \theta - \cos \theta \, \tan \phi'_d)] = 0 \tag{15.38}$$

Solving Eq. (15.38) gives the critical value of θ, or

$$\theta_{cr} = \frac{\beta + \phi'_d}{2} \tag{15.39}$$

Substitution of the value of $\theta = \theta_{cr}$ into Eq. (15.36) yields

$$c'_d = \frac{\gamma H}{4}\left[\frac{1 - \cos(\beta - \phi'_d)}{\sin \beta \, \cos \phi'_d}\right] \tag{15.40}$$

The preceding equation also can be written as

$$\frac{c'_d}{\gamma H} = m = \frac{1 - \cos(\beta - \phi'_d)}{4 \sin \beta \, \cos \phi'_d} \tag{15.41}$$

where m = stability number.

The maximum height of the slope for which critical equilibrium occurs can be obtained by substituting $c'_d = c'$ and $\phi'_d = \phi'$ into Eq. (15.40). Thus,

$$H_{cr} = \frac{4c'}{\gamma}\left[\frac{\sin \beta \, \cos \phi'}{1 - \cos(\beta - \phi')}\right] \tag{15.42}$$

Example 15.3

A cut is to be made in a soil having $\gamma = 16.5$ kN/m^3, $c' = 28.75$ kN/m^2, and $\phi' = 15°$. The side of the cut slope will make an angle of $45°$ with the horizontal. What should be the depth of the cut slope that will have a factor of safety (F_s) of 3?

Solution

Given: $\phi' = 15°$; $c' = 28.75$ kN/m^2. If $F_s = 3$, then $F_{c'}$ and $F_{\phi'}$ should both be equal to 3.

$$F_{c'} = \frac{c'}{c'_d}$$

or

$$c'_d = \frac{c'}{F_{c'}} = \frac{c'}{F_s} = \frac{28.75}{3} = 9.58 \text{ kN/m}^2$$

Similarly,

$$F_{\phi'} = \frac{\tan \phi'}{\tan \phi'_d}$$

$$\tan \phi'_d = \frac{\tan \phi'}{F_{\phi'}} = \frac{\tan \phi'}{F_s} = \frac{\tan 15}{3}$$

or

$$\phi'_d = \tan^{-1}\left[\frac{\tan 15}{3}\right] = 5.1°$$

Substituting the preceding values of c'_d and ϕ'_d in Eq. (15.40),

$$H = \frac{4c'_d}{\gamma}\left[\frac{\sin\beta \cdot \cos\phi'_d}{1 - \cos(\beta - \phi'_d)}\right]$$

$$= \frac{4 \times 9.58}{16.5}\left[\frac{\sin 45 \cdot \cos 5.1}{1 - \cos(45 - 5.1)}\right]$$

$$= \textbf{7.03 m}$$

15.7 Analysis of Finite Slopes with Circular Failure Surfaces—General

Modes of Failure

In general, finite slope failure occurs in one of the following modes (Figure 15.11):

1. When the failure occurs in such a way that the surface of sliding intersects the slope at or above its toe, it is called a *slope failure* (Figure 15.11a). The failure circle is referred to as a *toe circle* if it passes through the toe of the slope and as a *slope circle* if it passes above the toe of the slope. Under certain circumstances, a *shallow slope failure* can occur, as shown in Figure 15.11b.
2. When the failure occurs in such a way that the surface of sliding passes at some distance below the toe of the slope, it is called a *base failure* (Figure 15.11c). The failure circle in the case of base failure is called a *midpoint circle*.

Types of Stability Analysis Procedures

Various procedures of stability analysis may, in general, be divided into two major classes:

1. *Mass procedure:* In this case, the mass of the soil above the surface of sliding is taken as a unit. This procedure is useful when the soil that forms the slope is assumed to be homogeneous, although this is not the case in most natural slopes.
2. *Method of slices:* In this procedure, the soil above the surface of sliding is divided into a number of vertical parallel slices. The stability of each slice is calculated separately. This is a versatile technique in which the nonhomogeneity of the soils and pore water pressure can be taken into consideration. It also accounts for the variation of the normal stress along the potential failure surface.

The fundamentals of the analysis of slope stability by mass procedure and method of slices are given in the following sections.

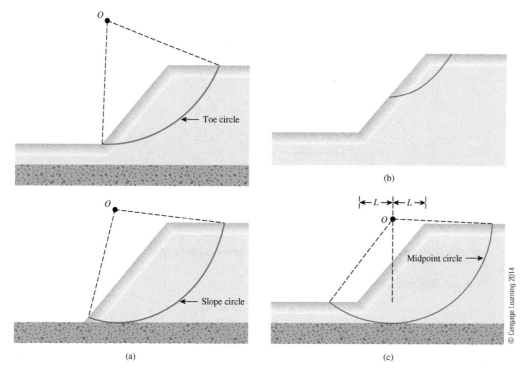

Figure 15.11 Modes of failure of finite slope: (a) slope failure; (b) shallow slope failure; (c) base failure

15.8 Mass Procedure—Slopes in Homogeneous Clay Soil with $\phi = 0$

Figure 15.12 shows a slope in a homogeneous soil. The undrained shear strength of the soil is assumed to be constant with depth and may be given by $\tau_f = c_u$. To perform the stability analysis, we choose a trial potential curve of sliding, *AED,* which is an arc of a circle that has a radius *r.* The center of the circle is located at *O.* Considering a unit length perpendicular to the section of the slope, we can give the weight of the soil above the curve *AED* as $W = W_1 + W_2$, where

$$W_1 = (\text{Area of } FCDEF)(\gamma)$$

and

$$W_2 = (\text{Area of } ABFEA)(\gamma)$$

Failure of the slope may occur by sliding of the soil mass. The moment of the driving force about *O* to cause slope instability is

$$M_d = W_1 l_1 - W_2 l_2 \tag{15.43}$$

where l_1 and l_2 are the moment arms.

The resistance to sliding is derived from the cohesion that acts along the potential surface of sliding. If c_d is the cohesion that needs to be developed, the moment of the resisting forces about *O* is

$$M_R = c_d(\widehat{AED})(1)(r) = c_d r^2 \theta \tag{15.44}$$

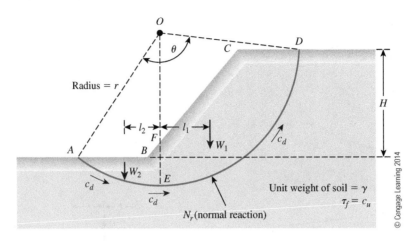

Figure 15.12 Stability analysis of slope in homogeneous saturated clay soil ($\phi=0$)

For equilibrium, $M_R = M_d$; thus,

$$c_d r^2 \theta = W_1 l_1 - W_2 l_2$$

or

$$c_d = \frac{W_1 l_1 - W_2 l_2}{r^2 \theta} \tag{15.45}$$

The factor of safety against sliding may now be found.

$$F_s = \frac{\tau_f}{c_d} = \frac{c_u}{c_d} \tag{15.46}$$

Note that the potential curve of sliding, *AED*, was chosen arbitrarily. The critical surface is that for which the ratio of c_u to c_d is a minimum. In other words, c_d is maximum. To find the critical surface for sliding, one must make a number of trials for different trial circles. The minimum value of the factor of safety thus obtained is the factor of safety against sliding for the slope, and the corresponding circle is the critical circle.

Stability problems of this type have been solved analytically by Fellenius (1927) and Taylor (1937). For the case of *critical circles*, the developed cohesion can be expressed by the relationship

$$c_d = \gamma H m$$

or

$$\frac{c_d}{\gamma H} = m \tag{15.47}$$

Note that the term m on the right-hand side of the preceding equation is nondimensional and is referred to as the *stability number*. The critical height (i.e., $F_s = 1$) of the slope can be evaluated by substituting $H = H_{cr}$ and $c_d = c_u$ (full mobilization of the undrained shear strength) into the preceding equation. Thus,

$$H_{cr} = \frac{c_u}{\gamma m} \tag{15.48}$$

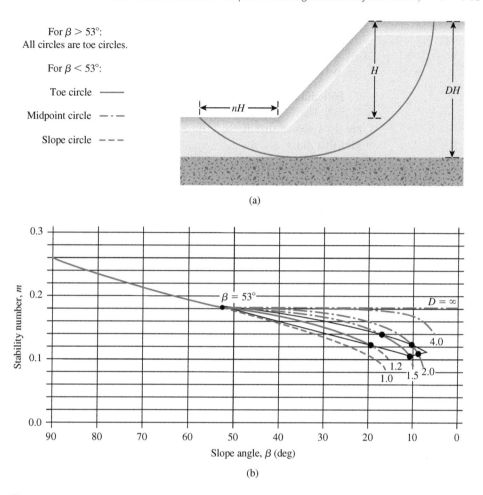

Figure 15.13 (a) Definition of parameters for midpoint circle type of failure; (b) plot of stability number against slope angle (*Adapted from Terzaghi and Peck, 1967. With permission of John Wiley & Sons, Inc.*)

Values of the stability number, m, for various slope angles, β, are given in Figure 15.13. Terzaghi used the term $\gamma H/c_d$, the reciprocal of m and called it the *stability factor.* Readers should be careful in using Figure 15.13 and note that it is valid for slopes of saturated clay and is applicable to only undrained conditions ($\phi = 0$).

In reference to Figure 15.13, the following must be pointed out:

1. For a slope angle β greater than 53°, the critical circle is always a toe circle. The location of the center of the critical toe circle may be found with the aid of Figure 15.14.

2. For $\beta < 53°$, the critical circle may be a toe, slope, or midpoint circle, depending on the location of the firm base under the slope. This is called the *depth function,* which is defined as

$$D = \frac{\text{Vertical distance from top of slope to firm base}}{\text{Height of slope}} \qquad (15.49)$$

3. When the critical circle is a midpoint circle (i.e., the failure surface is tangent to the firm base), its position can be determined with the aid of Figure 15.15.

4. The maximum possible value of the stability number for failure as a midpoint circle is 0.181.

Fellenius (1927) also investigated the case of critical toe circles for slopes with $\beta < 53°$. The location of these can be determined with the use of Figure 15.16 and Table 15.1. Note that these critical toe circles are not necessarily the most critical circles that exist.

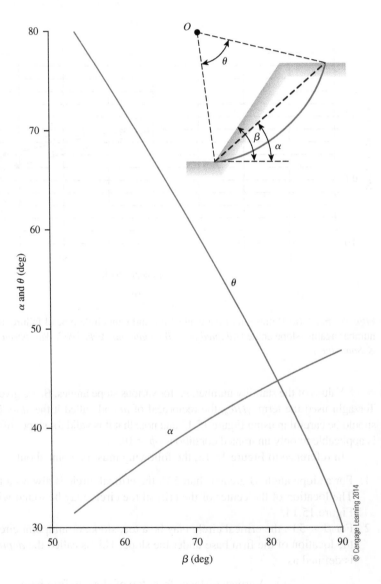

Figure 15.14 Location of the center of critical circles for $\beta > 53°$

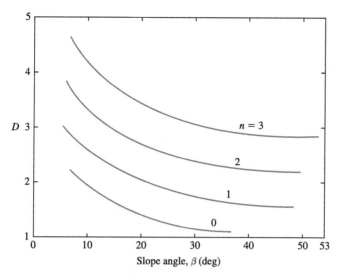

Figure 15.15 Location of midpoint circle (*Based on Fellenius, 1927; and Terzaghi and Peck, 1967*)

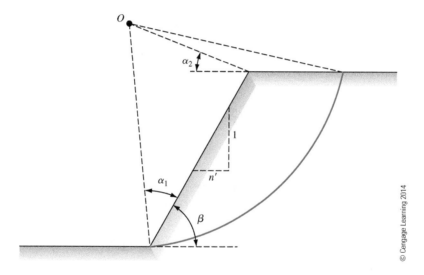

© Cengage Learning 2014

Figure 15.16 Location of the center of critical toe circles for $\beta < 53°$

Table 15.1 Location of the Center of Critical Toe Circles ($\beta < 53°$)

n'	β (deg)	α_1 (deg)	α_2 (deg)
1.0	45	28	37
1.5	33.68	26	35
2.0	26.57	25	35
3.0	18.43	25	35
5.0	11.32	25	37

© Cengage Learning 2014

Note: For notations of n', β, α_1, and α_2, see Figure 15.16.

Example 15.4

A cut slope is to be made in a soft saturated clay with its sides rising at an angle of 60° to the horizontal (Figure 15.17).
Given: $c_u = 40$ kN/m^2 and $\gamma = 17.5$ kN/m^3.
a. Determine the maximum depth up to which the excavation can be carried out.
b. Find the radius, r, of the critical circle when the factor of safety is equal to 1 (Part a).
c. Find the distance $\overline{BC}$.

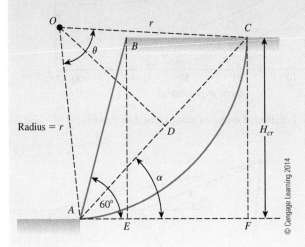

Figure 15.17

Solution
Part a

Since the slope angle $\beta = 60° > 53°$, the critical circle is a toe circle. From Figure 15.13, for $\beta = 60°$, the stability number $= 0.195$.

$$H_{cr} = \frac{c_u}{\gamma m} = \frac{40}{17.5 \times 0.195} = \textbf{11.72 m}$$

Part b

From Figure 15.17,

$$r = \frac{\overline{DC}}{\sin \dfrac{\theta}{2}}$$

But

$$\overline{DC} = \frac{\overline{AC}}{2} = \frac{\left(\dfrac{H_{cr}}{\sin \alpha}\right)}{2}$$

so,

$$r = \frac{H_{cr}}{2 \sin \alpha \ \sin \dfrac{\theta}{2}}$$

From Figure 15.14, for $\beta = 60°$, $\alpha = 35°$ and $\theta = 72.5°$. Substituting these values into the equation for r, we get

$$r = \frac{H_{cr}}{2 \sin \alpha \ \sin \dfrac{\theta}{2}}$$

$$= \frac{11.72}{2(\sin 35)(\sin 36.25)} = \textbf{17.28 m}$$

Part c

$$\overline{BC} = \overline{EF} = \overline{AF} - \overline{AE}$$
$$= H_{cr}(\cot \alpha - \cot 75°)$$
$$= 11.72(\cot 35 - \cot 60) = \textbf{9.97 m}$$

Example 15.5

A cut slope was excavated in a saturated clay. The slope made an angle of 40° with the horizontal. Slope failure occurred when the cut reached a depth of 7 m. Previous soil explorations showed that a rock layer was located at a depth of 10.5 m below the ground surface. Assuming an undrained condition and $\gamma_{sat} = 18$ kN/m³, find the following.

a. Determine the undrained cohesion of the clay (use Figure 15.13).
b. What was the nature of the critical circle?
c. With reference to the toe of the slope, at what distance did the surface of sliding intersect the bottom of the excavation?

Solution
Part a
Referring to Figure 15.13,

$$D = \frac{10.5}{7} = 1.5$$

$$\gamma_{sat} = 18 \text{ kN/m}^3$$

$$H_{cr} = \frac{c_u}{\gamma m}$$

From Figure 15.13, for $\beta = 40°$ and $D = 1.5$, $m = 0.175$. So,

$$c_u = (H_{cr})(\gamma)(m) = (7)(18)(0.175) = \mathbf{22.05\ kN/m^2}$$

Part b

Midpoint circle.

Part c

Again, from Figure 15.15, for $D = 1.5$, $\beta = 40°$; $n = 0.9$. So,

$$\text{Distance} = (n)(H_{cr}) = (0.9)(7) = \mathbf{6.3\ m}$$

15.9 Recent Developments on Critical Circle of Clay Slopes ($\phi = 0$)

More recently, Steward, Sivakugan, Shukla and Das (2011) made hundreds of runs using SLOPE/W to locate the critical circles of several slopes with different geometry and soil properties. According to this study, there appears to be five types of critical circles. (They are Figure 15.18):

- Compound slope circle
- Compound toe circle

a. Compound slope circle

d. Touch midpoint circle

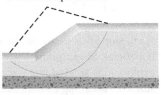

b. Compound toe circle

e. Shallow toe circle

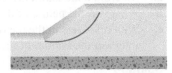

c. Compound midpoint circle

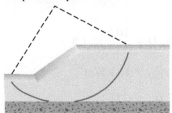

Figure 15.18 Nature of critical circles observed by Steward, Sivakugan, Shukla, and Das (2011)

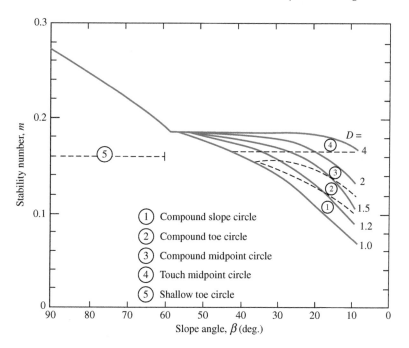

Figure 15.19 Analysis of Steward et al. (2011)—plot of m vs. β with zones of critical circles shown in Figure 15.18

- Compound midpoint circle
- Touch midpoint circle
- Shallow toe circle

Also, when $\beta > 58°$ the failure circle is a shallow toe circle. This is slightly different to the angle $53°$ shown in Figure 15.13. Based on this analysis, Figure 15.19 shows the plot of stability number (m) against β along with zones in which the above-stated five types of critical circles are encountered.

15.10 Mass Procedure–Slopes in Homogeneous $c' - \phi'$ Soil

A slope in a homogeneous soil is shown in Figure 15.20a. The shear strength of the soil is given by

$$\tau_f = c' + \sigma' \tan \phi'$$

The pore water pressure is assumed to be zero. $\overarc{AC}$ is a trial circular arc that passes through the toe of the slope, and O is the center of the circle. Considering a unit length perpendicular to the section of the slope, we find

$$\text{Weight of soil wedge } ABC = W = (\text{Area of } ABC)(\gamma)$$

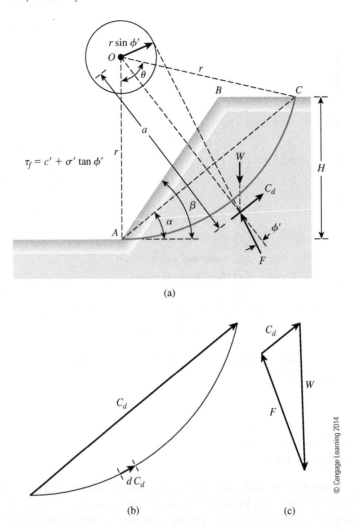

Figure 15.20 Stability analysis of slope in homogeneous $c' - \phi'$ soil

For equilibrium, the following other forces are acting on the wedge:

- C_d—resultant of the cohesive force that is equal to the cohesion per unit area developed times the length of the cord $\overline{AC}$. The magnitude of C_d is given by the following (Figure 15.20b).

$$C_d = c'_d(\overline{AC}) \tag{15.50}$$

C_d acts in a direction parallel to the cord $\overline{AC}$ (see Figure 15.20 b) and at a distance a from the center of the circle O such that

$$C_d(a) = c'_d(\widehat{AC})r$$

or

$$a = \frac{c'_d(\widehat{AC})r}{C_d} = \frac{\widehat{AC}}{\overline{AC}}r \qquad (15.51)$$

- F—the resultant of the normal and frictional forces along the surface of sliding. For equilibrium, the line of action of F will pass through the point of intersection of the line of action of W and C_d.

Now, if we assume that full friction is mobilized ($\phi'_d = \phi'$ or $F_{\phi'}=1$), the line of action of F will make an angle of ϕ' with a normal to the arc and thus will be a tangent to a circle with its center at O and having a radius of $r \sin \phi'$. This circle is called the *friction circle*. Actually, the radius of the friction circle is a little larger than $r \sin \phi'$.

Because the directions of W, C_d, and F are known and the magnitude of W is known, a force polygon, as shown in Figure 15.20c, can be plotted. The magnitude of C_d can be determined from the force polygon. So the cohesion per unit area developed can be found.

$$c'_d = \frac{C_d}{\overline{AC}}$$

Determination of the magnitude of c'_d described previously is based on a trial surface of sliding. Several trials must be made to obtain the most critical sliding surface, along which the developed cohesion is a maximum. Thus, we can express the maximum cohesion developed along the critical surface as

$$c'_d = \gamma H[f(\alpha, \beta, \theta, \phi')] \qquad (15.52)$$

For critical equilibrium—that is, $F_{c'} = F_{\phi'} = F_s = 1$—we can substitute $H = H_{cr}$ and $c'_d = c'$ into Eq. (15.52) and write

$$c' = \gamma H_{cr}[f(\alpha, \beta, \theta, \phi')]$$

or

$$\frac{c'}{\gamma H_{cr}} = f(\alpha, \beta, \theta, \phi') = m \qquad (15.53)$$

where m = stability number. The values of m for various values of ϕ' and β are given in Figure 15.21, which is based on Taylor (1937). This can be used to determine the factor of safety, F_s, of the homogeneous slope. The procedure to do the analysis is given as

Step 1: Determine c', ϕ', γ, β and H.
Step 2: Assume several values of ϕ'_d (Note: $\phi'_d \leq \phi'$, such as $\phi'_{d(1)}, \phi'_{d(2)}$. . . . (Column 1 of Table 15.2).

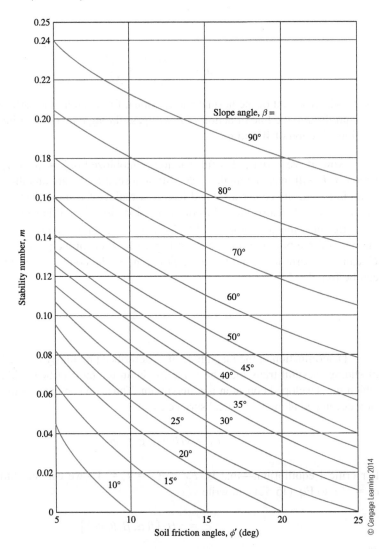

Figure 15.21 Taylor's stability number

Table 15.2 Determination of F_s by Friction Circle Method

ϕ'_d	$F_{\phi'} = \dfrac{\tan \phi'}{\tan \phi'_d}$	m	c'_d	$F_{c'}$
(1)	(2)	(3)	(4)	(5)
$\phi'_{d(1)}$	$\dfrac{\tan \phi'}{\tan \phi'_{d(1)}}$	m_1	$m_1 \gamma H = c'_{d(1)}$	$\dfrac{c'}{c'_{d(1)}} = F_{c'(1)}$
$\phi'_{d(2)}$	$\dfrac{\tan \phi'}{\tan \phi'_{d(2)}}$	m_2	$m_2 \gamma H = c'_{d(2)}$	$\dfrac{c'}{c'_{d(2)}} = F_{c'(2)}$

Step 3: Determine $F_{\phi'}$, for each assumed value of ϕ'_d as (Column 2, Table 15.2)

$$F_{\phi'(1)} = \frac{\tan \phi'}{\tan \phi'_{d(1)}}$$

$$F_{\phi'(2)} = \frac{\tan \phi'}{\tan \phi'_{d(2)}}$$

Step 4: For each assumed value of ϕ'_d and β, determine m (that is, $m_1, m_2, m_3, \ldots$) from Figure 15.21 (Column 3, Table 15.2).

Step 5: Determine the developed cohesion for each value of m as (Column 4, Table 15.2)

$$c'_{d(1)} = m_1 \gamma H$$

$$c'_{d(2)} = m_2 \gamma H$$

Step 6: Calculate $F_{c'}$ for each value of c'_d (Column 5, Table 15.2), or

$$F_{c'(1)} = \frac{c'}{c'_{d(1)}}$$

$$F_{c'(2)} = \frac{c'}{c'_{d(2)}}$$

Step 7: Plot a graph of $F_{\phi'}$ versus the corresponding $F_{c'}$ (Figure 15.22) and determine $F_s = F_{\phi'} = F_{c'}$.

An example of determining F_s using the procedure just described is given in Example 15.6.

Using Taylor's friction circle method of slope stability (as shown in Example 15.6) Singh (1970) provided graphs of equal factors of safety, F_s, for various slopes. This is shown in Figure 15.23.

Calculations have shown that for $\phi > \sim 3°$, the critical circles are all *toe circles*.

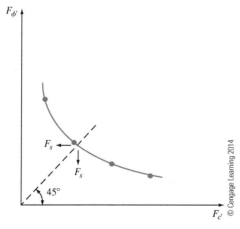

Figure 15.22 Plot of $F_{\phi'}$ versus $F_{c'}$ to determine F_s

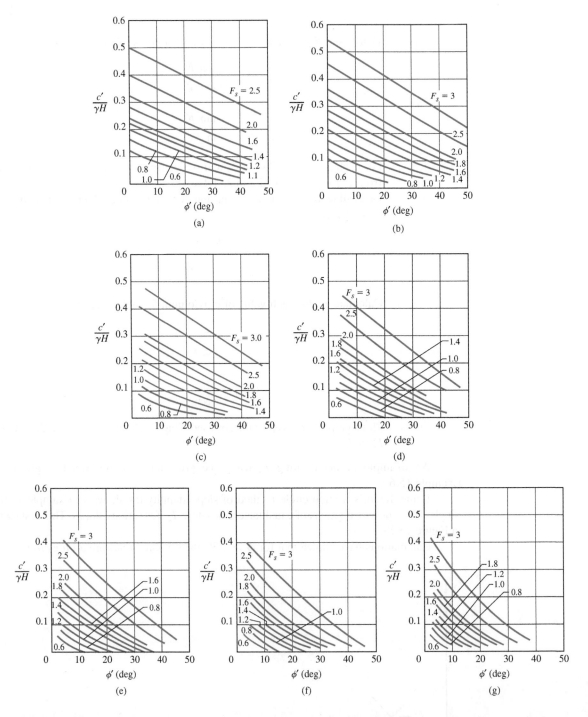

Figure 15.23 Contours of equal factors of safety: (a) slope − 1 vertical to 0.5 horizontal; (b) slope − 1 vertical to 0.75 horizontal; (c) slope − 1 vertical to 1 horizontal; (d) slope − 1 vertical to 1.5 horizontal; (e) slope − 1 vertical to 2 horizontal; (f) slope − 1 vertical to 2.5 horizontal; (g) slope − 1 vertical to 3 horizontal (*After Singh, 1970. With permission from ASCE.*)

Analysis of Michalowski (2002)

Michalowski (2002) made a stability analysis of simple slopes using the kinematic approach of limit analysis applied to a rigid rotational collapse mechanism. The failure surface in soil assumed in this study is an arc of a logarithmic spiral (Figure 15.24). The results of this study are summarized in Figure 15.25, from which F_s can be obtained directly (See Example 15.7).

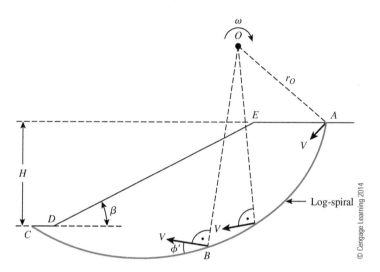

Figure 15.24 Stability analysis using rotational collapse mechanism

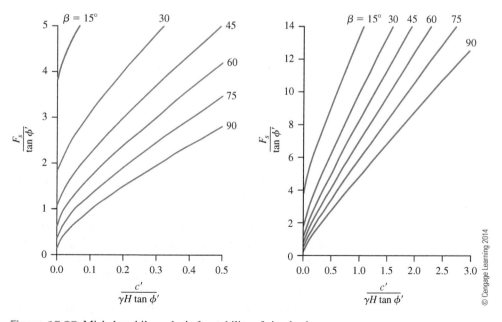

Figure 15.25 Michalowski's analysis for stability of simple slopes

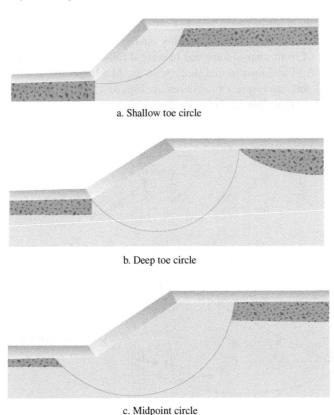

a. Shallow toe circle

b. Deep toe circle

c. Midpoint circle

Figure 15.26 Failure circles observed by Steward et al. (2011) for slopes in $c' - \phi'$ soil

Analysis of Steward, Sivakuga, Shukla, and Das (2011)

Steward et al. (2011) made hundreds of runs using SLOPE/W to locate the critical circles of slopes with $c' - \phi'$ soil. According to this study, the failure circles are mostly toe circles. However, in a few cases, as shown in Figure 15.26, they can be midpoint circles. Based on their study, a design chart has been developed and is shown in Figure 15.27.

15.11 Ordinary Method of Slices

Stability analysis by using the method of slices can be explained with the use of Figure 15.29a on page 611, in which AC is an arc of a circle representing the trial failure surface. The soil above the trial failure surface is divided into several vertical slices. The width of each slice need not be the same. Considering a unit length perpendicular to the cross section shown, the forces that act on a typical slice (nth slice) are shown in Figure 15.29b. W_n is the weight of the slice. The forces N_r and T_r, respectively, are the normal and tangential components of the reaction R. P_n and P_n+1 are the normal forces that act on the sides of the slice. Similarly, the shearing forces that act on the sides of the slice are T_n and T_n+1. For simplicity, the pore water pressure is assumed to be zero. The forces P_n, P_n+1, T_n, and T_n+1 are difficult to determine. However, we can make an approximate assumption that the resultants of P_n and T_n are equal in magnitude to the resultants of P_n+1 and T_n+1 and that their lines of action coincide.

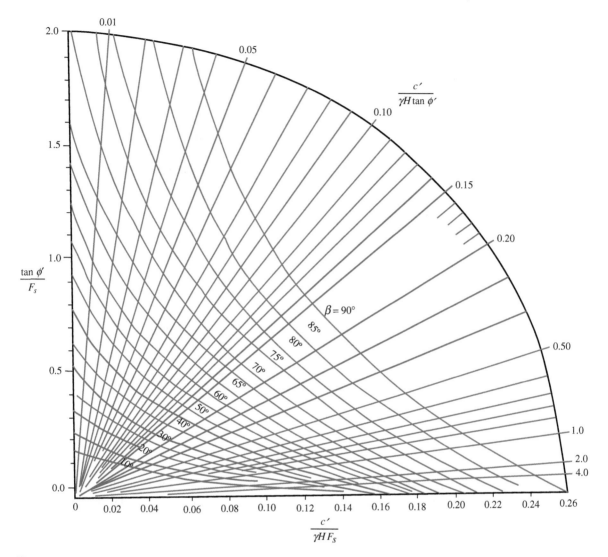

Figure 15.27 Analysis of Steward et al. (2011)—Design chart to estimate F_s

---

Example 15.6

A slope with $\beta = 45°$ is to be constructed with a soil that has $\phi' = 20°$ and $c' = 24$ kN/m². The unit weight of the compacted soil will be 18.9 kN/m³.

a. Find the critical height of the slope.
b. If the height of the slope is 10 m, determine the factor of safety with respect to strength.

Solution

Part a

We have

$$m = \frac{c'}{\gamma H_{cr}}$$

From Figure 15.21, for $\beta = 45°$ and $\phi' = 20°$, $m = 0.06$. So

$$H_{cr} = \frac{c'}{\gamma m} = \frac{24}{(18.9)(0.06)} = \textbf{21.1 m}$$

Part b

If we assume that full friction is mobilized, then, referring to Figure 15.21 (for $\beta = 45°$ and $\phi'_d = \phi' = 20°$), we have

$$m = 0.06 = \frac{c'_d}{\gamma H}$$

or

$$c'_d = (0.06)(18.9)(10) = 11.34 \text{ kN/m}^2$$

Thus,

$$F_{\phi'} = \frac{\tan \phi'}{\tan \phi'_d} = \frac{\tan 20}{\tan 20} = 1$$

and

$$F_{c'} = \frac{c'}{c'_d} = \frac{24}{11.34} = 2.12$$

Since $F_{c'} \neq F_{\phi'}$, this is not the factor of safety with respect to strength.

Now we can make another trial. Let the developed angle of friction, ϕ'_d, be equal to $15°$. For $\beta = 45°$ and the friction angle equal to $15°$, we find from Figure 15.21.

$$m = 0.083 = \frac{c'_d}{\gamma H}$$

or

$$c'_d = (0.083)(18.9)(10) = 15.69 \text{ kN/m}^2$$

For this trial,

$$F_{\phi'} = \frac{\tan \phi'}{\tan \phi'_d} = \frac{\tan 20}{\tan 15} = 1.36$$

and

$$F_{c'} = \frac{c'}{c'_d} = \frac{24}{15.69} = 1.53$$

Similar calculations of $F_{\phi'}$ and $F_{c'}$ for various assumed values of ϕ'_d are given in the following table.

ϕ'_d	$\tan \phi'_d$	$F_{\phi'}$	m	c'_d (kN/m²)	$F_{c'}$
20	0.364	1.0	0.06	11.34	2.12
15	0.268	1.36	0.083	15.69	1.53
10	0.176	2.07	0.105	19.85	1.21
5	0.0875	4.16	0.136	25.70	0.93

© Cengage Learning 2014

The values of $F_{\phi'}$ are plotted against their corresponding values of $F_{c'}$ in Figure 15.28, from which we find

$$F_{c'} = F_{\phi'} = F_s = \textbf{1.42}$$

Note: We could have found the value of F_s from Figure 15.23c. Since $\beta = 45°$, it is a slope of 1V:1H. For this slope

$$\frac{c'}{\gamma H} = \frac{24}{(18.9)(10)} = 0.127$$

From Figure 15.23c, for' $c'/\gamma H = 0.127$, the value of $F_s \approx \textbf{1.4}$.

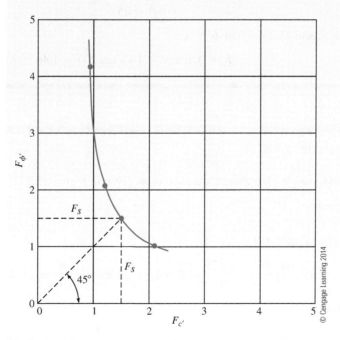

Figure 15.28

Example 15.7

Solve Example 15.6 using Michalowski's solution.

Solution
Part a

For critical height (H_{cr}), $F_s = 1$. Thus,

$$\frac{c'}{\gamma H \tan \phi'} = \frac{24}{(18.9)(H_{cr})(\tan 20)} = \frac{3.49}{H_{cr}}$$

$$\frac{F_s}{\tan \phi'} = \frac{1}{\tan 20} = 2.747$$

$$\beta = 45°$$

From Figure 15.25, for $\beta = 45°$ and $F_s/\tan \phi' = 2.747$, the value of $c'/\gamma H \tan \phi' \approx 0.17$. So

$$\frac{3.49}{H_{cr}} = 0.17; \quad H_{cr} = \mathbf{20.5 \ m}$$

Part b

$$\frac{c'}{\gamma H \tan \phi'} = \frac{24}{(18.9)(10)(\tan 20)} = 0.349$$

$$\beta = 45°$$

From Figure 15.25, $F_s/\tan \phi' = 4$.

$$F_s = 4 \ \tan \phi' = (4)(\tan 20) = \mathbf{1.46}$$

Example 15.8

Solve Example 15.6 using the design chart given in Figure 15.27.

Solution
Part a

For critical height H_{cr}, $F_s = 1$.

For $\beta = 45°$, $\dfrac{\tan \phi'}{F_s} = \dfrac{\tan 20}{1} = 0.364$, the value of $\dfrac{c'}{\gamma H_{cr} \tan \phi'} \approx 0.145$ (Figure 15.27).

Hence

$$H_{cr} = \frac{c'}{0.145\gamma \tan \phi'} = \frac{24}{(0.145)(18.9)(\tan 20)} = \mathbf{24.06 \ m}$$

Part b

For $\beta = 45°$ and $\dfrac{c'}{\gamma H \tan \phi'} = \dfrac{24}{(18.9)(10)(\tan 20)} = 0.349$, the value of $\dfrac{\tan \phi'}{F_s}$ is about 0.25. Hence

$$F_s = \frac{\tan 20}{0.25} = \mathbf{1.46}$$

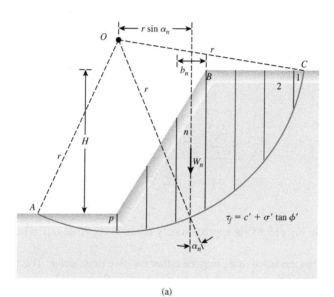

(a)

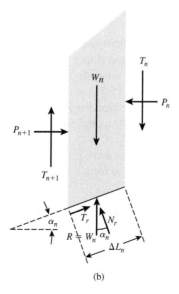

(b)

© Cengage Learning 2014

Figure 15.29
Stability analysis by ordinary method of slices: (a) trial failure surface; (b) forces acting on nth slice

For equilibrium consideration,

$$N_r = W_n \cos \alpha_n$$

The resisting shear force can be expressed as

$$T_r = \tau_d(\Delta L_n) = \frac{\tau_f(\Delta L_n)}{F_s} = \frac{1}{F_s}[c' + \sigma' \tan \phi']\Delta L_n \tag{15.54}$$

The normal stress, σ', in Eq. (15.54) is equal to

$$\frac{N_r}{\Delta L_n} = \frac{W_n \cos \alpha_n}{\Delta L_n}$$

For equilibrium of the trial wedge ABC, the moment of the driving force about O equals the moment of the resisting force about O, or

$$\sum_{n=1}^{n=p} W_n r \sin \alpha_n = \sum_{n=1}^{n=p} \frac{1}{F_s}\left(c' + \frac{W_n \cos \alpha_n}{\Delta L_n} \tan \phi'\right)(\Delta L_n)(r)$$

or

$$F_s = \frac{\sum\limits_{n=1}^{n=p} (c'\Delta L_n + W_n \cos \alpha_n \tan \phi')}{\sum\limits_{n=1}^{n=p} W_n \sin \alpha_n} \tag{15.55}$$

[*Note:* ΔL_n in Eq. (15.55) is approximately equal to $(b_n)/(\cos \alpha_n)$, where b_n = the width of the nth slice.]

Note that the value of α_n may be either positive or negative. The value of α_n is positive when the slope of the arc is in the same quadrant as the ground slope. To find the minimum factor of safety—that is, the factor of safety for the critical circle—one must make several trials by changing the center of the trial circle. This method generally is referred to as the *ordinary method of slices.*

For convenience, a slope in a homogeneous soil is shown in Figure 15.29. However, the method of slices can be extended to slopes with layered soil, as shown in Figure 15.30. The general procedure of stability analysis is the same. However, some minor points should be kept in mind. When Eq. (15.55) is used for the factor of safety calculation, the values of ϕ' and c' will not be the same for all slices. For example, for slice No. 3 (see Figure 15.30), we have to use a friction angle of $\phi' = \phi'_3$ and cohesion $c' = c'_3$; similarly, for slice No. 2, $\phi' = \phi'_2$ and $c' = c'_2$.

It is of interest to note that if total shear strength parameters (that is, $\tau_f = c + \tan \phi$) were used, Eq. (15.55) would take the form

$$F_s = \frac{\sum\limits_{n=1}^{n=p} (c\Delta L_n + W_n \cos \alpha_n \tan \phi)}{\sum\limits_{n=1}^{n=p} W_n \sin \alpha_n} \tag{15.56}$$

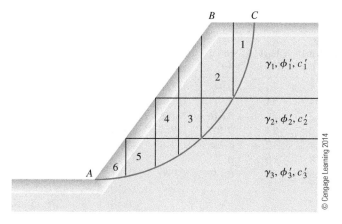

Figure 15.30 Stability analysis, by ordinary method of slices, for slope in layered soils

Example 15.9

For the slope shown in Figure 15.31, find the factor of safety against sliding for the trial slip surface AC. Use the ordinary method of slices.

Solution

The sliding wedge is divided into seven slices. Now the following table can be prepared:

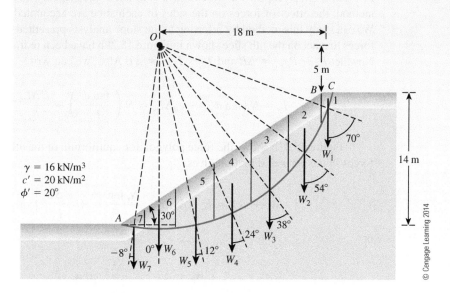

Figure 15.31 Stability analysis of a slope by ordinary method of slices

Slice no. (1)	W (kN/m) (2)	α_n (deg) (3)	$\sin \alpha_n$ (4)	$\cos \alpha_n$ (5)	ΔL_n (m) (6)	$W_n \sin \alpha_n$ (kN/m) (7)	$W_n \cos \alpha_n$ (kN/m) (8)
1	22.4	70	0.94	0.342	2.924	21.1	7.66
2	294.4	54	0.81	0.588	6.803	238.5	173.1
3	435.2	38	0.616	0.788	5.076	268.1	342.94
4	435.2	24	0.407	0.914	4.376	177.1	397.8
5	390.4	12	0.208	0.978	4.09	81.2	381.8
6	268.8	0	0	1	4	0	268.8
7	66.58	−8	−0.139	0.990	3.232	−9.25	65.9
					Σ Col. 6 = 30.501 m	Σ Col. 7 = 776.75 kN/m	Σ Col. 8 = 1638 kN/m

$$F_s = \frac{(\Sigma \text{ Col. } 6)(c') + (\Sigma \text{ Col. } 8) \tan \phi'}{\Sigma \text{ Col. } 7}$$

$$= \frac{(30.501)(20) + (1638)(\tan 20)}{776.75} = \textbf{1.55}$$

15.12 Bishop's Simplified Method of Slices

In 1955, Bishop proposed a more refined solution to the ordinary method of slices. In this method, the effect of forces on the sides of each slice are accounted for to some degree. We can study this method by referring to the slope analysis presented in Figure 15.29. The forces that act on the nth slice shown in Figure 15.29b have been redrawn in Figure 15.32a. Now, let $P_n - P_{n+1} = \Delta P$ and $T_n - T_{n+1} = \Delta T$. Also, we can write

$$T_r = N_r(\tan \phi'_d) + c'_d \Delta L_n = N_r \left(\frac{\tan \phi'}{F_s} \right) + \frac{c' \Delta L_n}{F_s} \tag{15.57}$$

Figure 15.32b shows the force polygon for equilibrium of the nth slice. Summing the forces in the vertical direction gives

$$W_n + \Delta T = N_r \cos \alpha_n + \left[\frac{N_r \tan \phi'}{F_s} + \frac{c' \Delta L_n}{F_s} \right] \sin \alpha_n$$

or

$$N_r = \frac{W_n + \Delta T - \dfrac{c' \Delta L_n}{F_s} \sin \alpha_n}{\cos \alpha_n + \dfrac{\tan \phi' \sin \alpha_n}{F_s}} \tag{15.58}$$

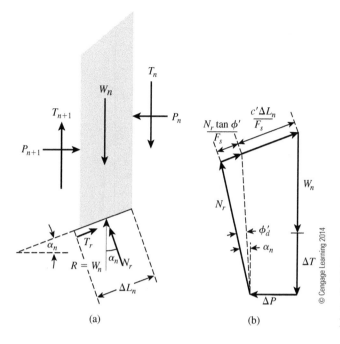

© Cengage Learning 2014

Figure 15.32
Bishop's simplified method of slices: (a) forces acting on the nth slice; (b) force polygon for equilibrium

For equilibrium of the wedge ABC (Figure 15.29a), taking the moment about O gives

$$\sum_{n=1}^{n=p} W_n r \sin \alpha_n = \sum_{n=1}^{n=p} T_r r \qquad (15.59)$$

where

$$T_r = \frac{1}{F_s}(c' + \sigma' \tan \phi')\Delta L_n$$

$$= \frac{1}{F_s}(c'\Delta L_n + N_r \tan \phi') \qquad (15.60)$$

Substitution of Eqs. (15.58) and (15.60) into Eq. (15.59) gives

$$F_s = \frac{\displaystyle\sum_{n=1}^{n=p} (c'b_n + W_n \tan \phi' + \Delta T \tan \phi')\dfrac{1}{m_{\alpha(n)}}}{\displaystyle\sum_{n=1}^{n=p} W_n \sin \alpha_n} \qquad (15.61)$$

where

$$m_{\alpha(n)} = \cos \alpha_n + \frac{\tan \phi' \sin \alpha_n}{F_s} \qquad (15.62)$$

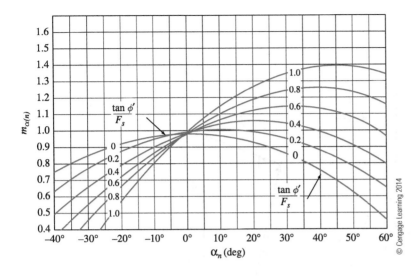

Figure 15.33 Variation of $m_{\alpha(n)}$ with α_n and $\tan \phi'/F_s$ [Eq. (15.62)]

Figure 15.33 shows the variation of $m_{\alpha(n)}$ with α_n and $\tan \phi'/F_s$. For simplicity, if we let $\Delta T = 0$, Eq. (15.61) becomes

$$F_s = \frac{\sum\limits_{n=1}^{n=p} (c'b_n + W_n \tan \phi')\dfrac{1}{m_{\alpha(n)}}}{\sum\limits_{n=1}^{n=p} W_n \sin \alpha_n} \qquad (15.63)$$

Note that the term F_s is present on both sides of Eq. (15.63). Hence, we must adopt a trial-and-error procedure to find the value of F_s. As in the method of ordinary slices, a number of failure surfaces must be investigated so that we can find the critical surface that provides the minimum factor of safety.

Bishop's simplified method is probably the most widely used. When incorporated into computer programs, it yields satisfactory results in most cases. The ordinary method of slices is presented in this chapter as a learning tool only. It is used rarely now because it is too conservative.

15.13 Stability Analysis by Method of Slices for Steady-State Seepage

The fundamentals of the ordinary method of slices and Bishop's simplified method of slices were presented in Sections 15.11 and 15.12, respectively, and we assumed the pore water pressure to be zero. However, for steady-state seepage through slopes, as is the situation in many practical cases, the pore water pressure must be considered when effective shear strength parameters are used. So we need to modify Eqs. (15.55) and (15.63) slightly.

Figure 15.34 shows a slope through which there is steady-state seepage. For the nth slice, the average pore water pressure at the bottom of the slice is equal to $u_n = h_n \gamma_w$. The total force caused by the pore water pressure at the bottom of the nth slice is equal to $u_n \Delta L_n$.

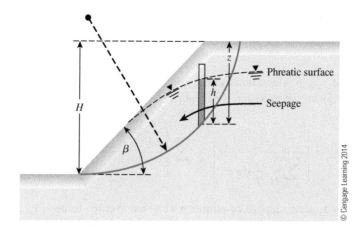

Figure 15.34
Stability analysis of slope
with steady-state seepage

Thus, Eq. (15.55) for the ordinary method of slices will be modified to read as follows.

$$F_s = \frac{\sum_{n=1}^{n=p} \left[c'\Delta L_n + (W_n \cos \alpha_n - u_n \Delta L_n) \right] \tan \phi'}{\sum_{n=1}^{n=p} W_n \sin \alpha_n} \qquad (15.64)$$

Similarly, Eq. (15.63) for Bishop's simplified method of slices will be modified to the form

$$F_s = \frac{\sum_{n=1}^{n=p} \left[c'b_n + (W_n - u_n b_n) \tan \phi' \right] \dfrac{1}{m_{(\alpha)n}}}{\sum_{n=1}^{n=p} W_n \sin \alpha_n} \qquad (15.65)$$

Note that W_n in Eqs. (15.64) and (15.65) is the *total weight* of the slice.

15.14 Solutions for Steady State Seepage

Bishop and Morgenstern Solution

Using Eq. (15.65), Bishop and Morgenstern (1960) developed tables for the calculation of F_s for simple slopes. The principles of these developments can be explained as follows. In Eq. (15.65),

$$W_n = \text{total weight of the } n\text{th slice} = \gamma b_n z_n \qquad (15.66)$$

where z_n = average height of the nth slice. Also in Eq. (15.65),

$$u_n = h_n \gamma_w$$

So, we can let

$$r_{u(n)} = \frac{u_n}{\gamma z_n} = \frac{h_n \gamma_w}{\gamma z_n}$$

(15.67)

Note that $r_{u(n)}$ is a nondimensional quantity. Substituting Eqs. (15.66) and (15.67) into Eq. (15.65) and simplifying, we obtain

$$F_s = \left[\frac{1}{\displaystyle\sum_{n=1}^{n=p} \frac{b_n}{H} \frac{z_n}{H} \sin \alpha_n} \right] \times \sum_{n=1}^{n=p} \left\{ \frac{\dfrac{c'}{\gamma H} \dfrac{b_n}{H} + \dfrac{b_n}{H} \dfrac{z_n}{H}[1 - r_{u(n)}] \tan \phi'}{m_{\alpha(n)}} \right\}$$

(15.68)

For a steady-state seepage condition, a weighted average value of $r_{u(n)}$ can be taken, which is a constant. Let the weighted averaged value of $r_{u(n)}$ be r_u. For most practical cases, the value of r_u may range up to 0.5. Thus,

$$F_s = \left[\frac{1}{\displaystyle\sum_{n=1}^{n=p} \frac{b_n}{H} \frac{z_n}{H} \sin \alpha_n} \right] \times \sum_{n=1}^{n=p} \left\{ \frac{\left[\dfrac{c'}{\gamma H} \dfrac{b_n}{H} + \dfrac{b_n}{H} \dfrac{z_n}{H} (1 - r_u) \tan \phi' \right]}{m_{\alpha(n)}} \right\}$$

(15.69)

The factor of safety based on the preceding equation can be solved and expressed in the form

$$F_s = m' - n' r_u$$

(15.70)

where m' and $n' = $ stability coefficients. Table 15.3 gives the values of m' and n' for various combinations of $c'/\gamma H$, D, ϕ', and β.

To determine F_s from Table 15.3, we must use the following step-by-step procedure:

Step 1:   Obtain ϕ', β, and $c'/\gamma H$.
Step 2:   Obtain r_u (weighted average value).
Step 3:   From Table 15.3, obtain the values of m' and n' for $D = 1$, 1.25, and 1.5 (for the required parameters ϕ', β, r_u, and $c'/\gamma H$).
Step 4:   Determine F_s, using the values of m' and n' for each value of D.
Step 5:   The required value of F_s is the smallest one obtained in Step 4.

Spencer's Solution

Bishop's simplified method of slices described in Sections 15.12, 15.13 and 15.14 satisfies the equations of equilibrium with respect to the moment but not with respect to the forces. Spencer (1967) has provided a method to determine the factor of safety (F_s) by taking into account the interslice forces (P_n, T_n, P_{n+1}, T_{n+1}, as shown in Figure 15.32), which does satisfy the equations of equilibrium with respect to moment and forces. The details of this method of analysis are beyond the scope of this text; however, the final results of Spencer's work are summarized in this section in Figure 15.35. Note that r_u, as shown in Figure 15.35, is the same as that given in by Eq. (15.69).

Table 15.3 Values of m' and n' [Eq. (15.70)]

a. Stability coefficients m' and n' for $c'/\gamma H = 0$

	Stability coefficients for earth slopes							
	Slope 2:1		Slope 3:1		Slope 4:1		Slope 5:1	
ϕ'	m'	n'	m'	n'	m'	n'	m'	n'
10.0	0.353	0.441	0.529	0.588	0.705	0.749	0.882	0.917
12.5	0.443	0.554	0.665	0.739	0.887	0.943	1.109	1.153
15.0	0.536	0.670	0.804	0.893	1.072	1.139	1.340	1.393
17.5	0.631	0.789	0.946	1.051	1.261	1.340	1.577	1.639
20.0	0.728	0.910	1.092	1.213	1.456	1.547	1.820	1.892
22.5	0.828	1.035	1.243	1.381	1.657	1.761	2.071	2.153
25.0	0.933	1.166	1.399	1.554	1.865	1.982	2.332	2.424
27.5	1.041	1.301	1.562	1.736	2.082	2.213	2.603	2.706
30.0	1.155	1.444	1.732	1.924	2.309	2.454	2.887	3.001
32.5	1.274	1.593	1.911	2.123	2.548	2.708	3.185	3.311
35.0	1.400	1.750	2.101	2.334	2.801	2.977	3.501	3.639
37.5	1.535	1.919	2.302	2.558	3.069	3.261	3.837	3.989
40.0	1.678	2.098	2.517	2.797	3.356	3.566	4.196	4.362

b. Stability coefficients m' and n' for $c'/\gamma H = 0.025$ and $D = 1.00$

	Stability coefficients for earth slopes							
	Slope 2:1		Slope 3:1		Slope 4:1		Slope 5:1	
ϕ'	m'	n'	m'	n'	m'	n'	m'	n'
10.0	0.678	0.534	0.906	0.683	1.130	0.846	1.367	1.031
12.5	0.790	0.655	1.066	0.849	1.337	1.061	1.620	1.282
15.0	0.901	0.776	1.224	1.014	1.544	1.273	1.868	1.534
17.5	1.012	0.898	1.380	1.179	1.751	1.485	2.121	1.789
20.0	1.124	1.022	1.542	1.347	1.962	1.698	2.380	2.050
22.5	1.239	1.150	1.705	1.518	2.177	1.916	2.646	2.317
25.0	1.356	1.282	1.875	1.696	2.400	2.141	2.921	2.596
27.5	1.478	1.421	2.050	1.882	2.631	2.375	3.207	2.886
30.0	1.606	1.567	2.235	2.078	2.873	2.622	3.508	3.191
32.5	1.739	1.721	2.431	2.285	3.127	2.883	3.823	3.511
35.0	1.880	1.885	2.635	2.505	3.396	3.160	4.156	3.849
37.5	2.030	2.060	2.855	2.741	3.681	3.458	4.510	4.209
40.0	2.190	2.247	3.090	2.993	3.984	3.778	4.885	4.592

(continued)

Table 15.3 (*continued*)

c. **Stability coefficients** m' **and** n' **for** $c'/\gamma H = 0.025$ **and** $D = 1.25$

| | Stability coefficients for earth slopes | | | | | | | |
| | Slope 2:1 | | Slope 3:1 | | Slope 4:1 | | Slope 5:1 | |
ϕ'	m'	n'	m'	n'	m'	n'	m'	n'
10.0	0.737	0.614	0.901	0.726	1.085	0.867	1.285	1.014
12.5	0.878	0.759	1.076	0.908	1.299	1.098	1.543	1.278
15.0	1.019	0.907	1.253	1.093	1.515	1.311	1.803	1.545
17.5	1.162	1.059	1.433	1.282	1.736	1.541	2.065	1.814
20.0	1.309	1.216	1.618	1.478	1.961	1.775	2.334	2.090
22.5	1.461	1.379	1.808	1.680	2.194	2.017	2.610	2.373
25.0	1.619	1.547	2.007	1.891	2.437	2.269	2.879	2.669
27.5	1.783	1.728	2.213	2.111	2.689	2.531	3.196	2.976
30.0	1.956	1.915	2.431	2.342	2.953	2.806	3.511	3.299
32.5	2.139	2.112	2.659	2.686	3.231	3.095	3.841	3.638
35.0	2.331	2.321	2.901	2.841	3.524	3.400	4.191	3.998
37.5	2.536	2.541	3.158	3.112	3.835	3.723	4.563	4.379
40.0	2.753	2.775	3.431	3.399	4.164	4.064	4.958	4.784

d. **Stability coefficients** m' **and** n' **for** $c'/\gamma H = 0.05$ **and** $D = 1.00$

| | Stability coefficients for earth slopes | | | | | | | |
| | Slope 2:1 | | Slope 3:1 | | Slope 4:1 | | Slope 5:1 | |
ϕ'	m'	n'	m'	n'	m'	n'	m'	n'
10.0	0.913	0.563	1.181	0.717	1.469	0.910	1.733	1.069
12.5	1.030	0.690	1.343	0.878	1.688	1.136	1.995	1.316
15.0	1.145	0.816	1.506	1.043	1.904	1.353	2.256	1.567
17.5	1.262	0.942	1.671	1.212	2.117	1.565	2.517	1.825
20.0	1.380	1.071	1.840	1.387	2.333	1.776	2.783	2.091
22.5	1.500	1.202	2.014	1.568	2.551	1.989	3.055	2.365
25.0	1.624	1.338	2.193	1.757	2.778	2.211	3.336	2.651
27.5	1.753	1.480	1.380	1.952	3.013	2.444	3.628	2.948
30.0	1.888	1.630	2.574	2.157	3.261	2.693	3.934	3.259
32.5	2.029	1.789	2.777	2.370	3.523	2.961	4.256	3.585
35.0	2.178	1.958	2.990	2.592	3.803	3.253	4.597	3.927
37.5	2.336	2.138	3.215	2.826	4.103	3.574	4.959	4.288
40.0	2.505	2.332	3.451	3.071	4.425	3.926	5.344	4.668

Table 15.3 (*continued*)

e. Stability coefficients m' and n' for $c'/\gamma H = 0.05$ and $D = 1.25$

	Stability coefficients for earth slopes							
	Slope 2:1		Slope 3:1		Slope 4:1		Slope 5:1	
ϕ'	m'	n'	m'	n'	m'	n'	m'	n'
10.0	0.919	0.633	1.119	0.766	1.344	0.886	1.594	1.042
12.5	1.065	0.792	1.294	0.941	1.563	1.112	1.850	1.300
15.0	1.211	0.950	1.471	1.119	1.782	1.338	2.109	1.562
17.5	1.359	1.108	1.650	1.303	2.004	1.567	2.373	1.831
20.0	1.509	1.266	1.834	1.493	2.230	1.799	2.643	2.107
22.5	1.663	1.428	2.024	1.690	2.463	2.038	2.921	2.392
25.0	1.822	1.595	2.222	1.897	2.705	2.287	3.211	2.690
27.5	1.988	1.769	2.428	2.113	2.957	2.546	3.513	2.999
30.0	2.161	1.950	2.645	2.342	3.221	2.819	3.829	3.324
32.5	2.343	2.141	2.873	2.583	3.500	3.107	4.161	3.665
35.0	2.535	2.344	3.114	2.839	3.795	3.413	4.511	4.025
37.5	2.738	2.560	3.370	3.111	4.109	3.740	4.881	4.405
40.0	2.953	2.791	3.642	3.400	4.442	4.090	5.273	4.806

f. Stability coefficients m' and n' for $c'/\gamma H = 0.05$ and $D = 1.50$

	Stability coefficients for earth slopes							
	Slope 2:1		Slope 3:1		Slope 4:1		Slope 5:1	
ϕ'	m'	n'	m'	n'	m'	n'	m'	n'
10.0	1.022	0.751	1.170	0.828	1.343	0.974	1.547	1.108
12.5	1.202	0.936	1.376	1.043	1.589	1.227	1.829	1.399
15.0	1.383	1.122	1.583	1.260	1.835	1.480	2.112	1.690
17.5	1.565	1.309	1.795	1.480	2.084	1.734	2.398	1.983
20.0	1.752	1.501	2.011	1.705	2.337	1.993	2.690	2.280
22.5	1.943	1.698	2.234	1.937	2.597	2.258	2.990	2.585
25.0	2.143	1.903	2.467	2.179	2.867	2.534	3.302	2.902
27.5	2.350	2.117	2.709	2.431	3.148	2.820	3.626	3.231
30.0	2.568	2.342	2.964	2.696	3.443	3.120	3.967	3.577
32.5	2.798	2.580	3.232	2.975	3.753	3.436	4.326	3.940
35.0	3.041	2.832	3.515	3.269	4.082	3.771	4.707	4.325
37.5	3.299	3.102	3.817	3.583	4.431	4.128	5.112	4.735
40.0	3.574	3.389	4.136	3.915	4.803	4.507	5.543	5.171

(*continued*)

Table 15.3 (*continued*)

g. Stability coefficients m' and n' for $c'/\gamma H = 0.075$ and toe circles

	Stability coefficients for earth slopes							
	Slope 2:1		Slope 3:1		Slope 4:1		Slope 5:1	
ϕ'	m'	n'	m'	n'	m'	n'	m'	n'
20	1.593	1.158	2.055	1.516	2.498	1.903	2.934	2.301
25	1.853	1.430	2.426	1.888	2.980	2.361	3.520	2.861
30	2.133	1.730	2.826	2.288	3.496	2.888	4.150	3.461
35	2.433	2.058	3.253	2.730	4.055	3.445	4.846	4.159
40	2.773	2.430	3.737	3.231	4.680	4.061	5.609	4.918

h. Stability coefficients m' and n' for $c'/\gamma H = 0.075$ and $D = 1.00$

	Stability coefficients for earth slopes							
	Slope 2:1		Slope 3:1		Slope 4:1		Slope 5:1	
ϕ'	m'	n'	m'	n'	m'	n'	m'	n'
20	1.610	1.100	2.141	1.443	2.664	1.801	3.173	2.130
25	1.872	1.386	2.502	1.815	3.126	2.259	3.742	2.715
30	2.142	1.686	2.884	2.201	3.623	2.758	4.357	3.331
35	2.443	2.030	3.306	2.659	4.177	3.331	5.024	4.001
40	2.772	2.386	3.775	3.145	4.785	3.945	5.776	4.759

i. Stability coefficients m' and n' for $c'/\gamma H = 0.075$ and $D = 1.25$

	Stability coefficients for earth slopes							
	Slope 2:1		Slope 3:1		Slope 4:1		Slope 5:1	
ϕ'	m'	n'	m'	n'	m'	n'	m'	n'
20	1.688	1.285	2.071	1.543	2.492	1.815	2.954	2.173
25	2.004	1.641	2.469	1.957	2.972	2.315	3.523	2.730
30	2.352	2.015	2.888	2.385	3.499	2.857	4.149	3.357
35	2.728	2.385	3.357	2.870	4.079	3.457	4.831	4.043
40	3.154	2.841	3.889	3.428	4.729	4.128	5.603	4.830

j. Stability coefficients m' and n' for $c'/\gamma H = 0.075$ and $D = 1.50$

	Stability coefficients for earth slopes							
	Slope 2:1		Slope 3:1		Slope 4:1		Slope 5:1	
ϕ'	m'	n'	m'	n'	m'	n'	m'	n'
20	1.918	1.514	2.199	1.728	2.548	1.985	2.931	2.272
25	2.308	1.914	2.660	2.200	3.083	2.530	3.552	2.915
30	2.735	2.355	3.158	2.714	3.659	3.128	4.128	3.585
35	3.211	2.854	3.708	3.285	4.302	3.786	4.961	4.343
40	3.742	3.397	4.332	3.926	5.026	4.527	5.788	5.185

Table 15.3 (*continued*)

k. Stability coefficients m' and n' for $c'/\gamma H = 0.100$ and toe circles

| | Stability coefficients for earth slopes | | | | | | | |
| | Slope 2:1 | | Slope 3:1 | | Slope 4:1 | | Slope 5:1 | |
ϕ'	m'	n'	m'	n'	m'	n'	m'	n'
20	1.804	2.101	2.286	1.588	2.748	1.974	3.190	2.361
25	2.076	1.488	2.665	1.945	3.246	2.459	3.796	2.959
30	2.362	1.786	3.076	2.359	3.770	2.961	4.442	3.576
35	2.673	2.130	3.518	2.803	4.339	3.518	5.146	4.249
40	3.012	2.486	4.008	3.303	4.984	4.173	5.923	5.019

l. Stability coefficients m' and n' for $c'/\gamma H = 0.100$ and $D = 1.00$

| | Stability coefficients for earth slopes | | | | | | | |
| | Slope 2:1 | | Slope 3:1 | | Slope 4:1 | | Slope 5:1 | |
ϕ'	m'	n'	m'	n'	m'	n'	m'	n'
20	1.841	1.143	2.421	1.472	2.982	1.815	3.549	2.157
25	2.102	1.430	2.785	1.845	3.458	2.303	4.131	2.743
30	2.378	1.714	3.183	2.258	3.973	2.830	4.751	3.372
35	2.692	2.086	3.612	2.715	4.516	3.359	5.426	4.059
40	3.025	2.445	4.103	3.230	5.144	4.001	6.187	4.831

m. Stability coefficients m' and n' for $c'/\gamma H = 0.100$ and $D = 1.25$

| | Stability coefficients for earth slopes | | | | | | | |
| | Slope 2:1 | | Slope 3:1 | | Slope 4:1 | | Slope 5:1 | |
ϕ'	m'	n'	m'	n'	m'	n'	m'	n'
20	1.874	1.301	2.283	1.558	2.751	1.843	3.253	2.158
25	2.197	1.642	2.681	1.972	3.233	2.330	3.833	2.758
30	2.540	2.000	3.112	2.415	3.753	2.858	4.451	3.372
35	2.922	2.415	3.588	2.914	4.333	3.458	5.141	4.072
40	3.345	2.855	4.119	3.457	4.987	4.142	5.921	4.872

n. Stability coefficients m' and n' for $c'/\gamma H = 0.100$ and $D = 1.50$

| | Stability coefficients for earth slopes | | | | | | | |
| | Slope 2:1 | | Slope 3:1 | | Slope 4:1 | | Slope 5:1 | |
ϕ'	m'	n'	m'	n'	m'	n'	m'	n'
20	2.079	1.528	2.387	1.742	2.768	2.014	3.158	2.285
25	2.477	1.942	2.852	2.215	3.297	2.542	3.796	2.927
30	2.908	2.385	3.349	2.728	3.881	3.143	4.468	3.614
35	3.385	2.884	3.900	3.300	4.520	3.800	5.211	4.372
40	3.924	3.441	4.524	3.941	5.247	4.542	6.040	5.200

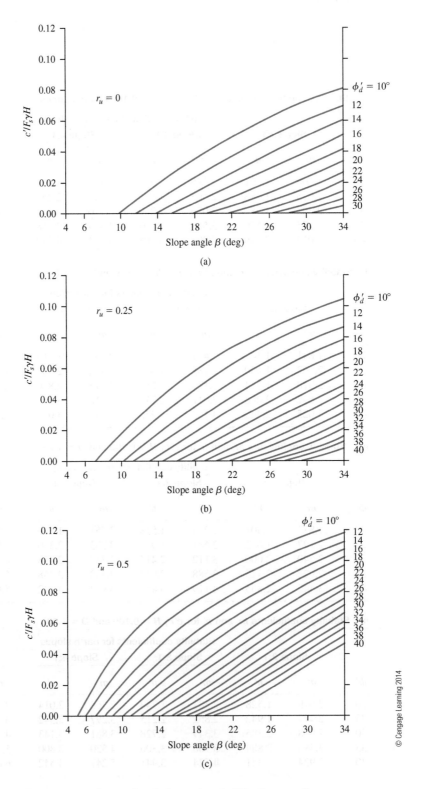

Figure 15.35 Spencer's solution—plot of $c'/F_s\gamma H$ versus β

In order to use the charts given in Figure 15.35 and to determine the required value of F_s, the following step-by-step procedure needs to be used.

Step 1: Determine c', γ, H, β, ϕ', and r_u for the given slope.

Step 2: Assume a value of F_s.

Step 3: Calculate $c'/[F_{s(assumed)}\ \gamma H]$.

$$\uparrow$$
$$\text{Step 2}$$

Step 4: With the value of $c'/F_s\gamma H$ calculated in step 3 and the slope angle β, enter the proper chart in Figure 15.35 to obtain ϕ'_d. Note that Figures 15.35 a, b, and c, are, respectively, for r_u of 0, 0.25, and 0.5, respectively.

Step 5: Calculate $F_s = \tan \phi' / \tan \phi'_d$.

$$\uparrow$$
$$\text{Step 4}$$

Step 6: If the values of F_s as assumed in step 2 are not the same as those calculated in step 5, repeat steps 2, 3, 4, and 5 until they are the same.

Michalowski's Solution

Michalowski (2002) used the kinematic approach of limit analysis similar to that shown in Figures 15.24 and 15.25 to analyze slopes with steady-state seepage. The results of this analysis are summarized in Figure 15.36 for $r_u = 0.25$ and $r_u = 0.5$. Note that Figure 15.25 is applicable for the $r_u = 0$ condition.

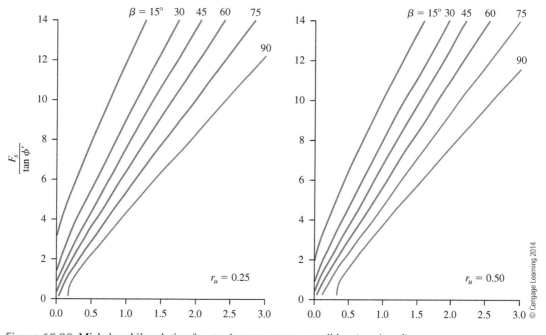

Figure 15.36 Michalowski's solution for steady-state seepage condition *(continued)*

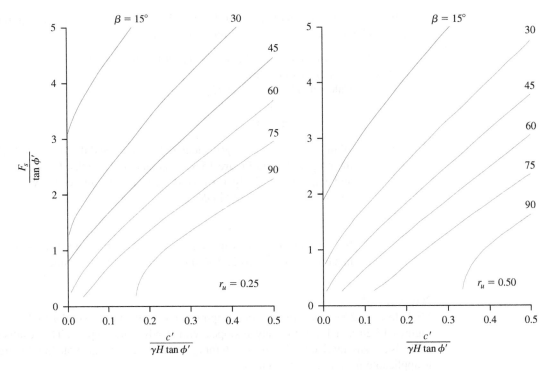

Figure 15.36 *(continued)*

Example 15.10

A given slope under steady-state seepage has the following: $H = 21.62$ m, $\phi' = 25°$, slope: 2H:1V, $c' = 20$ kN/m², $\gamma = 18.5$ kN/m³, $r_u = 0.25$. Determine the factor of safety, F_s. Use Table 15.3.

Solution

$$\beta = \tan^{-1}\left(\frac{1}{2}\right) = 26.57°$$

$$\frac{c'}{\gamma H} = \frac{20}{(18.5)(21.62)} = 0.05$$

Now the following table can be prepared.

β (deg)	ϕ' (deg)	$c'/\gamma H$	D	m'^a	n'^b	$F_s = m' - n'r_u{}^c$
26.57	25	0.05	1.00	1.624[d]	1.338[d]	1.29
26.57	25	0.05	1.25	1.822[e]	1.595[e]	1.423
26.57	25	0.05	1.5	2.143[f]	1.903[f]	1.667

[a]From Table 15.3
[b]From Table 15.3

^{c}Eq. (15.70); $r_u = 0.25$
^{d}Table 15.3d
^{e}Table 15.3e
^{f}Table 15.3f

So,

$$F_s \simeq \mathbf{1.29}$$

Example 15.11

Solve Example 15.10 using Spencer's solution (Figure 15.35).

Solution
Given: $H = 21.62$ m, $\beta = 26.57°$, $c' = 20$ kN/m^2, $\gamma = 18.5$ kN/m^3, $\phi' = 25°$, and $r_u = 0.25$. Now the following table can be prepared.

β (deg)	$F_{s(assumed)}$	$\dfrac{c'}{F_{s(assumed)}\gamma H}$	$\phi_d'{}^a$(deg)	$F_{s(calculated)} = \dfrac{\tan \phi'}{\tan \phi_d'}$
26.57	1.1	0.0455	18	1.435
26.57	1.2	0.0417	19	1.354
26.57	1.3	0.0385	20	1.281
26.57	1.4	0.0357	21	1.215

^{a} From Figure 15.35b

Figure 15.37 shows a plot of $F_{s(assumed)}$ against $F_{s(calculated)}$, from which $F_s \simeq \mathbf{1.3}$.

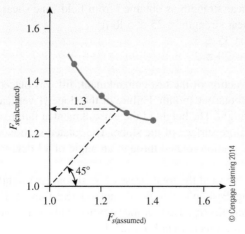

Figure 15.37

© Cengage Learning 2014

Example 15.12

Solve Example 15.10 using Michalowski's solution (Figure 15.36).

Solution

$$\frac{c'}{\gamma H \tan \phi'} = \frac{20}{(18.5)(21.62)(\tan 25)} = 0.107$$

For $r_u = 0.25$, from Figure 15.36, $\dfrac{F_s}{\tan \phi'} \approx 3.1$ So,

$$F_s = (3.1)(\tan 25) = 1.45$$

15.15 A Case History of Slope Failure

Ladd (1972) reported the results of a study of the failure of a slope that had been constructed over a sensitive clay. The study was conducted in relation to a major improvement program of Interstate Route 95 in Portsmouth, New Hampshire, which is located 50 miles north of Boston on the coast. To study the stability of the slope, a test embankment was built to failure during the spring of 1968. The test embankment was heavily instrumented. The general subsoil condition at the test site, the section of the test embankment, and the instruments placed to monitor the performance of the test section are shown in Figure 15.38.

The ground water level at the test section was at an elevation of +20 ft (mean sea level). The general physical properties of the soft to very soft gray silty clay layer as shown in Figure 15.38 are as follows:

Natural moisture content = 50 ± 5%
Undrained shear strength as obtained from field vane shear tests = 250 ± 50 lb/ft²
Remolded shear strength = 25 ± 5 lb/ft²
Liquid limit = 35 ± 5
Plastic limit = 20 ± 2

During construction of the test embankment, fill was placed at a fairly uniform rate within a period of about one month. Failure of the slope (1 vertical: 4 horizontal) occurred on June 6, 1968, at night. The height of the embankment at failure was 21.5 ft. Figure 15.39 shows the actual failure surface of the slope. The rotated section shown in Figure 15.39 is the "before failure" section rotated through an angle of 13 degrees about a point W 45 ft, El. 51 ft.

Ladd (1972) reported the total stress ($\phi = 0$ concept) stability analysis of the slope that failed by using Bishop's simplified method (Section 15.12). The variation of the undrained shear strengths (c_u) used for the stability analysis is given below. Note that these values have not been corrected via Eq. (12.51).

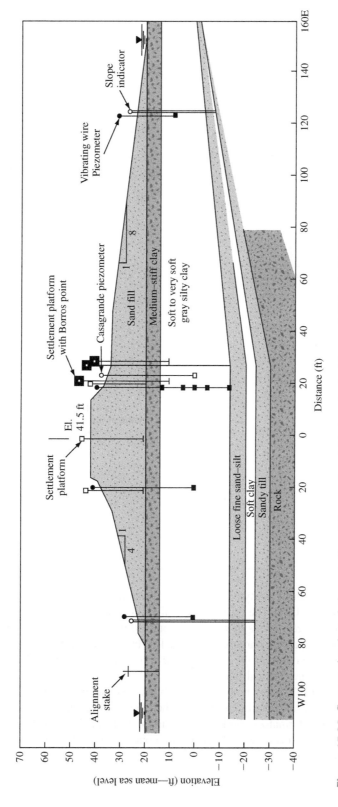

Figure 15.38 Cross section through the centerline of the experimental test section looking north (*After Ladd, 1972. With permission from ASCE.*)

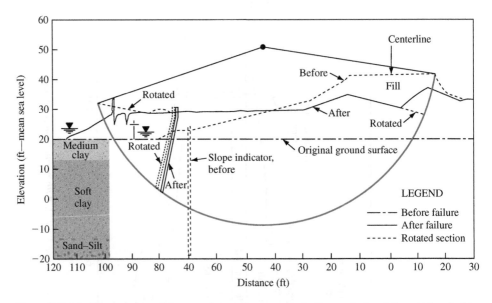

Figure 15.39 Cross section of the experimental test section before and after failure (*After Ladd, 1972. With permission from ASCE.*)

Elevation (ft—mean sea level)	c_u as obtained from vane shear strength tests (lb/ft²)
20 to 15	1000
15 to 10	400
10 to 5	240
5 to 0	250
0 to −2.5	300
−2 to −5	235
−5 to −10	265
−10 to −13	300

The factor of safety (F_s) as obtained from the stability analysis for the critical circle of sliding was 0.88. The critical circle of sliding is shown in Figure 15.40. The factor of safety for *actual surface of sliding* as obtained by using Bishop's simplified method was 0.92. For comparison purposes, the actual surface of sliding is also shown in Figure 15.40. Note that the bottom of the actual failure surface is about 3 ft (0.91 m) above the theoretically determined critical failure surface.

Ladd (1972) also reported the stability analysis of the slope based on the average undrained shear strength variation of the clay layer as determined by using the Stress History And Normalized Soil Engineering Properties (SHANSEP). The details of obtaining c_u by this procedure are beyond the scope of this text. However, the final results are given in the following table.

Elevation (ft—mean sea level)	Average c_u as obtained from SHANSEP (lb/ft²)
20 to 15	1000
15 to 10	335
10 to 5	230
5 to 0	260
0 to −2.5	300
−2.5 to −5	320
−5 to −10	355
−10 to −13	400

© Cengage Learning 2014

Using the preceding average value of c_u, Bishop's simplified method of stability analysis yields the following results:

Failure surface	Factor of safety, F_s
actual failure surface	1.02
critical failure surface	1.01

© Cengage Learning 2014

Figure 15.40 also shows the critical failure surface as determined by using the values of c_u obtained from SHANSEP.

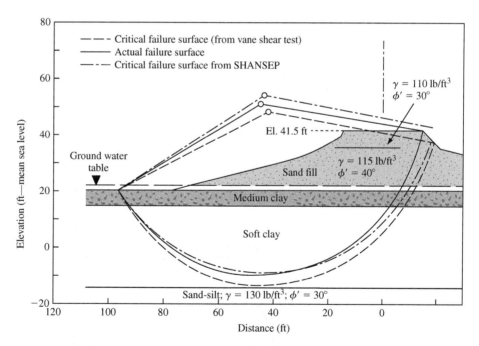

Figure 15.40 Results of total stress stability analysis (*After Ladd, 1972. With permission from ASCE.*) (*Note*: SHANSEP = Stress History And Normalized Soil Engineering Properties)

Based on the preceding results, we can draw the following conclusions:

a. The actual failure surface of a slope with limited height is an arc of a circle.
b. The disagreement between the predicted critical failure surface and the actual failure surface is primarily due to the shear strength assumptions. The c_u values obtained from SHANSEP give an $F_s \simeq 1$, and the critical failure surface is practically the same as the actual failure surface.

This case study is another example that demonstrates the importance of proper evaluation of soil parameters for prediction of the stability of various structures.

15.16 Morgenstern's Method of Slices for Rapid Drawdown Condition

Morgenstern (1963) used Bishop's method of slices (Section 15.12) to determine the factor of safety, F_s, during rapid draw-down. In preparing the solution, Morgenstern used the following notation (Figure 15.41):

- L = height of drawdown
- H = height of embankment
- β = angle that the slope makes with the horizontal

Morgenstern also assumed that

1. The embankment is made of homogeneous material and rests on an impervious base.
2. Initially, the water level coincides with the top of the embankment.
3. During draw-down, pore water pressure does not dissipate.
4. The unit weight of saturated soil $(\gamma_{sat}) = 2\gamma_w$ (γ_w = unit weight of water).

Figures 15.42 through 15.44 provide the draw-down stability charts developed by Morgenstern.

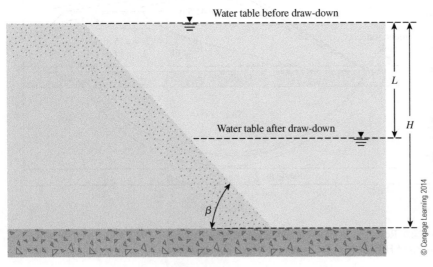

Water table before draw-down

Water table after draw-down

L

H

β

© Cengage Learning 2014

Impervious base

Figure 15.41 Stability analysis for rapid draw-down condition

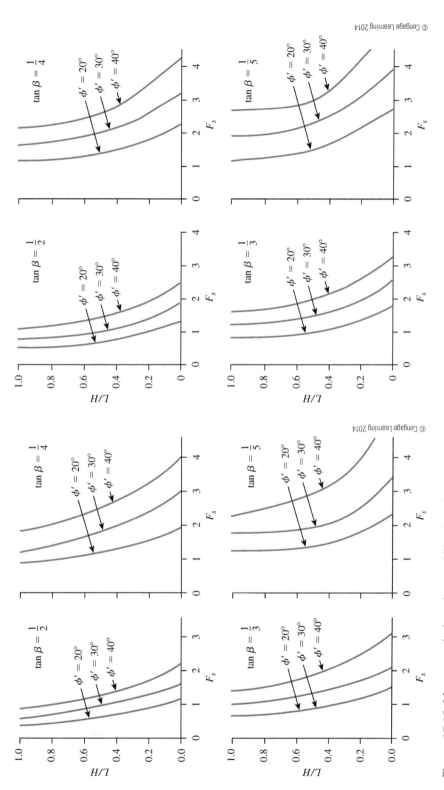

Figure 15.43 Morgenstern's draw-down stability chart for $c'/\gamma H = 0.025$

Figure 15.42 Morgenstern's draw-down stability chart for $c'/\gamma H = 0.0125$

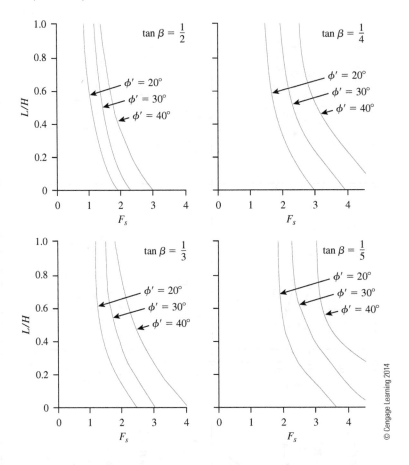

Figure 15.44 Morgenstern's draw-down stability chart for $c'/\gamma H = 0.05$

15.17 Fluctuation of Factor of Safety of Slopes in Clay Embankment on Saturated Clay

Figure 15.45a shows a clay embankment constructed on a *saturated soft clay*. Let P be a point on a potential failure surface APB that is an arc of a circle. Before construction of the embankment, the pore water pressure at P can be expressed as

$$u = h\gamma_w \tag{15.71}$$

Under ideal conditions, let us assume that the height of the fill needed for the construction of the embankment is placed uniformly, as shown in Figure 15.45b. At time $t = t_1$, the embankment height is equal to H, and it remains constant thereafter (that is, $t > t_1$). The average shear stress increase, τ, on the potential failure surface caused by the construction of the embankment also is shown in Figure 15.45b. The value of τ will increase linearly with time up to time $t = t_1$ and remain constant thereafter.

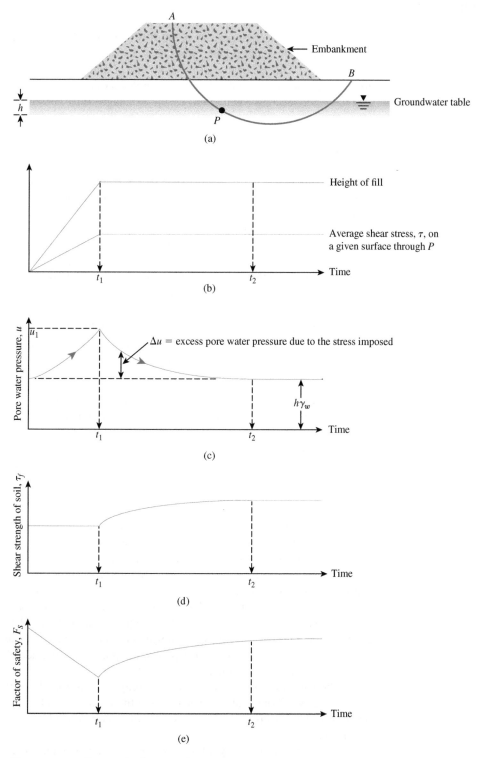

Figure 15.45 Factor of safety variation with time for embankment on soft clay (*Redrawn after Bishop and Bjerrum, 1960. With permission from ASCE.*)

The pore water pressure at point P (Figure 15.45a) will continue to increase as construction of the embankment progresses, as shown in Figure 15.45c. At time $t = t_1$, $u = u_1 > h\gamma_w$. This is because of the slow rate of drainage from the clay layer. However, after construction of the embankment is completed (that is, $t > t_1$), the pore water pressure gradually will decrease with time as the drainage (thus consolidation) progresses. At time $t \simeq t_2$,

$$u = h\gamma_w$$

For simplicity, if we assume that the embankment construction is rapid and that practically no drainage occurs during the construction period, the average *shear strength* of the clay will remain constant from $t = 0$ to $t = t_1$, or $\tau_f = c_u$ (undrained shear strength). This is shown in Figure 15.45d. For time $t > t_1$, as consolidation progresses, the magnitude of the shear strength, τ_f, will gradually increase. At time $t \geq t_2$—that is, after consolidation is completed—the average shear strength of the clay will be equal to $\tau_f = c' + \sigma' \tan \phi'$ (drained shear strength) (Figure 15.45d). The factor of safety of the embankment along the potential surface of sliding can be given as

$$F_s = \frac{\text{Average shear strength of clay, } \tau_f, \text{ along sliding surface (Figure 14.45d)}}{\text{Average shear stress, } \tau, \text{ along sliding surface (Figure 14.45b)}}$$

$$(15.72)$$

The general nature of the variation of the factor of safety, F_s, with time is shown in Figure 15.45e. As we can see from this figure, the magnitude of F_s initially decreases with time. At the end of construction (time $t = t_1$), the value of the factor of safety is a minimum. Beyond this point, the value of F_s continues to increase with drainage up to time $t = t_2$.

Cuts in Saturated Clay

Figure 15.46a shows a cut slope in a saturated soft clay in which APB is a circular potential failure surface. During advancement of the cut, the average shear stress, τ, on the potential failure surface passing through P will increase. The maximum value of the average shear stress, τ, will be attained at the end of construction—that is, at time $t = t_1$. This property is shown in Figure 15.46b.

Because of excavation of the soil, the effective overburden pressure at point P will decrease, which will induce a reduction in the pore water pressure. The variation of the net change of pore water pressure, Δu, is shown in Figure 15.46c. After excavation is complete (time $t > t_1$), the net negative excess pore water pressure will gradually dissipate. At time $t \geq t_2$, the magnitude of Δu will be equal to 0.

The variation of the average shear strength, τ_f, of the clay with time is shown in Figure 15.46d. Note that the shear strength of the soil after excavation gradually decreases. This decrease occurs because of dissipation of the negative excess pore water pressure.

If the factor of safety of the cut slope, F_s, along the potential failure surface is defined by Eq. (15.72), its variation will be as shown in Figure 15.46e. Note that the magnitude of F_s decreases with time, and its minimum value is obtained at time $t \geq t_2$.

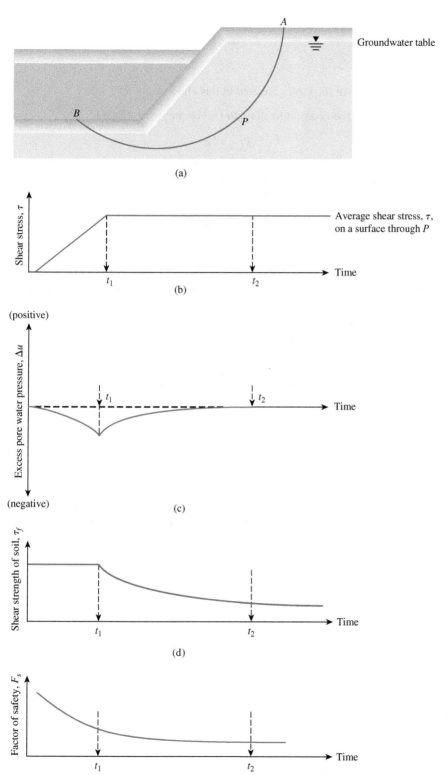

(a)

(b)

Average shear stress, τ, on a surface through P

(c)

(d)

(e)

Figure 15.46 Variation of factor of safety for cut slope in soft clay (*Redrawn after Bishop and Bjerrum, 1960. With permission from ASCE.*)

15.18 Summary

Following is a summary of the topics covered in this chapter:

- The factor of safety with respect to strength (F_s) occurs when [Eq. (15.7)]

$$F_s = F_{c'} = F_{\phi'}$$

- The factors of safety against sliding for infinite slopes for cases with and without seepage are given by Eqs. (15.15) and (15.28), respectively.
- The critical height of a finite slope with plane failure surface assumption can be given by Eq. (15.42).
- The modes of failure of finite slopes with circular failure surfaces can be categorized under (Section 15.7)
  - Slope failure
  - Base failure
- Stability analysis charts for clay slopes ($\phi = 0$ condition) are provided in Figures 15.13 and 15.19.
- Stability analysis charts for slopes with $c' - \phi'$ soil (pore water pressure equal to zero) are given in Figures 15.21, 15.23, 15.25, and 15.27.
- Determination of factor of safety with respect to strength using the method of slices with and without seepage is described in Sections 15.11 and 15.12.
- Stability analysis of slopes with circular failure surface under steady-state seepage is presented in Sections 15.13 and 15.14.

Problems

15.1 Refer to the infinite slope shown in Figure 15.47. Given: $\beta = 25°$, $\gamma = 17.8$ kN/m³, $\phi' = 28°$, and $c' = 31$ kN/m². Find the height, H, such that a factor of safety, F_s, of 2.75 is maintained against sliding along the soil–rock interface.

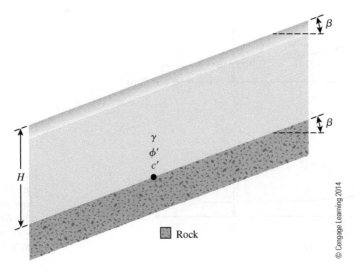

Figure 15.47

15.2 For the slope shown in Figure 15.47, determine the height, H, for critical equilibrium. Given: $\beta = 30°$, $\gamma = 115$ lb/ft³, $\phi' = 21°$, and $c' = 300$ lb/ft².

15.3 Determine the factor of safety, F_s, for the infinite slope shown in Figure 15.48, where seepage is occurring through the soil and the groundwater table coincides with the ground surface. Given: $H = 11$ m, $\beta = 18°$, $\gamma_{sat} = 19.2$ kN/m³, $\phi' = 22°$, and $c' = 46$ kN/m².

15.4 Figure 15.48 shows an infinite slope with $H = 27$ ft, and the groundwater table coinciding with the ground surface. If there is seepage through the soil, determine the factor of safety against sliding along the plane AB. The soil properties are as follows: $G_s = 2.73$, $e = 0.69$, $\beta = 28°$, $\phi' = 18°$, and $c' = 1000$ lb/ft².

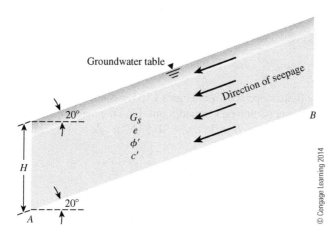

© Cengage Learning 2014

Figure 15.48

15.5 An infinite slope is shown in Figure 15.49. The shear strength parameters at the interface of soil and rock are $\phi' = 26°$, and $c' = 21$ kN/m². Given: $\rho = 1950$ kg/m³.

a. If $H = 5$ m and $\beta = 18°$, find the factor of safety against sliding on the rock surface.

b. If $\beta = 27°$, find the height, H, for which $F_s = 1.75$.

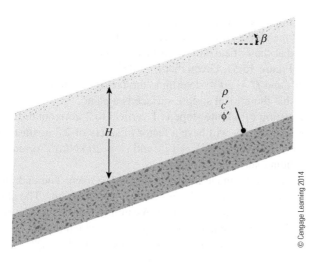

© Cengage Learning 2014

Figure 15.49

15.6 A slope is shown in Figure 15.50. If AC represents a trial failure plane, determine the factor of safety against sliding for the wedge ABC. Given: $\gamma = 115$ lb/ft^3, $\phi' = 25°$, and $c' = 400$ lb/ft^2.

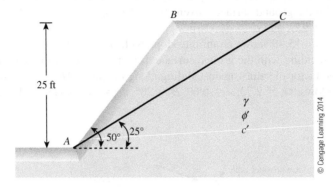

© Cengage Learning 2014

Figure 15.50

15.7 For the finite slope shown in Figure 15.51, assume that the slope failure would occur along a plane (Culmann's assumption). Find the height of the slope for critical equilibrium. Given: $\beta = 58°$, $\gamma = 16.5$ kN/m^3, $\phi' = 14°$, and $c' = 28$ kN/m^2.

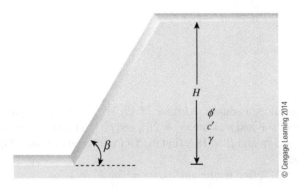

© Cengage Learning 2014

Figure 15.51

15.8 Refer to Figure 15.51. Using the soil parameters given in Problem 15.7, find the height of the slope, H, that will have a factor of safety of 2.5 against sliding. Assume that the critical sliding surface is a plane.

15.9 Refer to Figure 15.51. Given that $\beta = 45°$, $\gamma = 118$ lb/ft^3, $\phi' = 22°$, $c' = 700$ lb/ft^2, and $H = 30$ ft, determine the factor of safety with respect to sliding. Assume that the critical sliding surface is a plane.

15.10 The inclination of a finite slope is 1 vertical to 2 horizontal. Determine the slope height, H, that will have a factor of safety of 2.3 against sliding. Given: $\rho = 1800$ kg/m^3, $\phi' = 17°$, and $c' = 20$ kN/m^2. Assume that the critical sliding surface is a plane.

15.11 A cut slope is to be made in a saturated clay. Given: The undrained shear strength, $c_u = 26$ kN/m^2 ($\phi = 0$ condition), and $\gamma = 18.5$ kN/m^3. The slope makes an angle, $\beta = 55°$ with the horizontal. Assuming that the critical sliding surface is circular, determine the maximum depth up to which the cut could be made. What is the nature of the critical circle (toe, slope, or midpoint)?

15.12 For the cut slope described in Problem 15.11, how deep should the cut be made to ensure a factor of safety of 2.5 against sliding?

15.13 Using the graph shown in Figure 15.13, determine the height of a slope (1 vertical to 2 horizontal) in saturated clay with an undrained shear strength of 800 lb/ft^2 and a unit weight of 119 lb/ft^3. The desired factor of safety against sliding is 2.5. Given: $D = 1.50$.

15.14 Refer to Problem 15.13. What is the critical height of the slope? What is the nature of the critical circle?

15.15 A cut slope was excavated in a saturated clay with a slope angle, $\beta = 48°$ with the horizontal. Slope failure occurred when the cut reached a depth of 10 m. Previous soil explorations showed that a rock layer was located at a depth of 14 m below the ground surface. Assuming an undrained condition and $\gamma = 17$ kN/m^3:
  a. Determine the undrained cohesion of the clay (Figure 15.13).
  b. What was the nature of the critical circle?
  c. With reference to the top of the slope, at what distance did the surface of the sliding intersect the bottom of the excavation?

15.16 Refer to Figure 15.52. Using Michalowski's solution given in Figure 15.25 ($\phi' > 0$), determine the critical height of the slope for the following conditions.
  a. $n' = 2$, $\phi' = 12°$, $c' = 750$ lb/ft^2, and $\gamma = 118$ lb/ft^3
  b. $n' = 1$, $\phi' = 18°$, $c' = 30$ kN/m^2, and $\gamma = 17$ kN/m^3

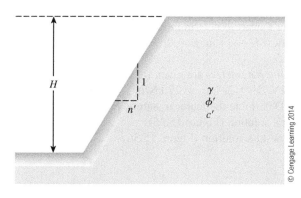

© Cengage Learning 2014

Figure 15.52

15.17 Refer to Figure 15.52. Using Taylor's stability chart (Figure 15.21), determine the factor of safety, F_s, against sliding for the slopes with the following characteristics: Slope: 2.5H:1V, $\gamma = 120$ lb/ft^3, $\phi' = 14°$, $H = 60$ ft, and $c' = 500$ lb/ft^2.

15.18 Repeat Problem 15.17 with the following data: Slope: 1H:1V, $\gamma = 18$ kN/m^3, $\phi' = 20°$, $H = 10$ m, and $c' = 32$ kN/m^2.

15.19 Repeat Problem 15.17 using the design chart given in Figure 15.27 (Steward, Sivakugan, Shukla, and Das, 2011).

15.20 Refer to Figure 15.53. Using the ordinary method of slices, find the factor of safety with respect to sliding for the following trial cases:
  a. $H = 50$ ft, $\beta = 45°$, $\alpha = 30°$, $\theta = 70°$, $\gamma = 121$ lb/ft^3, $\phi' = 18°$, and $c' = 650$ lb/ft^2
  b. $H = 8$ m, $\beta = 45°$, $\alpha = 30°$, $\theta = 80°$, $\gamma = 17$ kN/m^3, $\phi' = 20°$, and $c' = 27$ kN/m^2

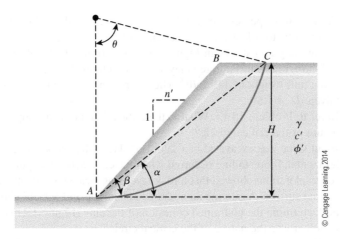

Figure 15.53

15.21 Determine the minimum factor of safety of a slope with the following parameters: $H = 14$ m, $\beta = 26.56°$, $\gamma = 19$ kN/m³, $\phi' = 25°$, $c' = 20$ kN/m², and $r_u = 0.5$. Use Bishop and Morgenstern's method.

15.22 Determine the minimum factor of safety of a slope with the following parameters: $H = 40$ ft, $\beta = 18.43°$, $\gamma = 118$ lb/ft³, $\phi' = 20°$, $c' = 475$ lb/ft², and $r_u = 0.5$. Use Bishop and Morgenstern's method.

15.23 Use Spencer's chart to determine the value of F_s for a slope with the following characteristics: $H = 17$ m, $\beta = 26°$, $\gamma = 19$ kN/m³, $\phi' = 21°$, $c' = 21$ kN/m², and $r_u = 0.5$.

15.24 The following parameters are given for a slope with steady-state seepage: Slope angle: 3H:1V, $\phi' = 24°$, $c' = 27$ kN/m², $\gamma = 17.5$ kN/m³, $H = 18$ m, and $r_u = 0.25$. Determine the factor of safety, F_s, using
  a. Spencer's solution (Figure 15.35)
  b. Michalowski's solution (Figure 15.36)

References

BISHOP, A. W. (1955). "The Use of Slip Circle in the Stability Analysis of Earth Slopes," *Geotechnique,* Vol. 5, No. 1, 7–17.

BISHOP, A. W., and BJERRUM, L. (1960). "The Relevance of the Triaxial Test to the Solution of Stability Problems," *Proceedings,* Research Conference on Shear Strength of Cohesive Soils, ASCE, 437–501.

BISHOP, A. W., and MORGENSTERN, N. R. (1960). "Stability Coefficients for Earth Slopes," *Geotechnique,* Vol. 10, No. 4, 129–147.

CRUDEN, D. M., and VARNES, D. J. (1996). "Landslide Types and Processes," *Special Report 247,* Transportation Research Board, 36–75.

CULMANN, C. (1875). *Die Graphische Statik,* Meyer and Zeller, Zurich.

FELLENIUS, W. (1927). *Erdstatische Berechnungen,* rev. ed., W. Ernst u. Sons, Berlin.

LADD, C. C. (1972). "Test Embankment on Sensitive Clay," *Proceedings,* Conference on Performance of Earth and Earth-Supported Structures, ASCE, Vol. 1, Part 1, 101–128.

MICHALOWSKI, R. L. (2002). "Stability Charts for Uniform Slopes," *Journal of Geotechnical and Geoenvironmental Engineering,* ASCE, Vol. 128, No. 4, 351–355.

MORGENSTERN, N. R. (1963). "Stability Charts for Earth Slopes During Rapid Drawdown," *Geotechnique,* Vol. 13, No. 2, 121–133.

SINGH, A. (1970). "Shear Strength and Stability of Man-Made Slopes," *Journal of the Soil Mechanics and Foundations Division,* ASCE, Vol. 96, No. SM6, 1879–1892.

SPENCER, E. (1967). "A Method of Analysis of the Stability of Embankments Assuming Parallel Inter-Slice Forces," *Geotechnique,* Vol. 17, No. 1, 11–26.

STEWARD, T., SIVAKUGAN, N., SHUKLA, S. K., and DAS, B. M. (2011). "Taylor's Slope Stability Charts Revisited," *International Journal of Geomechanics,* ASCE, Vol. 11, No. 4, 348–352.

TAYLOR, D. W. (1937). "Stability of Earth Slopes," *Journal of the Boston Society of Civil Engineers,* Vol. 24, 197–246.

TERZAGHI, K., and PECK, R. B. (1967). *Soil Mechanics in Engineering Practice,* 2nd ed., Wiley, New York.

4 Lateral Earth Pressure

12.1 Introduction

Vertical or near-vertical slopes of soil are supported by retaining walls, cantilever sheet-pile walls, sheet-pile bulkheads, braced cuts, and other, similar structures. The proper design of those structures requires an estimation of lateral earth pressure, which is a function of several factors, such as (a) the type and amount of wall movement, (b) the shear strength parameters of the soil, (c) the unit weight of the soil, and (d) the drainage conditions in the backfill. Figure 12.1 shows a retaining wall of height H. For similar types of backfill,

- **a.** The wall may be restrained from moving (Figure 12.1a). The lateral earth pressure on the wall at any depth is called the *at-rest earth pressure*.
- **b.** The wall may tilt away from the soil that is retained (Figure 12.1b). With sufficient wall tilt, a triangular soil wedge behind the wall will fail. The lateral pressure for this condition is referred to as *active earth pressure*.
- **c.** The wall may be pushed into the soil that is retained (Figure 12.1c). With sufficient wall movement, a soil wedge will fail. The lateral pressure for this condition is referred to as *passive earth pressure*.

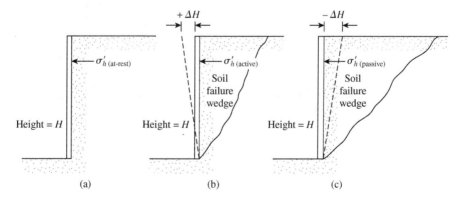

Figure 12.1 Nature of lateral earth pressure on a retaining wall

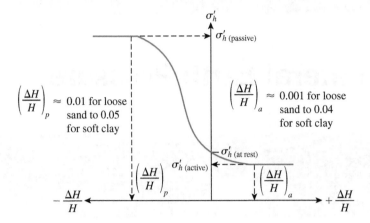

Figure 12.2 Nature of variation of lateral earth pressure at a certain depth

Figure 12.2 shows the nature of variation of the lateral pressure, σ_h', at a certain depth of the wall with the magnitude of wall movement.

In the sections that follow, we will discuss various relationships to determine the at-rest, active, and passive pressures on a retaining wall. It is assumed that the reader has studied lateral earth pressure in the past, so this chapter will serve as a review.

12.2 Lateral Earth Pressure at Rest

Consider a vertical wall of height H, as shown in Figure 12.3, retaining a soil having a unit weight of γ. A uniformly distributed load, q/unit area, is also applied at the ground surface. The shear strength of the soil is

$$s = c' + \sigma' \tan \phi'$$

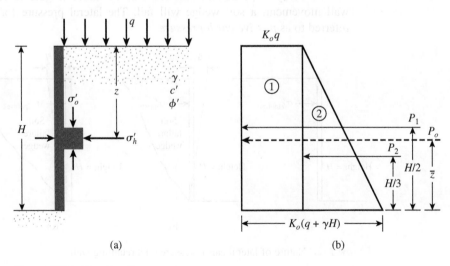

Figure 12.3 At-rest earth pressure

where

c' = cohesion
ϕ' = effective angle of friction
σ' = effective normal stress

At any depth z below the ground surface, the vertical subsurface stress is

$$\sigma'_o = q + \gamma z \tag{12.1}$$

If the *wall is at rest and is not allowed to move at all,* either away from the soil mass or into the soil mass (i.e., there is zero horizontal strain), the lateral pressure at a depth z is

$$\sigma_h = K_o \sigma'_o + u \tag{12.2}$$

where

u = pore water pressure
K_o = coefficient of at-rest earth pressure

For normally consolidated soil, the relation for K_o (Jaky, 1944) is

$$K_o \approx 1 - \sin \phi' \tag{12.3}$$

Equation (12.3) is an empirical approximation.

For overconsolidated soil, the at-rest earth pressure coefficient may be expressed as (Mayne and Kulhawy, 1982)

$$K_o = (1 - \sin \phi')\,\mathrm{OCR}^{\sin \phi'} \tag{12.4}$$

where OCR = overconsolidation ratio.

With a properly selected value of the at-rest earth pressure coefficient, Eq. (12.2) can be used to determine the variation of lateral earth pressure with depth z. Figure 12.3b shows the variation of σ'_h with depth for the wall depicted in Figure 12.3a. Note that if the surcharge $q = 0$ and the pore water pressure $u = 0$, the pressure diagram will be a triangle. The total force, P_o, *per unit length* of the wall given in Figure 12.3a can now be obtained from the area of the pressure diagram given in Figure 12.3b and is

$$P_o = P_1 + P_2 = qK_o H + \tfrac{1}{2}\gamma H^2 K_o \tag{12.5}$$

where

P_1 = area of rectangle 1
P_2 = area of triangle 2

The location of the line of action of the resultant force, P_o, can be obtained by taking the moment about the bottom of the wall. Thus,

$$\bar{z} = \frac{P_1\left(\dfrac{H}{2}\right) + P_2\left(\dfrac{H}{3}\right)}{P_o} \tag{12.6}$$

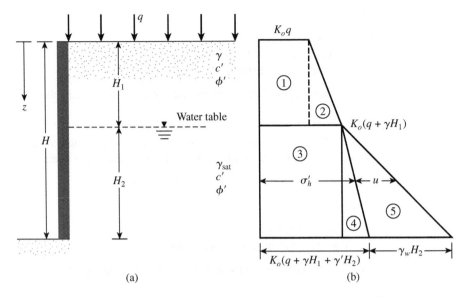

Figure 12.4 At-rest earth pressure with water table located at a depth $z < H$

If the water table is located at a depth $z < H$, the at-rest pressure diagram shown in Figure 12.3b will have to be somewhat modified, as shown in Figure 12.4. If the effective unit weight of soil below the water table equals γ' (i.e., $\gamma_{sat} - \gamma_w$), then

$$\text{at } z = 0, \qquad \sigma_h' = K_o \sigma_o' = K_o q$$
$$\text{at } z = H_1, \qquad \sigma_h' = K_o \sigma_o' = K_o(q + \gamma H_1)$$

and

$$\text{at } z = H_2, \qquad \sigma_h' = K_o \sigma_o' = K_o(q + \gamma H_1 + \gamma' H_2)$$

Note that in the preceding equations, σ_o' and σ_h' are effective vertical and horizontal pressures, respectively. Determining the total pressure distribution on the wall requires adding the hydrostatic pressure u, which is zero from $z = 0$ to $z = H_1$ and is $H_2 \gamma_w$ at $z = H_2$. The variation of σ_h' and u with depth is shown in Figure 12.4b. Hence, the total force per unit length of the wall can be determined from the area of the pressure diagram. Specifically,

$$P_o = A_1 + A_2 + A_3 + A_4 + A_5$$

where A = area of the pressure diagram.
So,

$$P_o = K_o q H_1 + \tfrac{1}{2} K_o \gamma H_1^2 + K_o(q + \gamma H_1) H_2 + \tfrac{1}{2} K_o \gamma' H_2^2 + \tfrac{1}{2} \gamma_w H_2^2 \qquad (12.7)$$

Example 12.1

For the retaining wall shown in Figure 12.5a, determine the lateral earth force at rest per unit length of the wall. Also determine the location of the resultant force. Assume OCR = 1.

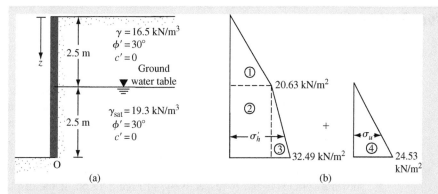

Figure 12.5

Solution

$$K_o = 1 - \sin \phi' = 1 - \sin 30° = 0.5$$

At $z = 0$, $\sigma_o' = 0$; $\sigma_h' = 0$

At $z = 2.5$ m, $\sigma_o' = (16.5)(2.5) = 41.25$ kN/m²;

$$\sigma_h' = K_o \sigma_o' = (0.5)(41.25) = 20.63 \text{ kN/m}^2$$

At $z = 5$ m, $\sigma_o' = (16.5)(2.5) + (19.3 - 9.81)2.5 = 64.98$ kN/m²;

$$\sigma_h' = K_o \sigma_o' = (0.5)(64.98) = 32.49 \text{ kN/m}^2$$

The hydrostatic pressure distribution is as follows:

From $z = 0$ to $z = 2.5$ m, $u = 0$. At $z = 5$ m, $u = \gamma_w(2.5) = (9.81)(2.5) = 24.53$ kN/m². The pressure distribution for the wall is shown in Figure 12.5b.

The total force per unit length of the wall can be determined from the area of the pressure diagram, or

$$P_o = \text{Area 1} + \text{Area 2} + \text{Area 3} + \text{Area 4}$$

$$= \tfrac{1}{2}(2.5)(20.63) + (2.5)(20.63) + \tfrac{1}{2}(2.5)(32.49 - 20.63)$$

$$+ \tfrac{1}{2}(2.5)(24.53) = \mathbf{122.85 \text{ kN/m}}$$

The location of the center of pressure measured from the bottom of the wall (point O) =

$$\bar{z} = \frac{(\text{Area 1})\left(2.5 + \dfrac{2.5}{3}\right) + (\text{Area 2})\left(\dfrac{2.5}{2}\right) + (\text{Area 3} + \text{Area 4})\left(\dfrac{2.5}{3}\right)}{P_o}$$

$$= \frac{(25.788)(3.33) + (51.575)(1.25) + (14.825 + 30.663)(0.833)}{122.85}$$

$$= \frac{85.87 + 64.47 + 37.89}{122.85} = \mathbf{1.53 \text{ m}}$$

Active Pressure

12.3 Rankine Active Earth Pressure

The lateral earth pressure described in Section 12.2 involves walls that do not yield at all. However, if a wall tends to move away from the soil a distance Δx, as shown in Figure 12.6a, the soil pressure on the wall at any depth will decrease. For a wall that is *frictionless,* the horizontal stress, σ'_h, at depth z will equal $K_o\sigma'_o$ $(=K_o\gamma z)$ when Δx is zero. However, with $\Delta x > 0$, σ'_h will be less than $K_o\sigma'_o$.

The Mohr's circles corresponding to wall displacements of $\Delta x = 0$ and $\Delta x > 0$ are shown as circles a and b, respectively, in Figure 12.6b. If the displacement of the wall, Δx, continues to increase, the corresponding Mohr's circle eventually will just touch the Mohr–Coulomb failure envelope defined by the equation

$$s = c' + \sigma' \tan \phi'$$

This circle, marked c in the figure, represents the failure condition in the soil mass; the horizontal stress then equals σ'_a, referred to as the *Rankine active pressure*. The *slip lines* (failure planes) in the soil mass will then make angles of $\pm(45 + \phi'/2)$ with the horizontal, as shown in Figure 12.6a.

Equation (2.91) relates the principal stresses for a Mohr's circle that touches the Mohr–Coulomb failure envelope:

$$\sigma'_1 = \sigma'_3 \tan^2\left(45 + \frac{\phi'}{2}\right) + 2c' \tan\left(45 + \frac{\phi'}{2}\right)$$

For the Mohr's circle c in Figure 12.6b,

Major principal stress, $\sigma'_1 = \sigma'_o$

and

Minor principal stress, $\sigma'_3 = \sigma'_a$

Thus,

$$\sigma'_o = \sigma'_a \tan^2\left(45 + \frac{\phi'}{2}\right) + 2c' \tan\left(45 + \frac{\phi'}{2}\right)$$

$$\sigma'_a = \frac{\sigma'_o}{\tan^2\left(45 + \dfrac{\phi'}{2}\right)} - \frac{2c'}{\tan\left(45 + \dfrac{\phi'}{2}\right)}$$

or

$$\sigma'_a = \sigma'_o \tan^2\left(45 - \frac{\phi'}{2}\right) - 2c' \tan\left(45 - \frac{\phi'}{2}\right)$$
$$= \sigma'_o K_a - 2c'\sqrt{K_a} \tag{12.8}$$

where $K_a = \tan^2(45 - \phi'/2) =$ Rankine active-pressure coefficient.

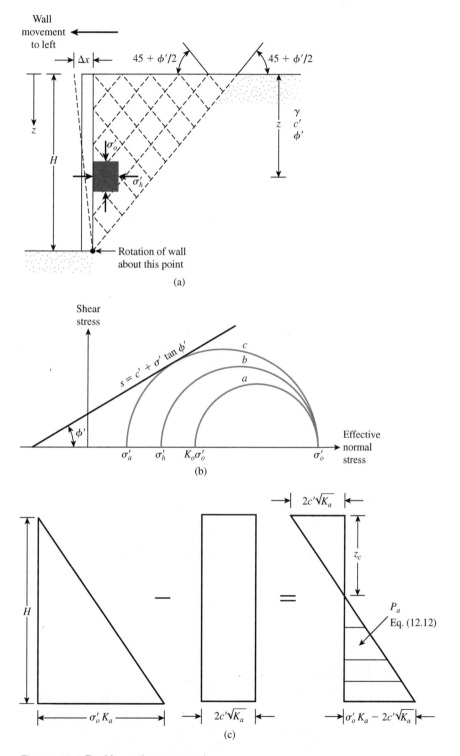

Figure 12.6 Rankine active pressure

The variation of the active pressure with depth for the wall shown in Figure 12.6a is given in Figure 12.6c. Note that $\sigma_o' = 0$ at $z = 0$ and $\sigma_o' = \gamma H$ at $z = H$. The pressure distribution shows that at $z = 0$ the active pressure equals $-2c'\sqrt{K_a}$, indicating a tensile stress that decreases with depth and becomes zero at a depth $z = z_c$, or

$$\gamma z_c K_a - 2c'\sqrt{K_a} = 0$$

and

$$z_c = \frac{2c'}{\gamma\sqrt{K_a}} \tag{12.9}$$

The depth z_c is usually referred to as the *depth of tensile crack,* because the tensile stress in the soil will eventually cause a crack along the soil–wall interface. Thus, the total Rankine active force per unit length of the wall before the tensile crack occurs is

$$P_a = \int_0^H \sigma_a' dz = \int_0^H \gamma z K_a dz - \int_0^H 2c'\sqrt{K_a} dz$$

$$= \tfrac{1}{2}\gamma H^2 K_a - 2c'H\sqrt{K_a} \tag{12.10}$$

After the tensile crack appears, the force per unit length on the wall will be caused only by the pressure distribution between depths $z = z_c$ and $z = H$, as shown by the hatched area in Figure 12.6c. This force may be expressed as

$$P_a = \tfrac{1}{2}(H - z_c)(\gamma H K_a - 2c'\sqrt{K_a}) \tag{12.11}$$

or

$$P_a = \frac{1}{2}\left(H - \frac{2c'}{\gamma\sqrt{K_a}}\right)\left(\gamma H K_a - 2c'\sqrt{K_a}\right) \tag{12.12}$$

However, it is important to realize that the active earth pressure condition will be reached only if the wall is allowed to "yield" sufficiently. The necessary amount of outward displacement of the wall is about $0.001H$ to $0.004H$ for granular soil backfills and about $0.01H$ to $0.04H$ for cohesive soil backfills.

Note further that if the *total stress* shear strength parameters (c, ϕ) were used, an equation similar to Eq. (12.8) could have been derived, namely,

$$\sigma_a = \sigma_o \tan^2\left(45 - \frac{\phi}{2}\right) - 2c \tan\left(45 - \frac{\phi}{2}\right)$$

Example 12.2

A 6-m-high retaining wall is to support a soil with unit weight $\gamma = 17.4$ kN/m^3, soil friction angle $\phi' = 26°$, and cohesion $c' = 14.36$ kN/m^2. Determine the Rankine active force per unit length of the wall both before and after the tensile crack occurs, and determine the line of action of the resultant in both cases.

Solution

For $\phi' = 26°$,

$$K_a = \tan^2\left(45 - \frac{\phi'}{2}\right) = \tan^2(45 - 13) = 0.39$$

$$\sqrt{K_a} = 0.625$$

$$\sigma_a' = \gamma H K_a - 2c'\sqrt{K_a}$$

From Figure 12.6c, at $z = 0$,

$$\sigma_a' = -2c'\sqrt{K_a} = -2(14.36)(0.625) = -17.95 \text{ kN/m}^2$$

and at $z = 6$ m,

$$\sigma_a' = (17.4)(6)(0.39) - 2(14.36)(0.625)$$
$$= 40.72 - 17.95 = 22.77 \text{ kN/m}^2$$

Active Force before the Tensile Crack Appeared: Eq. (12.10)

$$P_a = \tfrac{1}{2}\gamma H^2 K_a - 2c'H\sqrt{K_a}$$
$$= \tfrac{1}{2}(6)(40.72) - (6)(17.95) = 122.16 - 107.7 = \mathbf{14.46 \text{ kN/m}}$$

The line of action of the resultant can be determined by taking the moment of the area of the pressure diagrams about the bottom of the wall, or

$$P_a \bar{z} = (122.16)\left(\frac{6}{3}\right) - (107.7)\left(\frac{6}{2}\right)$$

Thus,

$$\bar{z} = \frac{244.32 - 323.1}{14.46} = \mathbf{-5.45 \text{ m}}.$$

Active Force after the Tensile Crack Appeared: Eq. (12.9)

$$z_c = \frac{2c'}{\gamma\sqrt{K_a}} = \frac{2(14.36)}{(17.4)(0.625)} = 2.64 \text{ m}$$

Using Eq. (12.11) gives

$$P_a = \tfrac{1}{2}(H - z_c)(\gamma H K_a - 2c'\sqrt{K_a}) = \tfrac{1}{2}(6 - 2.64)(22.77) = \mathbf{38.25 \text{ kN/m}}$$

Figure 12.6c indicates that the force $P_a = 38.25$ kN/m is the area of the hatched triangle. Hence, the line of action of the resultant will be located at a height $\bar{z} = (H - z_c)/3$ above the bottom of the wall, or

$$\bar{z} = \frac{6 - 2.64}{3} = \mathbf{1.12 \text{ m}}$$

∎

Example 12.3

Assume that the retaining wall shown in Figure 12.7a can yield sufficiently to develop an active state. Determine the Rankine active force per unit length of the wall and the location of the resultant line of action.

Solution

If the cohesion, c', is zero, then

$$\sigma_a' = \sigma_o' K_a$$

For the top layer of soil, $\phi_1' = 30°$, so

$$K_{a(1)} = \tan^2\left(45 - \frac{\phi_1'}{2}\right) = \tan^2(45 - 15) = \frac{1}{3}$$

Similarly, for the bottom layer of soil, $\phi_2' = 36°$, and it follows that

$$K_{a(2)} = \tan^2\left(45 - \frac{36}{2}\right) = 0.26$$

The following table shows the calculation of σ_a' and u at various depths below the ground surface.

Depth, z (ft)	σ_o' (lb/ft^2)	K_a	$\sigma_a' = K_a \sigma_o'$ (lb/ft^2)	u (lb/ft^2)
0	0	1/3	0	0
10$^-$	$(102)(10) = 1020$	1/3	340	0
10$^+$	1020	0.26	265.2	0
20	$(102)(10) + (121 - 62.4)(10) = 1606$	0.26	417.6	$(62.4)(10) = 624$

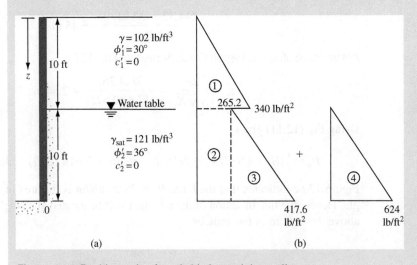

(a)   (b)

Figure 12.7 Rankine active force behind a retaining wall

The pressure distribution diagram is plotted in Figure 12.7b. The force per unit length is

$$P_a = \text{area 1} + \text{area 2} + \text{area 3} + \text{area 4}$$

$$= \tfrac{1}{2}(10)(340) + (265.2)(10) + \tfrac{1}{2}(417.6 - 265.2)(10) + \tfrac{1}{2}(624)(10)$$

$$= 1700 + 2652 + 762 + 3120 = \mathbf{8234\ lb/ft}$$

The distance of the line of action of the resultant force from the bottom of the wall can be determined by taking the moments about the bottom of the wall (point O in Figure 12.7a) and is

$$\bar{z} = \frac{(1700)\left(10 + \dfrac{10}{2}\right) + (2652)\left(\dfrac{10}{2}\right) + (762 + 3120)\left(\dfrac{10}{3}\right)}{8234} = \mathbf{5.93\ ft} \quad \blacksquare$$

12.4 A Generalized Case for Rankine Active Pressure—Granular Backfill

In Section 12.3, the relationship was developed for Rankine active pressure for a retaining wall with a vertical back and a horizontal backfill. That can be extended to general cases of frictionless walls with inclined backs and inclined backfills.

Figure 12.8 shows a retaining wall whose back is inclined at an angle θ with the vertical. The granular backfill is inclined at an angle α with the horizontal.

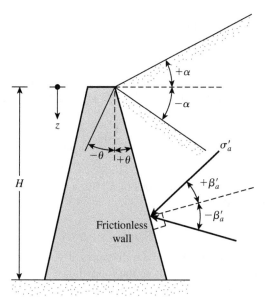

Figure 12.8 General case for a retaining wall with granular backfill

For a Rankine active case, the lateral earth pressure (σ'_a) at a depth z can be given as (Chu, 1991),

$$\sigma'_a = \frac{\gamma z \cos\alpha \sqrt{1 + \sin^2\phi' - 2\sin\phi' \cos\psi_a}}{\cos\alpha + \sqrt{\sin^2\phi' - \sin^2\alpha}} \tag{12.13}$$

where $\psi_a = \sin^{-1}\left(\dfrac{\sin\alpha}{\sin\phi'}\right) - \alpha + 2\theta.$ $\hspace{2cm}$ (12.14)

The pressure σ'_a will be inclined at an angle β'_a with the plane drawn at right angle to the backface of the wall, and

$$\beta'_a = \tan^{-1}\left(\frac{\sin\phi' \sin\psi_a}{1 - \sin\phi' \cos\psi_a}\right) \tag{12.15}$$

The active force P_a for unit length of the wall then can be calculated as

$$P_a = \frac{1}{2}\gamma H^2 K_a \tag{12.16}$$

where

$$K_{a(R)} = \frac{\cos(\alpha - \theta)\sqrt{1 + \sin^2\phi' - 2\sin\phi' \cos\psi_a}}{\cos^2\theta\left(\cos\alpha + \sqrt{\sin^2\phi' - \sin^2\alpha}\right)}$$
$$= \text{Rankine active earth-pressure coefficient for generalized case} \tag{12.17}$$

The location and direction of the resultant force P_a is shown in Figure 12.9. Also shown in this figure is the failure wedge, *ABC*. Note that *BC* will be inclined at an angle η. Or

$$\eta_a = \frac{\pi}{4} + \frac{\phi'}{2} + \frac{\alpha}{2} - \frac{1}{2}\sin^{-1}\left(\frac{\sin\alpha}{\sin\phi'}\right) \tag{12.18}$$

Tables 12.1 and 12.2 give the variations of K_a [Eq. (12.17)] and β'_a [Eq. (12.15)] for various values of α, θ, and ϕ'.

Granular Backfill with Vertical Back Face of Wall

As a special case, for a vertical back face of a wall (that is, $\theta = 0$), as shown in Figure 12.10, Eqs. (12.13), (12.16) and (12.17) simplify to the following.

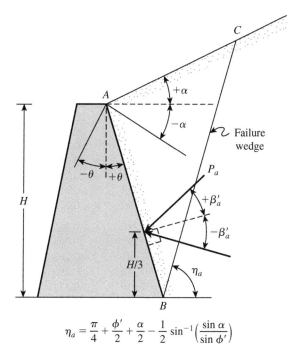

$$\eta_a = \frac{\pi}{4} + \frac{\phi'}{2} + \frac{\alpha}{2} - \frac{1}{2}\sin^{-1}\left(\frac{\sin\alpha}{\sin\phi'}\right)$$

Figure 12.9 Location and direction of Rankine active force

If the backfill of a frictionless retaining wall is a *granular soil* ($c' = 0$) and rises at an angle α with respect to the horizontal (see Figure 12.10), the *active earth-pressure coefficient* may be expressed in the form

$$K_a = \cos\alpha \frac{\cos\alpha - \sqrt{\cos^2\alpha - \cos^2\phi'}}{\cos\alpha + \sqrt{\cos^2\alpha - \cos^2\phi'}} \tag{12.19}$$

where ϕ' = angle of friction of soil.

At any depth z, the *Rankine active pressure* may be expressed as

$$\sigma'_a = \gamma z K_a \tag{12.20}$$

Also, the total force per unit length of the wall is

$$P_a = \tfrac{1}{2}\gamma H^2 K_a \tag{12.21}$$

Table 12.1 Variation of $K_{a(R)}$ [Eq. (12.17)]

| α (deg) | θ (deg) | $K_{a(R)}$ | | | | | | |
| | | ϕ' (deg) | | | | | | |
		28	30	32	34	36	38	40
0	0	0.361	0.333	0.307	0.283	0.260	0.238	0.217
	2	0.363	0.335	0.309	0.285	0.262	0.240	0.220
	4	0.368	0.341	0.315	0.291	0.269	0.248	0.228
	6	0.376	0.350	0.325	0.302	0.280	0.260	0.242
	8	0.387	0.362	0.338	0.316	0.295	0.276	0.259
	10	0.402	0.377	0.354	0.333	0.314	0.296	0.280
	15	0.450	0.428	0.408	0.390	0.373	0.358	0.345
5	0	0.366	0.337	0.311	0.286	0.262	0.240	0.219
	2	0.373	0.344	0.317	0.292	0.269	0.247	0.226
	4	0.383	0.354	0.328	0.303	0.280	0.259	0.239
	6	0.396	0.368	0.342	0.318	0.296	0.275	0.255
	8	0.412	0.385	0.360	0.336	0.315	0.295	0.276
	10	0.431	0.405	0.380	0.358	0.337	0.318	0.300
	15	0.490	0.466	0.443	0.423	0.405	0.388	0.373
10	0	0.380	0.350	0.321	0.294	0.270	0.246	0.225
	2	0.393	0.362	0.333	0.306	0.281	0.258	0.236
	4	0.408	0.377	0.348	0.322	0.297	0.274	0.252
	6	0.426	0.395	0.367	0.341	0.316	0.294	0.273
	8	0.447	0.417	0.389	0.363	0.339	0.317	0.297
	10	0.471	0.441	0.414	0.388	0.365	0.344	0.324
	15	0.542	0.513	0.487	0.463	0.442	0.422	0.404
15	0	0.409	0.373	0.341	0.311	0.283	0.258	0.235
	2	0.427	0.391	0.358	0.328	0.300	0.274	0.250
	4	0.448	0.411	0.378	0.348	0.320	0.294	0.271
	6	0.472	0.435	0.402	0.371	0.344	0.318	0.295
	8	0.498	0.461	0.428	0.398	0.371	0.346	0.323
	10	0.527	0.490	0.457	0.428	0.400	0.376	0.353
	15	0.610	0.574	0.542	0.513	0.487	0.463	0.442
20	0	0.461	0.414	0.374	0.338	0.306	0.277	0.250
	2	0.486	0.438	0.397	0.360	0.328	0.298	0.271
	4	0.513	0.465	0.423	0.386	0.353	0.323	0.296
	6	0.543	0.495	0.452	0.415	0.381	0.351	0.324
	8	0.576	0.527	0.484	0.446	0.413	0.383	0.355
	10	0.612	0.562	0.518	0.481	0.447	0.417	0.390
	15	0.711	0.660	0.616	0.578	0.545	0.515	0.488

Table 12.2 Variation of β_a' [Eq. (12.15)]

α (deg)	θ (deg)	β_a' ϕ' (deg)						
		28	30	32	34	36	38	40
0	0	0.000	0.000	0.000	0.000	0.000	0.000	0.000
	2	3.525	3.981	4.484	5.041	5.661	6.351	7.124
	4	6.962	7.848	8.821	9.893	11.075	12.381	13.827
	6	10.231	11.501	12.884	14.394	16.040	17.837	19.797
	8	13.270	14.861	16.579	18.432	20.428	22.575	24.876
	10	16.031	17.878	19.850	21.951	24.184	26.547	29.039
	15	21.582	23.794	26.091	28.464	30.905	33.402	35.940
5	0	5.000	5.000	5.000	5.000	5.000	5.000	5.000
	2	8.375	8.820	9.311	9.854	10.455	11.123	11.870
	4	11.553	12.404	13.336	14.358	15.482	16.719	18.085
	6	14.478	15.679	16.983	18.401	19.942	21.618	23.441
	8	17.112	18.601	20.203	21.924	23.773	25.755	27.876
	10	19.435	21.150	22.975	24.915	26.971	29.144	31.434
	15	23.881	25.922	28.039	30.227	32.479	34.787	37.140
10	0	10.000	10.000	10.000	10.000	10.000	10.000	10.000
	2	13.057	13.491	13.967	14.491	15.070	15.712	16.426
	4	15.839	16.657	17.547	18.519	19.583	20.751	22.034
	6	18.319	19.460	20.693	22.026	23.469	25.032	26.726
	8	20.483	21.888	23.391	24.999	26.720	28.559	30.522
	10	22.335	23.946	25.653	27.460	29.370	31.385	33.504
	15	25.683	27.603	29.589	31.639	33.747	35.908	38.114
15	0	15.000	15.000	15.000	15.000	15.000	15.000	15.000
	2	17.576	18.001	18.463	18.967	19.522	20.134	20.812
	4	19.840	20.631	21.485	22.410	23.417	24.516	25.719
	6	21.788	22.886	24.060	25.321	26.677	28.139	29.716
	8	23.431	24.778	26.206	27.722	29.335	31.052	32.878
	10	24.783	26.328	27.950	29.654	31.447	33.332	35.310
	15	27.032	28.888	30.793	32.747	34.751	36.802	38.894
20	0	20.000	20.000	20.000	20.000	20.000	20.000	20.000
	2	21.925	22.350	22.803	23.291	23.822	24.404	25.045
	4	23.545	24.332	25.164	26.054	27.011	28.048	29.175
	6	24.876	25.966	27.109	28.317	29.604	30.980	32.455
	8	25.938	27.279	28.669	30.124	31.657	33.276	34.989
	10	26.755	28.297	29.882	31.524	33.235	35.021	36.886
	15	27.866	29.747	31.638	33.552	35.498	37.478	39.491

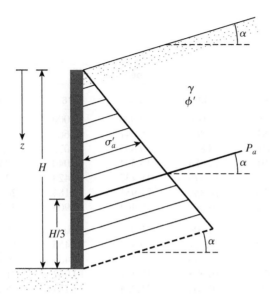

Figure 12.10 Notations for active pressure—Eqs. (12.19), (12.20), (12.21)

Note that, in this case, the direction of the resultant force P_a is *inclined at an angle α with the horizontal* and intersects the wall at a distance $H/3$ from the base of the wall. Table 12.3 presents the values of K_a (active earth pressure) for various values of α and ϕ'.

12.5 Rankine Active Pressure with Vertical Wall Backface and Inclined $c'-\phi'$ Soil Backfill

For a frictionless retaining wall with a vertical back face ($\theta = 0$) and inclined backfill of $c'-\phi'$ soil (see Figure 12.10), the active pressure at any depth z can be given as (Mazindrani and Ganjali, 1997)

$$\sigma_a' = \gamma z K_a = \gamma z K_a' \cos \alpha \tag{12.22}$$

where

$$K_a' = \frac{1}{\cos^2 \phi'} \left\{ \begin{array}{l} 2\cos^2 \alpha + 2\left(\dfrac{c'}{\gamma z}\right)\cos \phi' \sin \phi' \\[2mm] -\sqrt{\left[4\cos^2\alpha(\cos^2\alpha - \cos^2 \phi') + 4\left(\dfrac{c'}{\gamma z}\right)^2\cos^2\phi' + 8\left(\dfrac{c'}{\gamma z}\right)\cos^2 \alpha \sin \phi' \cos \phi'\right]} \end{array} \right\} - 1$$

$$\tag{12.23}$$

Table 12.3 Values of K_a [Eq. (12.19)]

α (deg)	ϕ' (deg) →												
↓	28	29	30	31	32	33	34	35	36	37	38	39	40
0	0.3610	0.3470	0.3333	0.3201	0.3073	0.2948	0.2827	0.2710	0.2596	0.2486	0.2379	0.2275	0.2174
1	0.3612	0.3471	0.3335	0.3202	0.3074	0.2949	0.2828	0.2711	0.2597	0.2487	0.2380	0.2276	0.2175
2	0.3618	0.3476	0.3339	0.3207	0.3078	0.2953	0.2832	0.2714	0.2600	0.2489	0.2382	0.2278	0.2177
3	0.3627	0.3485	0.3347	0.3214	0.3084	0.2959	0.2837	0.2719	0.2605	0.2494	0.2386	0.2282	0.2181
4	0.3639	0.3496	0.3358	0.3224	0.3094	0.2967	0.2845	0.2726	0.2611	0.2500	0.2392	0.2287	0.2186
5	0.3656	0.3512	0.3372	0.3237	0.3105	0.2978	0.2855	0.2736	0.2620	0.2508	0.2399	0.2294	0.2192
6	0.3676	0.3531	0.3389	0.3253	0.3120	0.2992	0.2868	0.2747	0.2631	0.2518	0.2409	0.2303	0.2200
7	0.3701	0.3553	0.3410	0.3272	0.3138	0.3008	0.2883	0.2761	0.2644	0.2530	0.2420	0.2313	0.2209
8	0.3730	0.3580	0.3435	0.3294	0.3159	0.3027	0.2900	0.2778	0.2659	0.2544	0.2432	0.2325	0.2220
9	0.3764	0.3611	0.3463	0.3320	0.3182	0.3049	0.2921	0.2796	0.2676	0.2560	0.2447	0.2338	0.2233
10	0.3802	0.3646	0.3495	0.3350	0.3210	0.3074	0.2944	0.2818	0.2696	0.2578	0.2464	0.2354	0.2247
11	0.3846	0.3686	0.3532	0.3383	0.3241	0.3103	0.2970	0.2841	0.2718	0.2598	0.2482	0.2371	0.2263
12	0.3896	0.3731	0.3573	0.3421	0.3275	0.3134	0.2999	0.2868	0.2742	0.2621	0.2503	0.2390	0.2281
13	0.3952	0.3782	0.3620	0.3464	0.3314	0.3170	0.3031	0.2898	0.2770	0.2646	0.2527	0.2412	0.2301
14	0.4015	0.3839	0.3671	0.3511	0.3357	0.3209	0.3068	0.2931	0.2800	0.2674	0.2552	0.2435	0.2322
15	0.4086	0.3903	0.3729	0.3564	0.3405	0.3253	0.3108	0.2968	0.2834	0.2705	0.2581	0.2461	0.2346
16	0.4165	0.3975	0.3794	0.3622	0.3458	0.3302	0.3152	0.3008	0.2871	0.2739	0.2612	0.2490	0.2373
17	0.4255	0.4056	0.3867	0.3688	0.3518	0.3356	0.3201	0.3053	0.2911	0.2776	0.2646	0.2521	0.2401
18	0.4357	0.4146	0.3948	0.3761	0.3584	0.3415	0.3255	0.3102	0.2956	0.2817	0.2683	0.2555	0.2433
19	0.4473	0.4249	0.4039	0.3842	0.3657	0.3481	0.3315	0.3156	0.3006	0.2862	0.2724	0.2593	0.2467
20	0.4605	0.4365	0.4142	0.3934	0.3739	0.3555	0.3381	0.3216	0.3060	0.2911	0.2769	0.2634	0.2504
21	0.4758	0.4498	0.4259	0.4037	0.3830	0.3637	0.3455	0.3283	0.3120	0.2965	0.2818	0.2678	0.2545
22	0.4936	0.4651	0.4392	0.4154	0.3934	0.3729	0.3537	0.3356	0.3186	0.3025	0.2872	0.2727	0.2590
23	0.5147	0.4829	0.4545	0.4287	0.4050	0.3832	0.3628	0.3438	0.3259	0.3091	0.2932	0.2781	0.2638
24	0.5404	0.5041	0.4724	0.4440	0.4183	0.3948	0.3731	0.3529	0.3341	0.3164	0.2997	0.2840	0.2692
25	0.5727	0.5299	0.4936	0.4619	0.4336	0.4081	0.3847	0.3631	0.3431	0.3245	0.3070	0.2905	0.2750

Table 12.4 Values of K'_a

ϕ' (deg)	α (deg)	$\dfrac{c'}{\gamma z}$ 0.025	0.05	0.1	0.5
15	0	0.550	0.512	0.435	−0.179
	5	0.566	0.525	0.445	−0.184
	10	0.621	0.571	0.477	−0.186
	15	0.776	0.683	0.546	−0.196
20	0	0.455	0.420	0.350	−0.210
	5	0.465	0.429	0.357	−0.212
	10	0.497	0.456	0.377	−0.218
	15	0.567	0.514	0.417	−0.229
25	0	0.374	0.342	0.278	−0.231
	5	0.381	0.348	0.283	−0.233
	10	0.402	0.366	0.296	−0.239
	15	0.443	0.401	0.321	−0.250
30	0	0.305	0.276	0.218	−0.244
	5	0.309	0.280	0.221	−0.246
	10	0.323	0.292	0.230	−0.252
	15	0.350	0.315	0.246	−0.263

Some values of K'_a are given in Table 12.4. For a problem of this type, the depth of tensile crack is given as

$$z_c = \frac{2c'}{\gamma}\sqrt{\frac{1 + \sin\phi'}{1 - \sin\phi'}} \tag{12.24}$$

For this case, the active pressure is inclined at an angle α with the horizontal.

Example 12.4

Refer to the retaining wall in Figure 12.9. The backfill is granular soil. Given:

$$\begin{aligned}
\text{Wall:} \quad & H = 10\text{ ft} \\
& \theta = +10° \\
\text{Backfill:} \quad & \alpha = 15° \\
& \phi' = 35° \\
& c' = 0 \\
& \gamma = 110\text{ lb/ft}^3
\end{aligned}$$

Determine the Rankine active force, P_a, and its location and direction.

Solution

From Table 12.1, for $\alpha = 15°$ and $\theta = +10°$, the value of $K_a \approx 0.42$. From Eq. (12.16),

$$P_a = \frac{1}{2}\gamma H^2 K_a = \left(\frac{1}{2}\right)(110)(10)^2(0.42) = \textbf{2310 lb/ft}$$

Again, from Table 12.2, for $\alpha = 15°$ and $\theta = +10°$, $\beta'_a \approx \textbf{30.5°}$.

The force P_a will act at a distance of $10/3 = 3.33$ ft above the bottom of the wall and will be inclined at an angle of $+30.5°$ to the normal drawn to the back face of the wall. ∎

Example 12.5

For the retaining wall shown in Figure 12.10, $H = 7.5$ m, $\gamma = 18$ kN/m³, $\phi' = 20°$, $c' = 13.5$ kN/m², and $\alpha = 10°$. Calculate the Rankine active force, P_a, per unit length of the wall and the location of the resultant force after the occurrence of the tensile crack.

Solution
From Eq. (12.24).

$$z_r = \frac{2c'}{\gamma}\sqrt{\frac{1 + \sin\phi'}{1 - \sin\phi'}} = \frac{(2)(13.5)}{18}\sqrt{\frac{1 + \sin 20}{1 - \sin 20}} = 2.14 \text{ m}$$

At $z = 7.5$ m,

$$\frac{c'}{\gamma z} = \frac{13.5}{(18)(7.5)} = 0.1$$

From Table 12.4, for $\phi' = 20°$, $c'/\gamma z = 0.1$, and $\alpha = 10°$, the value of K_a' is 0.377, so at $z = 7.5$ m,

$$\sigma_a' = \gamma z K_a' \cos\alpha = (18)(7.5)(0.377)(\cos 10) = 50.1 \text{ kN/m}^2$$

After the occurrence of the tensile crack, the pressure distribution on the wall will be as shown in Figure 12.11, so

$$P_a = \left(\frac{1}{2}\right)(50.1)(7.5 - 2.14) = \mathbf{134.3 \text{ kN/m}}$$

and

$$\bar{z} = \frac{7.5 - 2.14}{3} = \mathbf{1.79 \text{ m}}$$

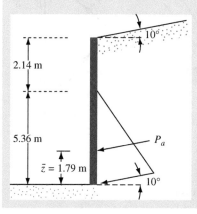

Figure 12.11 Calculation of Rankine active force, $c' - \phi'$ soil ∎

12.6 Coulomb's Active Earth Pressure

The Rankine active earth pressure calculations discussed in the preceding sections were based on the assumption that the wall is frictionless. In 1776, Coulomb proposed a theory for calculating the lateral earth pressure on a retaining wall with granular soil backfill. This theory takes wall friction into consideration.

To apply Coulomb's active earth pressure theory, let us consider a retaining wall with its back face inclined at an angle β with the horizontal, as shown in Figure 12.12a. The backfill is a granular soil that slopes at an angle α with the horizontal. Also, let δ' be the angle of friction between the soil and the wall (i.e., the angle of wall friction).

Under active pressure, the wall will move away from the soil mass (to the left in the figure). Coulomb assumed that, in such a case, the failure surface in the soil mass would be a plane (e.g., BC_1, BC_2, . . .). So, to find the active force, consider a possible soil failure wedge ABC_1. The forces acting on this wedge (per unit length at right angles to the cross section shown) are as follows:

1. The weight of the wedge, W.
2. The resultant, R, of the normal and resisting shear forces along the surface, BC_1. The force R will be inclined at an angle ϕ' to the normal drawn to BC_1.
3. The active force per unit length of the wall, P_a, which will be inclined at an angle δ' to the normal drawn to the back face of the wall.

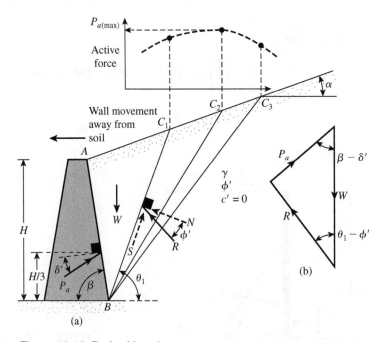

Figure 12.12 Coulomb's active pressure

For equilibrium purposes, a force triangle can be drawn, as shown in Figure 12.12b. Note that θ_1 is the angle that BC_1 makes with the horizontal. Because the magnitude of W, as well as the directions of all three forces, are known, the value of P_a can now be determined. Similarly, the active forces of other trial wedges, such as ABC_2, ABC_3, ..., can be determined. The maximum value of P_a thus determined is Coulomb's active force (see top part of Figure 12.12a), which may be expressed as

$$P_a = \tfrac{1}{2}K_a\gamma H^2 \tag{12.25}$$

where

$$K_a = \text{Coulomb's active earth pressure coefficient}$$

$$= \frac{\sin^2(\beta + \phi')}{\sin^2\beta \sin(\beta - \delta')\left[1 + \sqrt{\dfrac{\sin(\phi' + \delta')\sin(\phi' - \alpha)}{\sin(\beta - \delta')\sin(\alpha + \beta)}}\,\right]^2} \tag{12.26}$$

and H = height of the wall.

The values of the active earth pressure coefficient, K_a, for a vertical retaining wall ($\beta = 90°$) with horizontal backfill ($\alpha = 0°$) are given in Table 12.5. Note that the line of action of the resultant force (P_a) will act at a distance $H/3$ above the base of the wall and will be inclined at an angle δ' to the normal drawn to the back of the wall.

In the actual design of retaining walls, the value of the wall friction angle δ' is assumed to be between $\phi'/2$ and $\tfrac{2}{3}\phi'$. The active earth pressure coefficients for various values of ϕ', α, and β with $\delta' = \tfrac{1}{2}\phi'$ and $\tfrac{2}{3}\phi'$ are respectively given in Tables 12.6 and 12.7. These coefficients are very useful design considerations.

If a uniform surcharge of intensity q is located above the backfill, as shown in

Table 12.5 Values of K_a [Eq. (12.26)] for $\beta = 90°$ and $\alpha = 0°$

ϕ' (deg)	δ' (deg)					
	0	**5**	**10**	**15**	**20**	**25**
28	0.3610	0.3448	0.3330	0.3251	0.3203	0.3186
30	0.3333	0.3189	0.3085	0.3014	0.2973	0.2956
32	0.3073	0.2945	0.2853	0.2791	0.2755	0.2745
34	0.2827	0.2714	0.2633	0.2579	0.2549	0.2542
36	0.2596	0.2497	0.2426	0.2379	0.2354	0.2350
38	0.2379	0.2292	0.2230	0.2190	0.2169	0.2167
40	0.2174	0.2098	0.2045	0.2011	0.1994	0.1995
42	0.1982	0.1916	0.1870	0.1841	0.1828	0.1831

Table 12.6 Values of K_a [from Eq. (12.26)] for $\delta' = \frac{2}{3}\phi'$

α (deg)	φ' (deg)	β (deg) 90	85	80	75	70	65
0	28	0.3213	0.3588	0.4007	0.4481	0.5026	0.5662
	29	0.3091	0.3467	0.3886	0.4362	0.4908	0.5547
	30	0.2973	0.3349	0.3769	0.4245	0.4794	0.5435
	31	0.2860	0.3235	0.3655	0.4133	0.4682	0.5326
	32	0.2750	0.3125	0.3545	0.4023	0.4574	0.5220
	33	0.2645	0.3019	0.3439	0.3917	0.4469	0.5117
	34	0.2543	0.2916	0.3335	0.3813	0.4367	0.5017
	35	0.2444	0.2816	0.3235	0.3713	0.4267	0.4919
	36	0.2349	0.2719	0.3137	0.3615	0.4170	0.4824
	37	0.2257	0.2626	0.3042	0.3520	0.4075	0.4732
	38	0.2168	0.2535	0.2950	0.3427	0.3983	0.4641
	39	0.2082	0.2447	0.2861	0.3337	0.3894	0.4553
	40	0.1998	0.2361	0.2774	0.3249	0.3806	0.4468
	41	0.1918	0.2278	0.2689	0.3164	0.3721	0.4384
	42	0.1840	0.2197	0.2606	0.3080	0.3637	0.4302
5	28	0.3431	0.3845	0.4311	0.4843	0.5461	0.6190
	29	0.3295	0.3709	0.4175	0.4707	0.5325	0.6056
	30	0.3165	0.3578	0.4043	0.4575	0.5194	0.5926
	31	0.3039	0.3451	0.3916	0.4447	0.5067	0.5800
	32	0.2919	0.3329	0.3792	0.4324	0.4943	0.5677
	33	0.2803	0.3211	0.3673	0.4204	0.4823	0.5558
	34	0.2691	0.3097	0.3558	0.4088	0.4707	0.5443
	35	0.2583	0.2987	0.3446	0.3975	0.4594	0.5330
	36	0.2479	0.2881	0.3338	0.3866	0.4484	0.5221
	37	0.2379	0.2778	0.3233	0.3759	0.4377	0.5115
	38	0.2282	0.2679	0.3131	0.3656	0.4273	0.5012
	39	0.2188	0.2582	0.3033	0.3556	0.4172	0.4911
	40	0.2098	0.2489	0.2937	0.3458	0.4074	0.4813
	41	0.2011	0.2398	0.2844	0.3363	0.3978	0.4718
	42	0.1927	0.2311	0.2753	0.3271	0.3884	0.4625
10	28	0.3702	0.4164	0.4686	0.5287	0.5992	0.6834
	29	0.3548	0.4007	0.4528	0.5128	0.5831	0.6672
	30	0.3400	0.3857	0.4376	0.4974	0.5676	0.6516
	31	0.3259	0.3713	0.4230	0.4826	0.5526	0.6365
	32	0.3123	0.3575	0.4089	0.4683	0.5382	0.6219
	33	0.2993	0.3442	0.3953	0.4545	0.5242	0.6078
	34	0.2868	0.3314	0.3822	0.4412	0.5107	0.5942
	35	0.2748	0.3190	0.3696	0.4283	0.4976	0.5810
	36	0.2633	0.3072	0.3574	0.4158	0.4849	0.5682
	37	0.2522	0.2957	0.3456	0.4037	0.4726	0.5558
	38	0.2415	0.2846	0.3342	0.3920	0.4607	0.5437
	39	0.2313	0.2740	0.3231	0.3807	0.4491	0.5321
	40	0.2214	0.2636	0.3125	0.3697	0.4379	0.5207
	41	0.2119	0.2537	0.3021	0.3590	0.4270	0.5097
	42	0.2027	0.2441	0.2921	0.3487	0.4164	0.4990

Table 12.6 (*Continued*)

α (deg)	φ' (deg)	β (deg) 90	85	80	75	70	65
15	28	0.4065	0.4585	0.5179	0.5868	0.6685	0.7670
	29	0.3881	0.4397	0.4987	0.5672	0.6483	0.7463
	30	0.3707	0.4219	0.4804	0.5484	0.6291	0.7265
	31	0.3541	0.4049	0.4629	0.5305	0.6106	0.7076
	32	0.3384	0.3887	0.4462	0.5133	0.5930	0.6895
	33	0.3234	0.3732	0.4303	0.4969	0.5761	0.6721
	34	0.3091	0.3583	0.4150	0.4811	0.5598	0.6554
	35	0.2954	0.3442	0.4003	0.4659	0.5442	0.6393
	36	0.2823	0.3306	0.3862	0.4513	0.5291	0.6238
	37	0.2698	0.3175	0.3726	0.4373	0.5146	0.6089
	38	0.2578	0.3050	0.3595	0.4237	0.5006	0.5945
	39	0.2463	0.2929	0.3470	0.4106	0.4871	0.5805
	40	0.2353	0.2813	0.3348	0.3980	0.4740	0.5671
	41	0.2247	0.2702	0.3231	0.3858	0.4613	0.5541
	42	0.2146	0.2594	0.3118	0.3740	0.4491	0.5415
20	28	0.4602	0.5205	0.5900	0.6714	0.7689	0.8880
	29	0.4364	0.4958	0.5642	0.6445	0.7406	0.8581
	30	0.4142	0.4728	0.5403	0.6195	0.7144	0.8303
	31	0.3935	0.4513	0.5179	0.5961	0.6898	0.8043
	32	0.3742	0.4311	0.4968	0.5741	0.6666	0.7799
	33	0.3559	0.4121	0.4769	0.5532	0.6448	0.7569
	34	0.3388	0.3941	0.4581	0.5335	0.6241	0.7351
	35	0.3225	0.3771	0.4402	0.5148	0.6044	0.7144
	36	0.3071	0.3609	0.4233	0.4969	0.5856	0.6947
	37	0.2925	0.3455	0.4071	0.4799	0.5677	0.6759
	38	0.2787	0.3308	0.3916	0.4636	0.5506	0.6579
	39	0.2654	0.3168	0.3768	0.4480	0.5342	0.6407
	40	0.2529	0.3034	0.3626	0.4331	0.5185	0.6242
	41	0.2408	0.2906	0.3490	0.4187	0.5033	0.6083
	42	0.2294	0.2784	0.3360	0.4049	0.4888	0.5930

Table 12.7 Values of K_a [from Eq. (12.26)] for $\delta' = \phi'/2$

α (deg)	φ' (deg)	β (deg) 90	85	80	75	70	65
0	28	0.3264	0.3629	0.4034	0.4490	0.5011	0.5616
	29	0.3137	0.3502	0.3907	0.4363	0.4886	0.5492
	30	0.3014	0.3379	0.3784	0.4241	0.4764	0.5371
	31	0.2896	0.3260	0.3665	0.4121	0.4645	0.5253
	32	0.2782	0.3145	0.3549	0.4005	0.4529	0.5137
	33	0.2671	0.3033	0.3436	0.3892	0.4415	0.5025
	34	0.2564	0.2925	0.3327	0.3782	0.4305	0.4915
	35	0.2461	0.2820	0.3221	0.3675	0.4197	0.4807

(Continued)

Table 12.7 (*Continued*)

α (deg)	φ′ (deg)	β (deg) 90	85	80	75	70	65
	36	0.2362	0.2718	0.3118	0.3571	0.4092	0.4702
	37	0.2265	0.2620	0.3017	0.3469	0.3990	0.4599
	38	0.2172	0.2524	0.2920	0.3370	0.3890	0.4498
	39	0.2081	0.2431	0.2825	0.3273	0.3792	0.4400
	40	0.1994	0.2341	0.2732	0.3179	0.3696	0.4304
	41	0.1909	0.2253	0.2642	0.3087	0.3602	0.4209
	42	0.1828	0.2168	0.2554	0.2997	0.3511	0.4177
5	28	0.3477	0.3879	0.4327	0.4837	0.5425	0.6115
	29	0.3337	0.3737	0.4185	0.4694	0.5282	0.5972
	30	0.3202	0.3601	0.4048	0.4556	0.5144	0.5833
	31	0.3072	0.3470	0.3915	0.4422	0.5009	0.5698
	32	0.2946	0.3342	0.3787	0.4292	0.4878	0.5566
	33	0.2825	0.3219	0.3662	0.4166	0.4750	0.5437
	34	0.2709	0.3101	0.3541	0.4043	0.4626	0.5312
	35	0.2596	0.2986	0.3424	0.3924	0.4505	0.5190
	36	0.2488	0.2874	0.3310	0.3808	0.4387	0.5070
	37	0.2383	0.2767	0.3199	0.3695	0.4272	0.4954
	38	0.2282	0.2662	0.3092	0.3585	0.4160	0.4840
	39	0.2185	0.2561	0.2988	0.3478	0.4050	0.4729
	40	0.2090	0.2463	0.2887	0.3374	0.3944	0.4620
	41	0.1999	0.2368	0.2788	0.3273	0.3840	0.4514
	42	0.1911	0.2276	0.2693	0.3174	0.3738	0.4410
10	28	0.3743	0.4187	0.4688	0.5261	0.5928	0.6719
	29	0.3584	0.4026	0.4525	0.5096	0.5761	0.6549
	30	0.3432	0.3872	0.4368	0.4936	0.5599	0.6385
	31	0.3286	0.3723	0.4217	0.4782	0.5442	0.6225
	32	0.3145	0.3580	0.4071	0.4633	0.5290	0.6071
	33	0.3011	0.3442	0.3930	0.4489	0.5143	0.5920
	34	0.2881	0.3309	0.3793	0.4350	0.5000	0.5775
	35	0.2757	0.3181	0.3662	0.4215	0.4862	0.5633
	36	0.2637	0.3058	0.3534	0.4084	0.4727	0.5495
	37	0.2522	0.2938	0.3411	0.3957	0.4597	0.5361
	38	0.2412	0.2823	0.3292	0.3833	0.4470	0.5230
	39	0.2305	0.2712	0.3176	0.3714	0.4346	0.5103
	40	0.2202	0.2604	0.3064	0.3597	0.4226	0.4979
	41	0.2103	0.2500	0.2956	0.3484	0.4109	0.4858
	42	0.2007	0.2400	0.2850	0.3375	0.3995	0.4740
15	28	0.4095	0.4594	0.5159	0.5812	0.6579	0.7498
	29	0.3908	0.4402	0.4964	0.5611	0.6373	0.7284
	30	0.3730	0.4220	0.4777	0.5419	0.6175	0.7080
	31	0.3560	0.4046	0.4598	0.5235	0.5985	0.6884
	32	0.3398	0.3880	0.4427	0.5059	0.5803	0.6695
	33	0.3244	0.3721	0.4262	0.4889	0.5627	0.6513

Table 12.7 (*Continued*)

α (deg)	φ' (deg)	β (deg)					
		90	85	80	75	70	65
	34	0.3097	0.3568	0.4105	0.4726	0.5458	0.6338
	35	0.2956	0.3422	0.3953	0.4569	0.5295	0.6168
	36	0.2821	0.3282	0.3807	0.4417	0.5138	0.6004
	37	0.2692	0.3147	0.3667	0.4271	0.4985	0.5846
	38	0.2569	0.3017	0.3531	0.4130	0.4838	0.5692
	39	0.2450	0.2893	0.3401	0.3993	0.4695	0.5543
	40	0.2336	0.2773	0.3275	0.3861	0.4557	0.5399
	41	0.2227	0.2657	0.3153	0.3733	0.4423	0.5258
	42	0.2122	0.2546	0.3035	0.3609	0.4293	0.5122
20	28	0.4614	0.5188	0.5844	0.6608	0.7514	0.8613
	29	0.4374	0.4940	0.5586	0.6339	0.7232	0.8313
	30	0.4150	0.4708	0.5345	0.6087	0.6968	0.8034
	31	0.3941	0.4491	0.5119	0.5851	0.6720	0.7772
	32	0.3744	0.4286	0.4906	0.5628	0.6486	0.7524
	33	0.3559	0.4093	0.4704	0.5417	0.6264	0.7289
	34	0.3384	0.3910	0.4513	0.5216	0.6052	0.7066
	35	0.3218	0.3736	0.4331	0.5025	0.5851	0.6853
	36	0.3061	0.3571	0.4157	0.4842	0.5658	0.6649
	37	0.2911	0.3413	0.3991	0.4668	0.5474	0.6453
	38	0.2769	0.3263	0.3833	0.4500	0.5297	0.6266
	39	0.2633	0.3120	0.3681	0.4340	0.5127	0.6085
	40	0.2504	0.2982	0.3535	0.4185	0.4963	0.5912
	41	0.2381	0.2851	0.3395	0.4037	0.4805	0.5744
	42	0.2263	0.2725	0.3261	0.3894	0.4653	0.5582

Figure 12.13, the active force, P_a, can be calculated as

$$P_a = \tfrac{1}{2} K_a \gamma_{eq} H^2$$

$$\uparrow$$

Eq. (12.25)

(12.27)

where

$$\gamma_{eq} = \gamma + \left[\frac{\sin\beta}{\sin(\beta + \alpha)} \right]\left(\frac{2q}{H} \right)$$

(12.28)

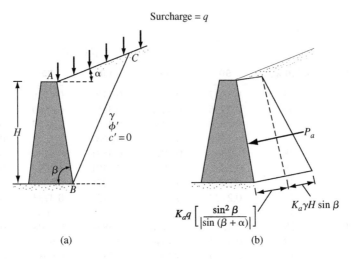

Figure 12.13 Coulomb's active pressure with a surcharge on the backfill

Example 12.6

Consider the retaining wall shown in Figure 12.12a. Given: $H = 5$ m; unit weight of soil $= 17.6$ kN/m³; angle of friction of soil $= 35°$; wall friction-angle, $\delta' = \frac{2}{3}\phi'$, soil cohesion, $c' = 0$; $\alpha = 0$, and $\beta = 90°$. Calculate the Coulomb's active force per unit length of the wall.

Solution
From Eq. (12.25)

$$P_a = \tfrac{1}{2}\gamma H^2 K_a$$

From Table 12.6, for $\alpha = 0°$, $\beta = 90°$, $\phi' = 35°$, and $\delta' = \frac{2}{3}\phi' = 23.33°$, $K_a = 0.2444$. Hence,

$$P_a = \tfrac{1}{2}(17.6)(5)^2(0.2444) = \mathbf{53.77\ kN/m}$$ ∎

Example 12.7

Refer to Figure 12.13a. Given: $H = 20$ ft, $\phi' = 30°$, $\delta' = 20°$, $\alpha = 5°$, $\beta = 85°$, $q = 2000$ lb/ft², and $\gamma = 115$ lb/ft³. Determine Coulomb's active force and the location of the line of action of the resultant P_a.

Solution

For $\beta = 85°$, $\alpha = 5°$, $\delta' = 20°$, $\phi' = 30°$, and $K_a = 0.3578$ (Table 12.6). From Eqs. (12.27) and (12.28),

$$P_a = \frac{1}{2}K_a\gamma_{eq}H^2 = \underbrace{\frac{1}{2}K_a\left[\gamma + \frac{2q}{H}\frac{\sin\beta}{\sin(\beta+\alpha)}\right]H^2}_{} = \underbrace{\frac{1}{2}K_a\gamma H^2}_{P_{a(1)}}$$

$$\underbrace{+ K_aHq\left[\frac{\sin\beta}{\sin(\beta+\alpha)}\right]}_{P_{a(2)}}$$

$$= (0.5)(0.3578)(115)(20)^2 + (0.3578)(20)(2000)\left[\frac{\sin 85}{\sin(85+5)}\right]$$

$$= 8229.4 + 14{,}257.5 = \mathbf{22{,}486.9\ lb/ft}$$

Location of the line of action of the resultant:

$$P_a\bar{z} = P_{a(1)}\frac{H}{3} + P_{a(2)}\frac{H}{2}$$

or

$$\bar{z} = \frac{(8229.4)\left(\dfrac{20}{3}\right) + (14{,}257.5)\left(\dfrac{20}{2}\right)}{22{,}486.9}$$

$$= \mathbf{8.78\ ft}\ \text{(measured vertically from the bottom of the wall)} \qquad \blacksquare$$

12.7 Lateral Earth Pressure Due to Surcharge

In several instances, the theory of elasticity is used to determine the lateral earth pressure on unyielding retaining structures caused by various types of surcharge loading, such as *line loading* (Figure 12.14a) and *strip loading* (Figure 12.14b).

According to the theory of elasticity, the stress at any depth, z, on a retaining structure caused by a line load of intensity q/unit length (Figure 12.14a) may be given as

$$\sigma = \frac{2q}{\pi H}\frac{a^2b}{(a^2+b^2)^2} \qquad (12.29)$$

where σ = horizontal stress at depth $z = bH$

(See Figure 12.14a for explanations of the terms a and b.)

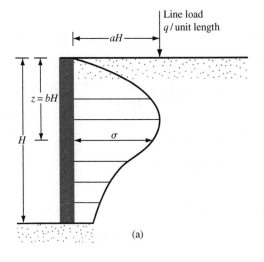

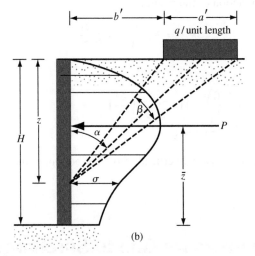

Figure 12.14 Lateral earth pressure caused by (a) line load and (b) strip load

However, because soil is not a perfectly elastic medium, some deviations from Eq. (12.29) may be expected. The modified forms of this equation generally accepted for use with soils are as follows:

$$\sigma = \frac{4a}{\pi H} \frac{a^2 b}{(a^2 + b^2)} \quad \text{for } a > 0.4 \tag{12.30}$$

and

$$\sigma = \frac{q}{H} \frac{0.203b}{(0.16 + b^2)^2} \quad \text{for } a \leq 0.4 \tag{12.31}$$

Figure 12.14b shows a strip load with an intensity of q/unit area located at a distance b' from a wall of height H. Based on the theory of elasticity, the horizontal stress, σ, at any depth z on a retaining structure is

$$\sigma = \frac{q}{\pi}(\beta - \sin \beta \cos 2\alpha) \tag{12.32}$$

(The angles α and β are defined in Figure 12.14b.)

However, in the case of soils, the right-hand side of Eq. (12.32) is doubled to account for the yielding soil continuum, or

$$\sigma = \frac{2q}{\pi}(\beta - \sin \beta \cos 2\alpha) \tag{12.33}$$

The total force per unit length (P) due to the *strip loading only* (Jarquio, 1981) may be expressed as

$$P = \frac{q}{90}[H(\theta_2 - \theta_1)] \tag{12.34}$$

where

$$\theta_1 = \tan^{-}\left(\frac{b'}{H}\right)(\text{deg}) \tag{12.35}$$

$$\theta_2 = \tan^{-1}\left(\frac{a' + b'}{H}\right)(\text{deg}) \tag{12.36}$$

The location $\bar{z}$ (see Figure 12.14b) of the resultant force, P, can be given as

$$\bar{z} = H - \left[\frac{H^2(\theta_2 - \theta_1) + (R - Q) - 57.3a'H}{2H(\theta_2 - \theta_1)}\right] \tag{12.37}$$

where

$$R = (a' + b')^2(90 - \theta_2) \tag{12.38}$$

$$Q = b'^2(90 - \theta_1) \tag{12.39}$$

EXAMPLE 12.8

Refer to Figure 12.14a which shows a line load surcharge. Given: $H = 6$ m, $a = 0.25$, amd $q = 3$ kN/m. Calculate the variation of the lateral stress σ on the retaining structure at $z = 1, 2, 3, 4, 5,$ and 6 m.

Solution

For $a = 0.25$, which is less than 0.4, we will use Eq. (12.31). Now the following table can be prepared.

z (m)	H (m)	b = z/H	a	σ (kN/m²)
1	6	0.167	0.25	0.48
2	6	0.333	0.25	0.46
3	6	0.5	0.25	0.302
4	6	0.667	0.25	0.185
5	6	0.833	0.25	0.116
6	6	1	0.25	0.073

Example 12.9

Refer to Figure 12.14b. Here, $a' = 2$ m, $b' = 1$ m, $q = 40$ kN/m², and $H = 6$ m. Determine the total force on the wall (kN/m) caused by the strip loading only.

Solution

From Eqs. (12.35) and (12.36),

$$\theta_1 = \tan^{-1}\left(\frac{1}{6}\right) = 9.46°$$

$$\theta_2 = \tan^{-1}\left(\frac{2+1}{6}\right) = 26.57°$$

From Eq. (12.34)

$$P = \frac{q}{90}[H(\theta_2 - \theta_1)] = \frac{40}{90}[6(26.57 - 9.46)] = \textbf{45.63 kN/m}$$ ∎

Example 12.10

Refer to Example 12.9. Determine the location of the resultant $\bar{z}$.

Solution

From Eqs. (12.38) and (12.39),

$$R = (a' + b')^2(90 - \theta_2) = (2 + 1)^2(90 - 26.57) = 570.87$$
$$Q = b'^2(90 - \theta_1) = (1)^2(90 - 9.46) = 80.54$$

From Eq. (12.37),

$$\bar{z} = H - \left[\frac{H^2(\theta_2 - \theta_1) + (R - Q) - 57.3a'H}{2H(\theta_2 - \theta_1)}\right]$$

$$= 6 - \left[\frac{(6)^2(26.57 - 9.46) + (570.87 - 80.54) - (57.3)(2)(6)}{(2)(6)(26.57 - 9.46)}\right] = \textbf{3.96 m}$$ ∎

12.8 Active Earth Pressure for Earthquake Conditions–Granular Backfill

Coulomb's active earth pressure theory (see Section 12.6) can be extended to take into account the forces caused by an earthquake. Figure 12.15 shows a condition of active pressure with a granular backfill ($c' = 0$). Note that the forces acting on the soil failure wedge in Figure 12.15 are essentially the same as those shown in Figure 12.12a with the addition of $k_h W$ and $k_v W$ in the horizontal and vertical direction respectively; k_h and k_v may be defined as

$$k_h = \frac{\text{horizontal earthquake acceleration component}}{\text{acceleration due to gravity, } g} \tag{12.40}$$

$$k_v = \frac{\text{vertical earthquake acceleration component}}{\text{acceleration due to gravity, } g} \tag{12.41}$$

As in Section 12.6, the relation for the active force per unit length of the wall (P_{ae}) can be determined as

$$P_{ae} = \tfrac{1}{2}\gamma H^2 (1 - k_v) K_{ae} \tag{12.42}$$

where

K_{ae} = active earth pressure coefficient

$$= \frac{\sin^2(\phi' + \beta - \theta')}{\cos\theta' \sin^2\beta \sin(\beta - \theta' - \delta')\left[1 + \sqrt{\dfrac{\sin(\phi' + \delta')\sin(\phi' - \theta' - \alpha)}{\sin(\beta - \delta' - \theta')\sin(\alpha + \beta)}}\right]^2}$$

$$\tag{12.43}$$

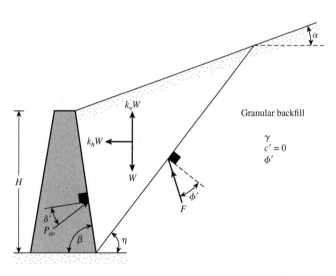

Figure 12.15 Derivation of Eq. (12.42)

$$\theta' = \tan^{-1}\left[\frac{k_h}{(1 - k_v)}\right]$$ (12.44)

Note that for no earthquake condition, $k_h = 0$, $k_v = 0$, and $\theta' = 0$. Hence $K_{ae} = K_a$ [as given by Eq. (12.26)].

Some values of K_{ae} for $\beta = 90°$ and $k_v = 0$ are given in Table 12.8.

The magnitude of P_{ae} as given in Eq. (12.42) also can be determined as (Seed and Whitman, 1970),

$$P_{ae} = \frac{1}{2}\gamma H^2(1 - k_v)[K_a(\beta',\alpha')]\left(\frac{\sin^2\beta'}{\cos\theta'\sin^2\beta}\right)$$ (12.45)

where

$$\beta' = \beta - \theta'$$ (12.46)
$$\alpha' = \theta' + \alpha$$ (12.47)

$K_a(\beta',\alpha')$ = Coulomb's active earth-pressure coefficient on a wall with a back face inclination of β' with the horizontal and with a back fill inclined at an angle α' with the horizontal (such as Tables 12.6 and 12.7)

Equation (12.42) is usually referred to as the *Mononobe–Okabe* solution. Unlike the case shown in Figure 12.12a, the resultant earth pressure in this situation, as calculated by Eq. (12.42) *does not act* at a distance of $H/3$ from the bottom of the wall. The following procedure may be used to obtain the location of the resultant force P_{ae}:

Step 1. Calculate P_{ae} by using Eq. (12.42)
Step 2. Calculate P_a by using Eq. (12.25)
Step 3. Calculate

$$\Delta P_{ae} = P_{ae} - P_a$$ (12.48)

Step 4. Assume that P_a acts at a distance of $H/3$ from the bottom of the wall (Figure 12.16)
Step 5. Assume that ΔP_{ae} acts at a distance of $0.6H$ from the bottom of the wall (Figure 12.16)
Step 6. Calculate the location of the resultant as

$$\bar{z} = \frac{(0.6H)(\Delta P_{ae}) + \left(\frac{H}{3}\right)(P_a)}{P_{ae}}$$ (12.49)

Table 12.8 Values of K_{ae} [Eq. (12.43)] for $\beta = 90°$ and $k_v = 0$

k_h	δ' (deg)	α (deg)	ϕ' (deg) 28	30	35	40	45
0.1	0	0	0.427	0.397	0.328	0.268	0.217
0.2			0.508	0.473	0.396	0.382	0.270
0.3			0.611	0.569	0.478	0.400	0.334
0.4			0.753	0.697	0.581	0.488	0.409
0.5			1.005	0.890	0.716	0.596	0.500
0.1	0	5	0.457	0.423	0.347	0.282	0.227
0.2			0.554	0.514	0.424	0.349	0.285
0.3			0.690	0.635	0.522	0.431	0.356
0.4			0.942	0.825	0.653	0.535	0.442
0.5			—	—	0.855	0.673	0.551
0.1	0	10	0.497	0.457	0.371	0.299	0.238
0.2			0.623	0.570	0.461	0.375	0.303
0.3			0.856	0.748	0.585	0.472	0.383
0.4			—	—	0.780	0.604	0.486
0.5			—	—	—	0.809	0.624
0.1	$\phi'/2$	0	0.396	0.368	0.306	0.253	0.207
0.2			0.485	0.452	0.380	0.319	0.267
0.3			0.604	0.563	0.474	0.402	0.340
0.4			0.778	0.718	0.599	0.508	0.433
0.5			1.115	0.972	0.774	0.648	0.522
0.1	$\phi'/2$	5	0.428	0.396	0.326	0.268	0.218
0.2			0.537	0.497	0.412	0.342	0.283
0.3			0.699	0.640	0.526	0.438	0.367
0.4			1.025	0.881	0.690	0.568	0.475
0.5			—	—	0.962	0.752	0.620
0.1	$\phi'/2$	10	0.472	0.433	0.352	0.285	0.230
0.2			0.616	0.562	0.454	0.371	0.303
0.3			0.908	0.780	0.602	0.487	0.400
0.4			—	—	0.857	0.656	0.531
0.5			—	—	—	0.944	0.722
0.1	$\frac{2}{3}\phi'$	0	0.393	0.366	0.306	0.256	0.212
0.2			0.486	0.454	0.384	0.326	0.276
0.3			0.612	0.572	0.486	0.416	0.357
0.4			0.801	0.740	0.622	0.533	0.462
0.5			1.177	1.023	0.819	0.693	0.600
0.1	$\frac{2}{3}\phi'$	5	0.427	0.395	0.327	0.271	0.224
0.2			0.541	0.501	0.418	0.350	0.294
0.3			0.714	0.655	0.541	0.455	0.386
0.4			1.073	0.921	0.722	0.600	0.509
0.5			—	—	1.034	0.812	0.679
0.1	$\frac{2}{3}\phi'$	10	0.472	0.434	0.354	0.290	0.237
0.2			0.625	0.570	0.463	0.381	0.317
0.3			0.942	0.807	0.624	0.509	0.423
0.4			—	—	0.909	0.699	0.573
0.5			—	—	—	1.037	0.800

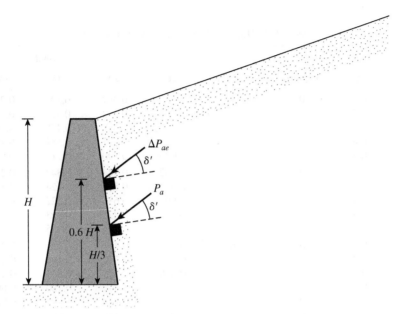

Figure 12.16 Determining the line of action of P_{ae}

Example 12.11

Refer to Figure 12.15. Given: $H = 10$ ft, $\beta = 85°$, $\alpha = 5°$, $k_v = 0$, and $k_h = 0.3$. For the backfill, $\phi' = 35°$, $\gamma = 105$ lb/ft^3, and $\delta' = 17.5°$. Determine:

    a. P_{ae} using Eq. (12.45)
    b. The location of the resultant, $\bar{z}$, from the bottom of the wall

Solution
Part a
    From Eq. (12.44),

$$\theta' = \tan^{-1}\left(\frac{k_h}{1 - k_v}\right) = \tan^{-1}\left(\frac{0.3}{1 - 0}\right) = 16.7°$$

From Eqs. (12.46) and (12.47),

$$\beta' = \beta - \theta' = 85 - 16.7 = 68.3°$$

$$\alpha' = \theta' + \alpha = 16.7 + 5 = 21.7°$$

$$\frac{\delta'}{\phi'} = \frac{17.5}{35} = 0.5$$

We will refer to Eq. (12.26). For $\phi' = 35°$, $\delta'/\phi' = 0.5$, $\beta' = 68.3°$, and $\alpha' = 21.7°$, the value of $K_a(\beta', \alpha') \approx 0.642$. Thus, from Eq. (12.45),

$$P_{ae} = \frac{1}{2}\gamma H^2(1 - k_v)[K_a(\beta', \alpha')]\left(\frac{\sin^2\beta'}{\cos\theta'\sin^2\beta}\right)$$

$$= \frac{1}{2}(105)(10^2)(1 - 0)(0.642)\left(\frac{\sin^2 68.3}{\cos 21.7 \sin^2 85}\right) = \textbf{3155.6 lb/ft}$$

Part b

From Eq. (12.25),

$$P_a = \frac{1}{2}\gamma H^2 K_a$$

From Eq. (12.26) with $\delta' = 17.5°$, $\beta = 85°$, and $\alpha = 5°$, $K_a \approx 0.2986$ (Table 12.7).

$$P_a = \frac{1}{2}(105)(10)^2(0.2986) = 1568 \text{ lb/ft}$$

$$\Delta P_{ae} = P_{ae} - P_a = 3155.6 - 1568 = 1587.6 \text{ lb/ft}$$

From Eq. (12.49),

$$\bar{z} = \frac{(0.6H)(\Delta P_{ae}) + (H/3)(P_a)}{P_{ae}}$$

$$= \frac{[(0.6)(10)](1587.6) + (10/3)(1568)}{3155.6} = \textbf{4.67 ft}$$ ∎

12.9 Active Earth Pressure for Earthquake Condition (Vertical Backface of Wall and $c'-\phi'$ Backfill)

Shukla et al. (2009) developed a procedure for estimation of P_{ae} for a retaining wall with a vertical back face and horizontal backfill with a $c'-\phi'$ soil (Figure 12.17a). In Figure 12.17a, *ABC* is the trail failure wedge. The following assumptions have been made in the analysis:

1. The effect of tensile crack is not taken into account.
2. The friction and adhesion between the back face of the wall and backfill are neglected.

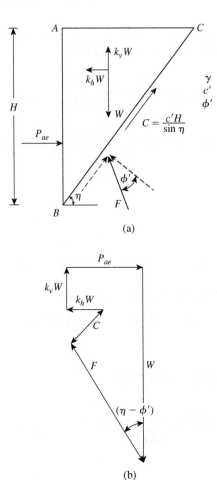

(a)

(b)

Figure 12.17 Estimation of P_{ae} with $c'-\phi'$ backfill: (a) trail failure wedge; (b) force polygon

Figure 12.17b shows the polygon for all the forces acting on the wedge *ABC*. The notations are similar to those shown in Figure 12.15. According to this analysis, the critical wedge angle $\eta = \eta_c$ for maximum value of P_{ae} can be given as

$$\tan \eta_c = \frac{\sin\phi'\sin(\phi' - \theta') + m\sin 2\phi' + \left[\begin{array}{c} \sin\phi'\sin(\phi' - \theta')\cos\theta' \\ + 4m^2\cos^2\phi' + 2m\cos\phi' \\ \{\sin\phi'\cos\theta' + \sin(\phi' - \theta')\} \end{array}\right]^{0.5}}{\sin\phi'\cos(\phi' - \theta') + 2m\cos^2\phi'} \quad (12.50)$$

where

$$m = \frac{c'\cos\theta'}{\gamma H(1 - k_v)} \quad (12.51)$$

For definition of θ', see Eq. (12.44)

Thus, the magnitude of P_{ae} can be expressed as

$$\frac{P_{ae}}{\gamma H^2} = P_{ae}^* = \frac{1}{2}(1 - k_v)\,K_{ae\gamma} - c^*K_{aec} \tag{12.52}$$

where

$$c^* = \frac{c'}{\gamma H} \tag{12.53}$$

$$K_{ae\gamma} = \frac{\cos(\phi' - \theta') - \dfrac{\sin(\phi' - \theta')}{\tan\eta_c}}{\cos\theta'\,(\cos\phi' + \tan\eta_c\,\sin\phi')} \tag{12.54}$$

$$K_{aec} = \frac{\cos\phi'\,(1 + \tan^2\eta_c)}{\tan\eta_c\,(\cos\phi' + \tan\eta_c\,\sin\phi')} \tag{12.55}$$

Figure 12.18 gives plots of P_{ae}^* against ϕ' for various of c^* and k_h ($k_v = 0$).

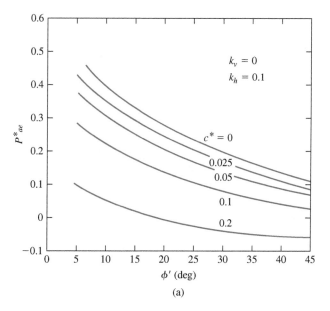

Figure 12.18 Plot of P_{ae}^* vs. ϕ' for various values of c^*: (a) $k_h = 0.1$; (b) $k_h = 0.2$; (c) $k_h = 0.3$; (d) $k_h = 0.4$ *(Note: $k_v = 0$)*

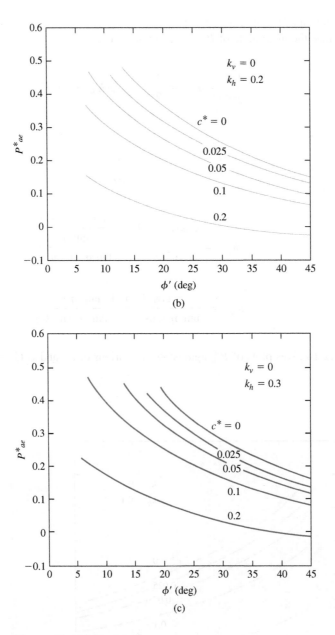

Figure 12.18 (*Continued*)

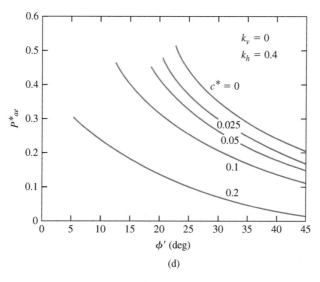

Figure 12.18 (*Continued*)

Example 12.12

For a retaining wall with a vertical back face and horizontal backfill, the following are given.

- $H = 28$ ft
- $\phi' = 20°$
- $c' = 165$ lb/ft^2

- $\gamma = 118$ lb/ft^3
- $k_h = 0.1$

Determine the magnitude of the active force, P_{ae},

Solution
From Eq. (12.53)

$$c^* = \frac{c'}{\gamma H} = \frac{165}{(118)(28)} = 0.0499 \approx 0.05$$

$$\phi' = 20°$$

From Figure 12.18a, for $\phi' = 20°$ and $c^* = 0.05$, the value of $P^*_{ae} = 0.207$. Hence

$$P_{ae} = P^*_{ae}\gamma H^2 = (0.207)(118)(28)^2 = \textbf{19,150 lb/ft}$$

Passive Pressure

12.10 Rankine Passive Earth Pressure

Figure 12.19a shows a vertical frictionless retaining wall with a horizontal backfill. At depth z, the effective vertical pressure on a soil element is $\sigma_o' = \gamma z$. Initially, if the wall does not yield at all, the lateral stress at that depth will be $\sigma_h' = K_o\sigma_o'$. This state of stress is illustrated by the Mohr's circle a in Figure 12.19b. Now, if the wall is pushed into the soil mass by an amount Δx, as shown in Figure 12.19a, the vertical stress at depth z will stay the same; however, the horizontal stress will increase. Thus, σ_h' will be greater than $K_o\sigma_o'$. The state of stress can now be represented by the Mohr's circle b in Figure 12.19b. If the wall moves farther inward (i.e., Δx is increased still more), the stresses at depth z will ultimately reach the state represented by Mohr's circle c. Note that this Mohr's circle touches the Mohr–Coulomb failure envelope, which implies that the soil behind the wall will fail by being pushed upward. The horizontal stress, σ_h', at this point is referred to as the *Rankine passive pressure,* or $\sigma_h' = \sigma_p'$.

For Mohr's circle c in Figure 12.19b, the major principal stress is σ_p', and the minor principal stress is σ_o'. Substituting these quantities into Eq. (2.91) yields

$$\sigma_p' = \sigma_o'\tan^2\left(45 + \frac{\phi'}{2}\right) + 2c'\tan\left(45 + \frac{\phi'}{2}\right) \tag{12.56}$$

Now, let

$$K_p = \text{Rankine passive earth pressure coefficient}$$
$$= \tan^2\left(45 + \frac{\phi'}{2}\right) \tag{12.57}$$

Then, from Eq. (12.56), we have

$$\sigma_p' = \sigma_o'K_p + 2c'\sqrt{K_p} \tag{12.58}$$

Equation (12.58) produces (Figure 12.19c), the passive pressure diagram for the wall shown in Figure 12.19a. Note that at $z = 0$,

$$\sigma_o' = 0 \quad \text{and} \quad \sigma_p' = 2c'\sqrt{K_p}$$

and at $z = H$,

$$\sigma_o' = \gamma H \quad \text{and} \quad \sigma_p' = \gamma H K_p + 2c'\sqrt{K_p}$$

The passive force per unit length of the wall can be determined from the area of the pressure diagram, or

$$P_p = \tfrac{1}{2}\gamma H^2 K_p + 2c'H\sqrt{K_p} \tag{12.59}$$

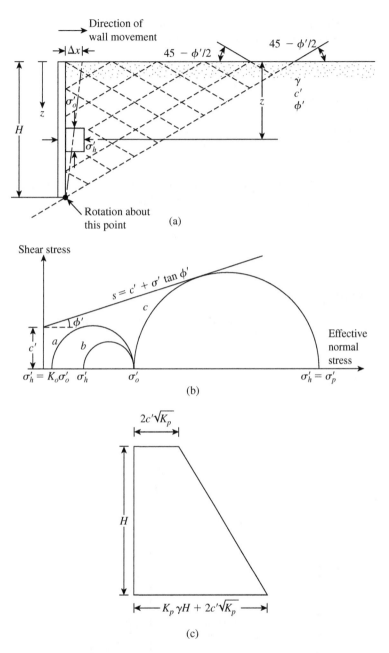

Figure 12.19 Rankine passive pressure

The approximate magnitudes of the wall movements, Δx, required to develop failure under passive conditions are as follows:

Soil type	Wall movement for passive condition, Δx
Dense sand	$0.005H$
Loose sand	$0.01H$
Stiff clay	$0.01H$
Soft clay	$0.05H$

If the backfill behind the wall is a granular soil (i.e., $c' = 0$), then, from Eq. (12.59), the passive force per unit length of the wall will be

$$P_p = \frac{1}{2} \gamma H^2 K_p \tag{12.60}$$

Example 12.13

A 3-m-high wall is shown in Figure 12.20a. Determine the Rankine passive force per unit length of the wall.

Solution
For the top layer

$$K_{p(1)} = \tan^2\left(45 + \frac{\phi_1'}{2}\right) = \tan^2(45 + 15) = 3$$

From the bottom soil layer

$$K_{p(2)} = \tan^2\left(45 + \frac{\phi_2'}{2}\right) = \tan^2(45 + 13) = 2.56$$

$$\sigma_p' = \sigma_o'K_p + 2c'\sqrt{K_p}$$

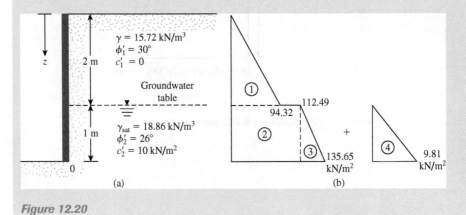

(a)

(b)

Figure 12.20

where

σ_o' = effective vertical stress

at $z = 0$, $\sigma_o' = 0$, $c_1' = 0$, $\sigma_p' = 0$

at $z = 2$ m, $\sigma_o' = (15.72)(2) = 31.44$ kN/m^2, $c_1' = 0$

So, for the top soil layer

$$\sigma_p' = 31.44K_{p(1)} + 2(0)\sqrt{K_{p(1)}} = 31.44(3) = 94.32 \text{ kN/m}^2$$

At this depth, that is $z = 2$ m, for the bottom soil layer

$$\sigma_p' = \sigma_o'K_{p(2)} + 2c_2'\sqrt{K_{p(2)}} = 31.44(2.56) + 2(10)\sqrt{2.56}$$

$$= 80.49 + 32 = 112.49 \text{ kN/m}^2$$

Again, at $z = 3$ m,

$$\sigma_o' = (15.72)(2) + (\gamma_{\text{sat}} - \gamma_w)(1)$$

$$= 31.44 + (18.86 - 9.81)(1) = 40.49 \text{ kN/m}^2$$

Hence,

$$\sigma_p' = \sigma_o'K_{p(2)} + 2c_2'\sqrt{K_{p(2)}} = 40.49(2.56) + (2)(10)(1.6)$$

$$= \mathbf{135.65 \text{ kN/m}^2}$$

Note that, because a water table is present, the hydrostatic stress, u, also has to be taken into consideration. For $z = 0$ to 2 m, $u = 0$; $z = 3$ m, $u = (1)(\gamma_w) = 9.81$ kN/m^2.

The passive pressure diagram is plotted in Figure 12.20b. The passive force per unit length of the wall can be determined from the area of the pressure diagram as follows:

Area no.	Area	
1	$\left(\frac{1}{2}\right)(2)(94.32)$	$= 94.32$
2	$(112.49)(1)$	$= 112.49$
3	$\left(\frac{1}{2}\right)(1)(135.65 - 112.49)$	$= 11.58$
4	$\left(\frac{1}{2}\right)(9.81)(1)$	$= 4.905$
		$P_P \approx \mathbf{223.3 \text{ kN/m}}$

12.11 Rankine Passive Earth Pressure—Vertical Backface and Inclined Backfill

Granular Soil

For a frictionless vertical retaining wall (Figure 12.10) with a *granular backfill* ($c' = 0$), the Rankine passive pressure at any depth can be determined in a manner similar to that done in the case of active pressure in Section 12.4. The pressure is

$$\sigma_p' = \gamma z K_p \tag{12.61}$$

Table 12.9 Passive Earth Pressure Coefficient K_p [from Eq. (12.63)]

	ϕ' (deg)→						
↓α (deg)	28	30	32	34	36	38	40
0	2.770	3.000	3.255	3.537	3.852	4.204	4.599
5	2.715	2.943	3.196	3.476	3.788	4.136	4.527
10	2.551	2.775	3.022	3.295	3.598	3.937	4.316
15	2.284	2.502	2.740	3.003	3.293	3.615	3.977
20	1.918	2.132	2.362	2.612	2.886	3.189	3.526
25	1.434	1.664	1.894	2.135	2.394	2.676	2.987

and the passive force is

$$P_p = \tfrac{1}{2}\gamma H^2 K_p \tag{12.62}$$

where

$$K_p = \cos \alpha \, \frac{\cos \alpha + \sqrt{\cos^2 \alpha - \cos^2 \phi'}}{\cos \alpha - \sqrt{\cos^2 \alpha - \cos^2 \phi'}} \tag{12.63}$$

As in the case of the active force, the resultant force, P_p, is inclined at an angle α with the horizontal and intersects the wall at a distance $H/3$ from the bottom of the wall. The values of K_p (the passive earth pressure coefficient) for various values of α and ϕ' are given in Table 12.9.

$c'-\phi'$ Soil

If the backfill of the frictionless vertical retaining wall is a $c-\phi'$ soil (see Figure 12.10), then (Mazindrani and Ganjali, 1997)

$$\sigma_p' = \gamma z K_p = \gamma z K_p' \cos \alpha \tag{12.64}$$

where

$$K_p' = \frac{1}{\cos^2 \phi'} \left\{ \frac{2\cos^2 \alpha + 2\left(\dfrac{c'}{\gamma z}\right)\cos \phi' \sin \phi'}{+ \sqrt{4\cos^2 \alpha(\cos^2 \alpha - \cos^2 \phi') + 4\left(\dfrac{c'}{\gamma z}\right)^2 \cos^2 \phi' + 8\left(\dfrac{c'}{\gamma z}\right)\cos^2 \alpha \sin \phi' \cos \phi'}} \right\} - 1 \tag{12.65}$$

The variation of K_p' with ϕ', α, and $c'/\gamma z$ is given in Table 12.10 (Mazindrani and Ganjali, 1997).

Table 12.10 Values of K_p'

ϕ' (deg)	α (deg)	$c'/\gamma z$			
		0.025	0.050	0.100	0.500
15	0	1.764	1.829	1.959	3.002
	5	1.716	1.783	1.917	2.971
	10	1.564	1.641	1.788	2.880
	15	1.251	1.370	1.561	2.732
20	0	2.111	2.182	2.325	3.468
	5	2.067	2.140	2.285	3.435
	10	1.932	2.010	2.162	3.339
	15	1.696	1.786	1.956	3.183
25	0	2.542	2.621	2.778	4.034
	5	2.499	2.578	2.737	3.999
	10	2.368	2.450	2.614	3.895
	15	2.147	2.236	2.409	3.726
30	0	3.087	3.173	3.346	4.732
	5	3.042	3.129	3.303	4.674
	10	2.907	2.996	3.174	4.579
	15	2.684	2.777	2.961	4.394

12.12 Coulomb's Passive Earth Pressure

Coulomb (1776) also presented an analysis for determining the passive earth pressure (i.e., when the wall moves *into* the soil mass) for walls possessing friction (δ' = angle of wall friction) and retaining a granular backfill material similar to that discussed in Section 12.6.

To understand the determination of Coulomb's passive force, P_p, consider the wall shown in Figure 12.21a. As in the case of active pressure, Coulomb assumed that the potential failure surface in soil is a plane. For a trial failure wedge of soil, such as ABC_1, the forces per unit length of the wall acting on the wedge are

1. The weight of the wedge, W
2. The resultant, R, of the normal and shear forces on the plane BC_1, and
3. The passive force, P_p

Figure 12.21b shows the force triangle at equilibrium for the trial wedge ABC_1. From this force triangle, the value of P_p can be determined, because the direction of all three forces and the magnitude of one force are known.

Similar force triangles for several trial wedges, such as $ABC_1, ABC_2, ABC_3, \ldots$, can be constructed, and the corresponding values of P_p can be determined. The top part of Figure 12.21a shows the nature of variation of the P_p values for different wedges. The *minimum value of P_p* in this diagram is *Coulomb's passive force,* mathematically expressed as

$$P_p = \tfrac{1}{2}\gamma H^2 K_p \qquad (12.66)$$

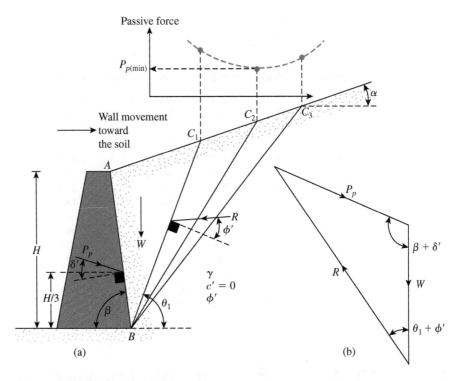

Figure 12.21 Coulomb's passive pressure

where

$$K_p = \text{Coulomb's passive pressure coefficient}$$
$$= \frac{\sin^2(\beta-\phi')}{\sin^2\beta \sin(\beta+\delta')\left[1 - \sqrt{\dfrac{\sin(\phi'+\delta')\sin(\phi'+\alpha)}{\sin(\beta+\delta')\sin(\beta+\alpha)}}\right]^2} \tag{12.67}$$

The values of the passive pressure coefficient, K_p, for various values of ϕ' and δ' are given in Table 12.11 ($\beta = 90°$, $\alpha = 0°$).

Note that the resultant passive force, P_p, will act at a distance $H/3$ from the bottom of the wall and will be inclined at an angle δ' to the normal drawn to the back face of the wall.

Table 12.11 Values of K_p [from Eq. (12.67)] for $\beta = 90°$ and $\alpha = 0°$

ϕ' (deg)	δ' (deg)				
	0	5	10	15	20
15	1.698	1.900	2.130	2.405	2.735
20	2.040	2.313	2.636	3.030	3.525
25	2.464	2.830	3.286	3.855	4.597
30	3.000	3.506	4.143	4.977	6.105
35	3.690	4.390	5.310	6.854	8.324
40	4.600	5.590	6.946	8.870	11.772

12.13 Comments on the Failure Surface Assumption for Coulomb's Pressure Calculations

Coulomb's pressure calculation methods for active and passive pressure have been discussed in Sections 12.6 and 12.12. The fundamental assumption in these analyses is the acceptance of *plane failure surface.* However, for walls with friction, this assumption does not hold in practice. The nature of *actual* failure surface in the soil mass for active and passive pressure is shown in Figure 12.22a and b, respectively (for a vertical wall with a horizontal backfill). Note that the failure surface BC is curved and that the failure surface CD is a plane.

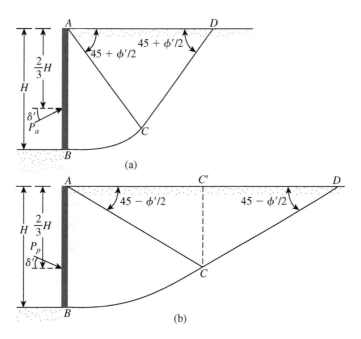

Figure 12.22 Nature of failure surface in soil with wall friction: (a) active pressure; (b) passive pressure

Although the actual failure surface in soil for the case of active pressure is somewhat different from that assumed in the calculation of the Coulomb pressure, the results are not greatly different. However, in the case of passive pressure, as the value of δ' increases, Coulomb's method of calculation gives increasingly erroneous values of P_p. This factor of error could lead to an unsafe condition because the values of P_p would become higher than the soil resistance.

Several studies have been conducted to determine the passive force P_p, assuming that the curved portion BC in Figure 12.22b is an arc of a circle, an ellipse, or a logarithmic spiral (e.g., Caquot and Kerisel, 1948; Terzaghi and Peck, 1967; Shields and Tolunay, 1973; Zhu and Qian, 2000). Section 12.14 presents the solution of Caquot and Kerisel (1948) which will suffice the purpose of this text.

12.14 Caquot and Kerisel Solution for Passive Earth Pressure (Granular Backfill)

Figure 12.23 shows a retaining wall with an inclined back and a horizontal backfill. For this case, the passive pressure per unit length of the wall can be calculated as

$$P_p = \tfrac{1}{2}\gamma H_1^2 K_p \tag{12.68}$$

where K_p = the passive pressure coefficient.

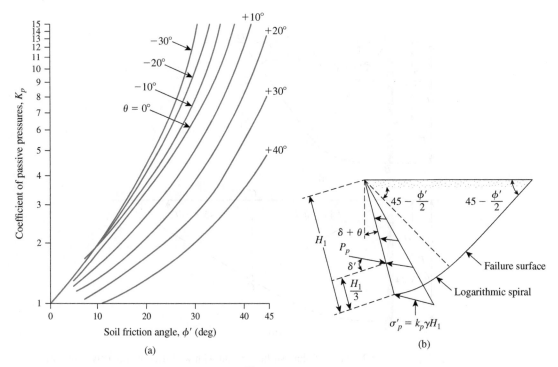

Figure 12.23 Caquot and Kerisel's solution for K_p [Eq.(12.68)]

For definition of H_1, refer to Figure 12.23b. The variation of K_p determined by Caquot and Kerisel (1948) also is shown in Figure 12.23a. It is important to note that the K_p values shown are for $\delta'/\phi' = 1$. If $\delta'/\phi' \neq 1$, the following procedure must be used to determine K_p.

1. Assume δ' and ϕ'.
2. Calculate δ'/ϕ'.
3. Using the ratio of δ'/ϕ' (step 2), determine the reduction factor, R', from Table 12.12.
4. Determine K_p from Figure 12.23 for $\delta'/\phi' = 1$
5. Calculate K_p for the required δ'/ϕ' as

$$K_p = (R')[K_{p(\delta'/\phi' = 1)}] \tag{12.69}$$

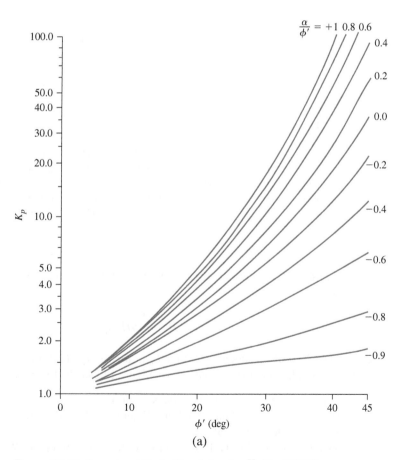

(a)

Figure 12.24 Caquot and Kerisel's solution for K_p [Eq.(12.70)]

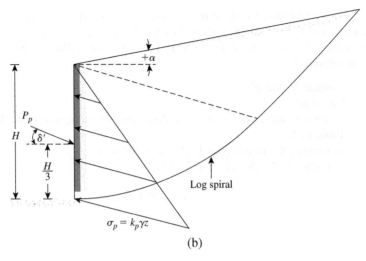

(b)

Figure 12.24 (*Continued*)

Table 12.12 Caquot and Kerisel's Reduction Factor, R', for Passive Pressure Calculation

ϕ'	δ'/ϕ'							
	0.7	**0.6**	**0.5**	**0.4**	**0.3**	**0.2**	**0.1**	**0.0**
10	0.978	0.962	0.946	0.929	0.912	0.898	0.881	0.864
15	0.961	0.934	0.907	0.881	0.854	0.830	0.803	0.775
20	0.939	0.901	0.862	0.824	0.787	0.752	0.716	0.678
25	0.912	0.860	0.808	0.759	0.711	0.666	0.620	0.574
30	0.878	0.811	0.746	0.686	0.627	0.574	0.520	0.467
35	0.836	0.752	0.674	0.603	0.536	0.475	0.417	0.362
40	0.783	0.682	0.592	0.512	0.439	0.375	0.316	0.262
45	0.718	0.600	0.500	0.414	0.339	0.276	0.221	0.174

Figure 12.24b shows a vertical retaining wall with an inclined granular backfill. For this case,

$$P_p = \frac{1}{2}\gamma H^2 K_p \tag{12.70}$$

Caquot and Kerisel's solution (1948) for K_p to use in Eq. (12.70) is given in Figure 12.24a for $\delta'/\phi' = 1$. In order to determine K_p via Figure 12.24a, the following steps are necessary:

Step 1. Determine α/ϕ' (note the sign of α).
Step 2. Knowing ϕ' and α/ϕ', use Figure 12.24a to determine K_p for $\delta'/\phi' = 1$.
Step 3. Calculate δ'/ϕ'.
Step 4. Go to Table 12.12 to determine the reduction factor, R'.
Step 5. $K_p = (R') [K_{p(\delta'/\phi' = 1)}]$. $\tag{12.71}$

Example 12.14

Consider a 3-m-high (H) retaining wall a vertical back ($\theta = 0°$) and a horizontal granular backfill Given: $\gamma = 15.7 \text{ kN/m}^3$, $\delta' = 15°$, and $\phi' = 30°$. Estimate the passive force, P_p, by using

    a. Coulomb's theory
    b. Caquot and Kerisel's theory

Solution
Part a
From Eq. (12.66)

$$P_p = \frac{1}{2}K_p\gamma H^2$$

From Table 12.11, for $\phi' = 30°$ and $\delta' = 15°$, the value of K_p is 4.977. Thus,

$$P_p = \left(\frac{1}{2}\right)(4.977)(15.7)(3)^2 = \textbf{351.6 kN/m}$$

Part b
From Eq. (12.68), with $\theta = 0, H_1 = H$,

$$P_p = \frac{1}{2}\gamma H^2 K_p$$

From Figure 12.23a, for $\phi' = 30°$ and $\delta'/\phi' = 1$, the value of $K_{p(\delta'/\phi' = 1)}$ is about 5.9. Also, from Table 12.12, with $\phi' = 30°$ and $\delta'/\phi' = 0.5$, the value of R' is 0.746.

Hence,

$$P_p = \frac{1}{2}\gamma H^2 K_p = \frac{1}{2}(15.7)(3)^2(0.746 \times 5.9) \approx \textbf{311 kN/m}$$

    ∎

12.15 Passive Pressure under Earthquake Conditions

The relationship for passive earth pressure on a retaining wall with a granular backfill and under earthquake conditions was evaluated by Subba Rao and Choudhury (2005) by the method of limit equilibrium using the pseudo-static approach. Figure 12.25 shows the

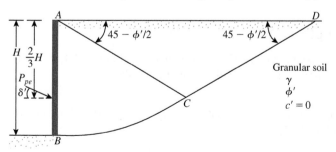

Figure 12.25 Nature of failure surface in soil considered in the analysis to determine P_{pe}

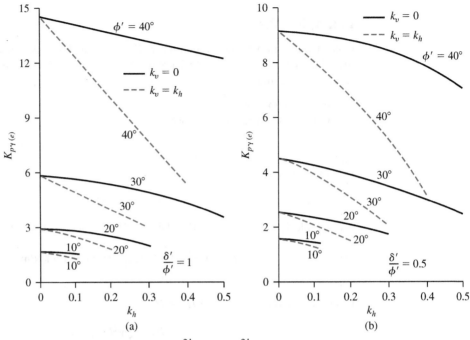

Figure 12.26 Variation of $K_{p\gamma(e)}$: (a) $\dfrac{\delta'}{\phi'} = 1$; (b) $\dfrac{\delta'}{\phi'} = 0.5$

nature of failure surface in soil considered in this analysis. The passive pressure, P_{pe}, can be expressed as

$$P_{pe} = \left[\tfrac{1}{2}\gamma H^2 K_{p\gamma(e)}\right]\frac{1}{\cos \delta'} \qquad (12.72)$$

where $K_{p\gamma(e)}$ = passive earth-pressure coefficient in the normal direction to the wall.

$K_{p\gamma(e)}$ is a function of k_h and k_v that are, respectively, coefficient of horizontal and vertical acceleration due to earthquake. The variations of $K_{p\gamma(e)}$ for $\delta'/\phi' = 0.5$ and 1 are shown in Figures 12.26a and b. The passive pressure P_{pe} will be inclined at an angle δ' to the back face of the wall and will act at a distance of $H/3$ above the bottom of the wall.

Example 12.15

Refer to Example 12.14. For an earthquake condition with $k_v = 0$ and $k_k = 0.2$, estimate the passive pressure P_{pe} per unit length of the wall.

Solution
For this case, $\phi' = 30°$, $\delta'/\phi' = 0.5$, $k_v = 0$, $k_k = 0.2$. From Figure 12.26b, $K_{p\gamma(e)} \approx 4$. From Eq. (12.72),

$$P_{pe} = \left[\frac{1}{2}\gamma H^2 K_{p\gamma(e)}\right]\frac{1}{\cos \delta'} = \left[\left(\frac{1}{2}\right)(15.7)(3)^2(4)\right]\frac{1}{\cos 15} = \textbf{292.6 kN/m} \qquad \blacksquare$$

Problems

12.1 Refer to Figure 12.3a. Given: $H = 12$ ft, $q = 0$, $\gamma = 108$ lb/ft^3, $c' = 0$, and $\phi' = 30°$. Determine the at-rest lateral earth force per foot length of the wall. Also, find the location of the resultant. Use Eq. (12.4) and OCR $= 2$.

12.2 Use Eq. (12.3), Figure P12.2, and the following values to determine the at-rest lateral earth force per unit length of the wall. Also find the location of the resultant. $H = 5$ m, $H_1 = 2$ m, $H_2 = 3$ m, $\gamma = 15.5$ kN/m^3, $\gamma_{sat} = 18.5$ kN/m^3, $\phi' = 34°$, $c' = 0$, $q = 20$ kN/m^2, and OCR $= 1$.

12.3 Refer to Figure 12.6a. Given the height of the retaining wall, H is 18 ft; the backfill is a saturated clay with $\phi = 0°$, $c = 500$ lb/ft^2, $\gamma_{sat} = 120$ lb/ft^3,
   a. Determine the Rankine active pressure distribution diagram behind the wall.
   b. Determine the depth of the tensile crack, z_c.
   c. Estimate the Rankine active force per foot length of the wall before and after the occurrence of the tensile crack.

12.4 A vertical retaining wall (Figure 12.6a) is 7 m high with a horizontal backfill. For the backfill, assume that $\gamma = 16.5$ kN/m^3, $\phi' = 26°$, and $c' = 18$ kN/m^2. Determine the Rankine active force per unit length of the wall after the occurrence of the tensile crack.

12.5 Refer to Problem 12.2. For the retaining wall, determine the Rankine active force per unit length of the wall and the location of the line of action of the resultant.

12.6 Refer to Figure 12.10. For the retaining wall, $H = 8$ m, $\phi' = 36°$, $\alpha = 10°$, $\gamma = 17$ kN/m^3, and $c' = 0$.
   a. Determine the intensity of the Rankine active force at $z = 2$ m, 4 m, and 6 m.
   b. Determine the Rankine active force per meter length of the wall and also the location and direction of the resultant.

12.7 Refer to Figure 12.10. Given: $H = 7$ m, $\gamma = 18$ kN/m^3, $\phi' = 25°$, $c' = 12$ kN/m^2, and $\alpha = 10°$. Calculate the Rankine active force per unit length of the wall after the occurrence of the tensile crack.

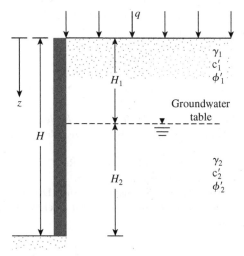

Figure P12.2

12.8 Refer to Figure 12.12a. Given: $H = 4$ m, $\gamma = 16.5$ kN/m^3, $\phi' = 30°$, $c' = 0$, and $\beta = 85°$. Determine the Coulomb's active force per meter length of the wall and the location and direction of the resultant for the following cases:
a. $\alpha = 10°$ and $\delta' = 20°$
b. $\alpha = 20°$ and $\delta' = 15°$

12.9 Refer to Figure 12.13a. Given $H = 4$ m, $\alpha = 0$, $\beta = 85°$, $\gamma = 17$ kN/m^3, $c' = 0$, $\phi' = 36°$, $\delta'/\phi' = 0.5$, and $q = 30$ kN/m^2. Determine the Coulomb's active force per unit length of the wall.

12.10 Refer to Figure 12.14b. Given $H = 3.3$ m, $a' = 1$ m, $b' = 1.5$ m, and $q = 25$ kN/m^2. Determine the lateral force per unit length of the unyielding wall caused by the surcharge loading only.

12.11 Refer to Figure 12.15. Here, $H = 5$ m, $\gamma = 18.2$ kN/m^3, $\phi' = 30°$, $\delta' = 20°$, $c' = 0$, $\alpha = 10°$, and $\beta = 85°$. Determine the Coulomb's active force for earthquake conditions (P_{ae}) per meter length of the wall and the location and direction of the resultant. Given $k_h = 0.2$ and $k_v = 0$.

12.12 For a retaining wall with a vertical back and horizontal backfill with a $c'-\phi'$ soil, the following are given:

$$H = 10 \text{ ft} \qquad \gamma = 115 \text{ lb/ft}^3$$
$$\phi' = 25° \qquad k_h = 0.2$$
$$c' = 113 \text{ lb/ft}^2 \qquad k_v = 0$$

Determine the magnitude of active force P_{ae} on the wall.

12.13 Refer to Problem 12.3.
a. Draw the Rankine passive pressure distribution diagram behind the wall.
b. Estimate the Rankine passive force per foot length of the wall and also the location of the resultant.

12.14 In Figure 12.24, which shows a vertical retaining wall with a granular backfill, let $H = 4$ m, $\alpha = 17.5°$, $\gamma = 16.5$ kN/m^3, $\phi' = 35°$, and $\delta' = 10°$. Based on Caquot and Kerisel's solution, what would be the passive force per meter length of the wall?

12.15 Consider a 4-m-high retaining wall with a vertical back and horizontal granular backfill, as shown in Figure 12.25. Given: $\gamma = 18$ kN/m^3, $\phi' = 40°$, $c' = 0$, $\delta' = 20°$, $k_v = 0$, and $k_h = 0.2$. Determine the passive force P_{pe} per unit length of the wall taking the earthquake effect into consideration.

References

CHU, S. C. (1991). "Rankine Analysis of Active and Passive Pressures on Dry Sand," *Soils and Foundations,* Vol. 31, No. 4, pp. 115–120.

COULOMB, C. A. (1776). *Essai sur une Application des Règles de Maximis et Minimum à quelques Problemes de Statique Relatifs à l'Architecture,* Mem. Acad. Roy. des Sciences, Paris, Vol. 3, p. 38.

JAKY, J. (1944). "The Coefficient of Earth Pressure at Rest," *Journal for the Society of Hungarian Architects and Engineers,* October, pp. 355–358.

JARQUIO, R. (1981). "Total Lateral Surcharge Pressure Due to Strip Load," *Journal of the Geotechnical Engineering Division,* American Society of Civil Engineers, Vol. 107, No. GT10, pp. 1424–1428.

Mayne, P. W. and Kulhawy, F. H. (1982). "K_o–*OCR* Relationships in Soil," *Journal of the Geotechnical Engineering Division,* ASCE, Vol. 108, No. GT6, pp. 851–872.

Mazindrani, Z. H. and Ganjali, M. H. (1997). "Lateral Earth Pressure Problem of Cohesive Backfill with Inclined Surface," *Journal of Geotechnical and Geoenvironmental Engineering,* ASCE, Vol. 123, No. 2, pp. 110–112.

Seed, H. B. and Whitman, R. V. (1970). "Design of Earth Retaining Structures for Dynamic Loads," *Proceedings, Specialty Conference on Lateral Stresses in the Ground and Design of Earth Retaining Structures*, American Society of Civil Engineers, pp. 103–147.

Shields, D. H. and Tolunay, A. Z. (1973). "Passive Pressure Coefficients by Method of Slices," *Journal of the Soil Mechanics and Foundations Division,* ASCE, Vol. 99, No. SM12, pp. 1043–1053.

Shukla, S. K., Gupta, S. K., and Sivakugan, N. (2009). "Active Earth Pressure on Retaining Wall for c–ϕ Soil Backfill Under Seismic Loading Condition," *Journal of Geotechnical and Geoenvironmental Engineering,* ASCE, Vol. 135, No. 5, 690–696.

Subba Rao, K. S. and Choudhury, D. (2005). "Seismic Passive Earth Pressures in Soil," *Journal of Geotechnical and Geoenvironmental Engineering*, ASCE, Vol. 131, No. 1, pp. 131–135.

Terzaghi, K. and Peck, R. B. (1967). *Soil Mechanics in Engineering Practice,* 2nd ed. New York.

Zhu, D. Y. and Qian, Q. (2000). "Determination of Passive Earth Pressure Coefficient by the Method of Triangular Slices," *Canadian Geotechnical Journal,* Vol. 37, No. 2, pp. 485–491.

5 Retaining Walls

13.1 Introduction

In Chapter 12, you were introduced to various theories of lateral earth pressure. Those theories will be used in this chapter to design various types of retaining walls. In general, retaining walls can be divided into two major categories: (a) conventional retaining walls and (b) mechanically stabilized earth walls.

Conventional retaining walls can generally be classified into four varieties:

1. Gravity retaining walls
2. Semigravity retaining walls
3. Cantilever retaining walls
4. Counterfort retaining walls

Gravity retaining walls (Figure 13.1a) are constructed with plain concrete or stone masonry. They depend for stability on their own weight and any soil resting on the masonry. This type of construction is not economical for high walls.

In many cases, a small amount of steel may be used for the construction of gravity walls, thereby minimizing the size of wall sections. Such walls are generally referred to as *semigravity walls* (Figure 13.1b).

Cantilever retaining walls (Figure 13.1c) are made of reinforced concrete that consists of a thin stem and a base slab. This type of wall is economical to a height of about 8 m (25 ft). Figure 13.2 shows a cantilever retaining wall under construction.

Counterfort retaining walls (Figure 13.1d) are similar to cantilever walls. At regular intervals, however, they have thin vertical concrete slabs known as *counterforts* that tie the wall and the base slab together. The purpose of the counterforts is to reduce the shear and the bending moments.

To design retaining walls properly, an engineer must know the basic parameters—the *unit weight, angle of friction,* and *cohesion*—of the soil retained behind the wall and the soil below the base slab. Knowing the properties of the soil behind the wall enables the engineer to determine the lateral pressure distribution that has to be designed for.

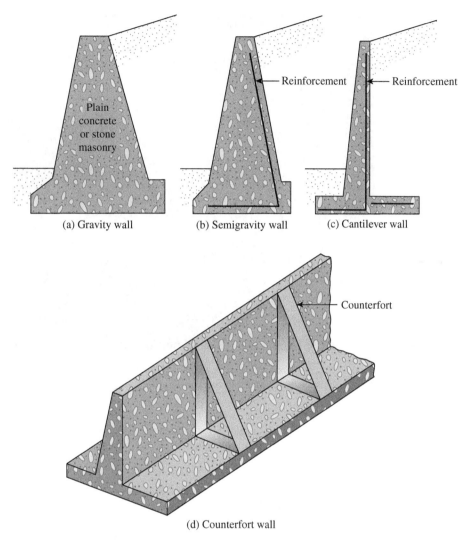

Plain concrete or stone masonry

(a) Gravity wall

Reinforcement

(b) Semigravity wall

Reinforcement

(c) Cantilever wall

Counterfort

(d) Counterfort wall

Figure 13.1 Types of retaining wall

There are two phases in the design of a conventional retaining wall. First, with the lateral earth pressure known, the structure as a whole is checked for *stability*. The structure is examined for possible *overturning, sliding,* and *bearing capacity* failures. Second, each component of the structure is checked for *strength,* and the *steel reinforcement* of each component is determined.

This chapter presents the procedures for determining the stability of the retaining wall. Checks for strength can be found in any textbook on reinforced concrete.

Some retaining walls have their backfills stabilized mechanically by including reinforcing elements such as metal strips, bars, welded wire mats, geotextiles, and

Figure 13.2 A cantilever retaining wall under construction (*Courtesy of Dharma Shakya, Geotechnical Solutions, Inc., Irvine, California*)

geogrids. These walls are relatively flexible and can sustain large horizontal and vertical displacements without much damage.

Gravity and Cantilever Walls

13.2   Proportioning Retaining Walls

In designing retaining walls, an engineer must assume some of their dimensions. Called *proportioning,* such assumptions allow the engineer to check trial sections of the walls for stability. If the stability checks yield undesirable results, the sections can be changed and rechecked. Figure 13.3 shows the general proportions of various retaining-wall components that can be used for initial checks.

Note that the top of the stem of any retaining wall should not be less than about 0.3 m. ($\approx$12 in.) for proper placement of concrete. The depth, D, to the bottom of the base slab should be a minimum of 0.6 m ($\approx$2 ft). However, the bottom of the base slab should be positioned below the seasonal frost line.

For counterfort retaining walls, the general proportion of the stem and the base slab is the same as for cantilever walls. However, the counterfort slabs may be about 0.3 m ($\approx$12 in.) thick and spaced at center-to-center distances of $0.3H$ to $0.7H$.

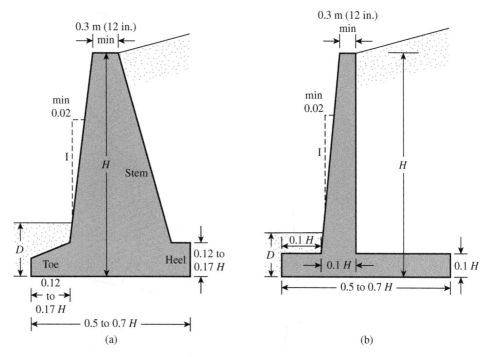

Figure 13.3 Approximate dimensions for various components of retaining wall for initial stability checks: (a) gravity wall; (b) cantilever wall

13.3 Application of Lateral Earth Pressure Theories to Design

The fundamental theories for calculating lateral earth pressure were presented in Chapter 12. To use these theories in design, an engineer must make several simple assumptions. In the case of cantilever walls, the use of the Rankine earth pressure theory for stability checks involves drawing a vertical line *AB* through point *A,* located at the edge of the heel of the base slab in Figure 13.4a. The Rankine active condition is assumed to exist along the vertical plane *AB*. Rankine active earth pressure equations may then be used to calculate the lateral pressure on the face *AB* of the wall. In the analysis of the wall's stability, the force $P_{a(\text{Rankine})}$, the weight of soil above the heel, and the weight W_c of the concrete all should be taken into consideration. The assumption for the development of Rankine active pressure along the soil face *AB* is theoretically correct if the shear zone bounded by the line *AC* is not obstructed by the stem of the wall. The angle, η, that the line *AC* makes with the vertical is

$$\eta = 45 + \frac{\alpha}{2} - \frac{\phi'}{2} - \frac{1}{2}\sin^{-1}\left(\frac{\sin\alpha}{\sin\phi'}\right) \qquad (13.1)$$

A similar type of analysis may be used for gravity walls, as shown in Figure 13.4b. However, *Coulomb's active earth pressure theory* also may be used, as shown in Figure 13.4c. If it is used, the only forces to be considered are $P_{a(\text{Coulomb})}$ and the weight of the wall, W_c.

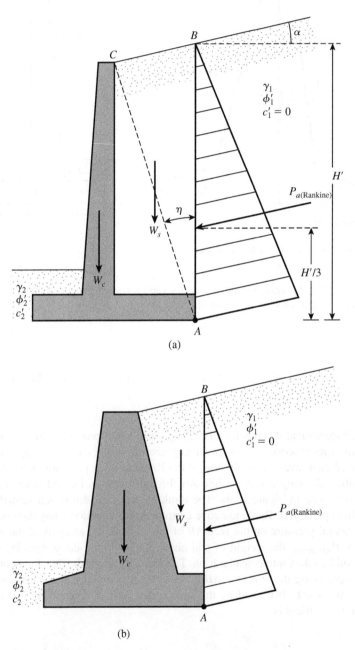

Figure 13.4 Assumption for the determination of lateral earth pressure: (a) cantilever wall; (b) and (c) gravity wall

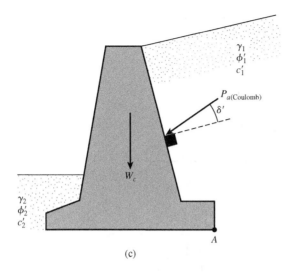

(c)

Figure 13.4 (*continued*)

If Coulomb's theory is used, it will be necessary to know the range of the wall friction angle δ' with various types of backfill material. Following are some ranges of wall friction angle for masonry or mass concrete walls:

Backfill material	Range of δ' (deg)
Gravel	27–30
Coarse sand	20–28
Fine sand	15–25
Stiff clay	15–20
Silty clay	12–16

In the case of ordinary retaining walls, water table problems and hence hydrostatic pressure are not encountered. Facilities for drainage from the soils that are retained are always provided.

13.4 Stability of Retaining Walls

A retaining wall may fail in any of the following ways:

- It may overturn about its toe. (See Figure 13.5a.)
- It may *slide* along its base. (See Figure 13.5b.)
- It may fail due to the loss of *bearing capacity* of the soil supporting the base. (See Figure 13.5c.)
- It may undergo deep-seated shear failure. (See Figure 13.5d.)
- It may go through excessive settlement.

The checks for stability against overturning, sliding, and bearing capacity failure will be described in Sections 13.5, 13.6, and 13.7. The principles used to estimate settlement were covered in Chapter 7 and will not be discussed further. When a weak soil layer is located at a shallow depth—that is, within a depth of 1.5 times the width

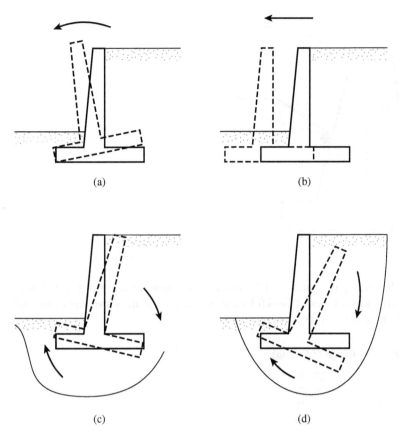

(a)

(b)

(c)

(d)

Figure 13.5 Failure of retaining wall:
(a) by overturning; (b) by sliding;
(c) by bearing capacity failure;
(d) by deep-seated shear failure

of the base slab of the retaining wall—the possibility of excessive settlement should be considered. In some cases, the use of lightweight backfill material behind the retaining wall may solve the problem.

Deep shear failure can occur along a cylindrical surface, such as *abc* shown in Figure 13.6, as a result of the existence of a weak layer of soil underneath the wall at a

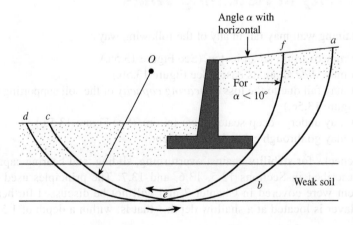

Angle α with horizontal

For $\alpha < 10°$

Weak soil

Figure 13.6 Deep-seated shear failure

depth of about 1.5 times the width of the base slab of the retaining wall. In such cases, the critical cylindrical failure surface *abc* has to be determined by trial and error, using various centers such as *O*. The failure surface along which the minimum factor of safety is obtained is the *critical surface of sliding*. For the backfill slope with α less than about 10°, the critical failure circle apparently passes through the edge of the heel slab (such as *def* in the figure). In this situation, the minimum factor of safety also has to be determined by trial and error by changing the center of the trial circle.

13.5 Check for Overturning

Figure 13.7 shows the forces acting on a cantilever and a gravity retaining wall, based on the assumption that the Rankine active pressure is acting along a vertical plane *AB* drawn through the heel of the structure. P_p is the Rankine passive pressure; recall that its magnitude is [from Eq. (12.59)].

$$P_p = \tfrac{1}{2} K_p \gamma_2 D^2 + 2 c_2' \sqrt{K_p} D$$

where

$\quad \gamma_2$ = unit weight of soil in front of the heel and under the base slab
$\quad K_p$ = Rankine passive earth pressure coefficient = $\tan^2(45 + \phi_2'/2)$
c_2', ϕ_2' = cohesion and effective soil friction angle, respectively

The factor of safety against overturning about the toe—that is, about point *C* in Figure 13.7—may be expressed as

$$FS_{(\text{overturning})} = \frac{\Sigma M_R}{\Sigma M_o} \tag{13.2}$$

where

ΣM_o = sum of the moments of forces tending to overturn about point *C*
ΣM_R = sum of the moments of forces tending to resist overturning about point *C*

The overturning moment is

$$\Sigma M_o = P_h \left(\frac{H'}{3} \right) \tag{13.3}$$

where $P_h = P_a \cos \alpha$.

To calculate the resisting moment, ΣM_R (neglecting P_p), a table such as Table 13.1 can be prepared. The weight of the soil above the heel and the weight of the concrete (or masonry) are both forces that contribute to the resisting moment. Note that the force P_v also contributes to the resisting moment. P_v is the vertical component of the active force P_a, or

$$P_v = P_a \sin \alpha$$

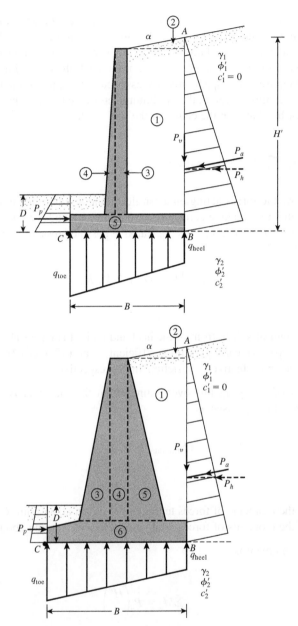

Figure 13.7 Check for overturning, assuming that the Rankine pressure is valid

The moment of the force P_v about C is

$$M_v = P_v B = P_a \sin \alpha B \tag{13.4}$$

where B = width of the base slab.

Once ΣM_R is known, the factor of safety can be calculated as

$$\text{FS}_{(\text{overturning})} = \frac{M_1 + M_2 + M_3 + M_4 + M_5 + M_6 + M_v}{P_a \cos \alpha (H'/3)} \tag{13.5}$$

Table 13.1 Procedure for Calculating ΣM_R

Section (1)	Area (2)	Weight/unit length of wall (3)	Moment arm measured from C (4)	Moment about C (5)
1	A_1	$W_1 = \gamma_1 \times A_1$	X_1	M_1
2	A_2	$W_2 = \gamma_1 \times A_2$	X_2	M_2
3	A_3	$W_3 = \gamma_c \times A_3$	X_3	M_3
4	A_4	$W_4 = \gamma_c \times A_4$	X_4	M_4
5	A_5	$W_5 = \gamma_c \times A_5$	X_5	M_5
6	A_6	$W_6 = \gamma_c \times A_6$	X_6	M_6
		P_v	B	M_v
		ΣV		ΣM_R

(*Note:* γ_l = unit weight of backfill
γ_c = unit weight of concrete
X_i = horizontal distance between C and the centroid of the section)

The usual minimum desirable value of the factor of safety with respect to overturning is 2 to 3.

Some designers prefer to determine the factor of safety against overturning with the formula

$$FS_{(\text{overturning})} = \frac{M_1 + M_2 + M_3 + M_4 + M_5 + M_6}{P_a \cos\alpha(H'/3) - M_v} \tag{13.6}$$

13.6 Check for Sliding along the Base

The factor of safety against sliding may be expressed by the equation

$$FS_{(\text{sliding})} = \frac{\Sigma F_{R'}}{\Sigma F_d} \tag{13.7}$$

where

$\Sigma F_{R'}$ = sum of the horizontal resisting forces
ΣF_d = sum of the horizontal driving forces

Figure 13.8 indicates that the shear strength of the soil immediately below the base slab may be represented as

$$s = \sigma' \tan\delta' + c_a'$$

where

δ' = angle of friction between the soil and the base slab
c_a' = adhesion between the soil and the base slab

Thus, the maximum resisting force that can be derived from the soil per unit length of the wall along the bottom of the base slab is

$$R' = s(\text{area of cross section}) = s(B \times 1) = B\sigma' \tan\delta' + Bc_a'$$

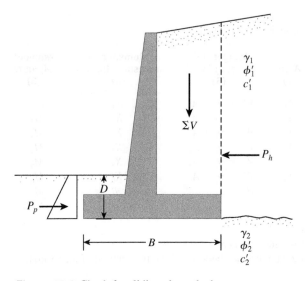

Figure 13.8 Check for sliding along the base

However,

$$B\sigma' = \text{sum of the vertical force} = \Sigma V (\text{see Table 13.1})$$

so

$$R' = (\Sigma V)\tan\delta' + Bc_a'$$

Figure 13.8 shows that the passive force P_p is also a horizontal resisting force. Hence,

$$\Sigma F_{R'} = (\Sigma V)\tan\delta' + Bc_a' + P_p \tag{13.8}$$

The only horizontal force that will tend to cause the wall to slide (a *driving force*) is the horizontal component of the active force P_a, so

$$\Sigma F_d = P_a \cos\alpha \tag{13.9}$$

Combining Eqs. (13.7), (13.8), and (13.9) yields

$$\text{FS}_{(\text{sliding})} = \frac{(\Sigma V)\tan\delta' + Bc_a' + P_p}{P_a\cos\alpha} \tag{13.10}$$

A minimum factor of safety of 1.5 against sliding is generally required.

In many cases, the passive force P_p is ignored in calculating the factor of safety with respect to sliding. In general, we can write $\delta' = k_1\phi_2'$ and $c_a' = k_2c_2'$. In most cases, k_1 and k_2 are in the range from $\frac{1}{2}$ to $\frac{2}{3}$. Thus,

$$\text{FS}_{(\text{sliding})} = \frac{(\Sigma V)\tan(k_1\phi_2') + Bk_2c_2' + P_p}{P_a\cos\alpha} \tag{13.11}$$

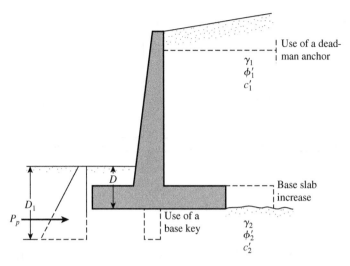

Figure 13.9 Alternatives for increasing the factor of safety with respect to sliding

If the desired value of $FS_{(sliding)}$ is not achieved, several alternatives may be investigated (see Figure 13.9):

- Increase the width of the base slab (i.e., the heel of the footing).
- Use a key to the base slab. If a key is included, the passive force per unit length of the wall becomes

$$P_p = \frac{1}{2}\gamma_2 D_1^2 K_p + 2c_2' D_1 \sqrt{K_p}$$

where $K_p = \tan^2\left(45 + \dfrac{\phi_2'}{2}\right)$.

- Use a *deadman anchor* at the stem of the retaining wall.
- Another possible way to increase the value of $FS_{(sliding)}$ is to consider reducing the value of P_a [see Eq. (13.11)]. One possible way to do so is to use the method developed by Elman and Terry (1988). The discussion here is limited to the case in which the retaining wall has a horizontal granular backfill (Figure 13.10). In Figure 13.10, the active force, P_a, is horizontal ($\alpha = 0$) so that

$$P_a \cos \alpha = P_h = P_a$$

and

$$P_a \sin \alpha = P_v = 0$$

However,

$$P_a = P_{a(1)} + P_{a(2)} \tag{13.12}$$

The magnitude of $P_{a(2)}$ can be reduced if the heel of the retaining wall is sloped as shown in Figure 13.10. For this case,

$$P_a = P_{a(1)} + AP_{a(2)} \tag{13.13}$$

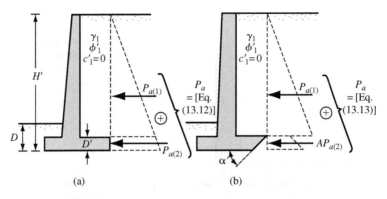

Figure 13.10 Retaining wall with sloped heel

The magnitude of A, as shown in Table 13.2, is valid for $\alpha' = 45°$. However note that in Figure 13.10a

$$P_{a(1)} = \frac{1}{2} \gamma_1 K_a (H' - D')^2$$

and

$$P_a = \frac{1}{2} \gamma_1 K_a H'^2$$

Hence,

$$P_{a(2)} = \frac{1}{2} \gamma_1 K_a [H'^2 - (H' - D')^2]$$

So, for the active pressure diagram shown in Figure 13.10b,

$$P_a = \frac{1}{2} \gamma_1 K_a (H' - D')^2 + \frac{A}{2} \gamma_1 K_a [H'^2 - (H' - D')^2] \tag{13.14}$$

Sloping the heel of a retaining wall can thus be extremely helpful in some cases.

Table 13.2 Variation of A with ϕ_1' (for $\alpha' = 45°$)

Soil friction angle, ϕ_1' (deg)	A
20	0.28
25	0.14
30	0.06
35	0.03
40	0.018

13.7 Check for Bearing Capacity Failure

The vertical pressure transmitted to the soil by the base slab of the retaining wall should be checked against the ultimate bearing capacity of the soil. The nature of variation of the vertical pressure transmitted by the base slab into the soil is shown in Figure 13.11. Note that q_{toe} and q_{heel} are the *maximum* and the *minimum* pressures occurring at the ends of the toe and heel sections, respectively. The magnitudes of q_{toe} and q_{heel} can be determined in the following manner:

The sum of the vertical forces acting on the base slab is ΣV (see column 3 of Table 13.1), and the horizontal force $\mathbf{P}_h$ is $P_a \cos \alpha$. Let

$$\mathbf{R} = \Sigma \mathbf{V} + \mathbf{P}_h \tag{13.15}$$

be the resultant force. The net moment of these forces about point C in Figure 13.11 is

$$M_{net} = \Sigma M_R - \Sigma M_o \tag{13.16}$$

Note that the values of ΣM_R and ΣM_o were previously determined. [See Column 5 of Table 13.1 and Eq. (13.3)]. Let the line of action of the resultant R intersect the base slab at E. Then the distance

$$\overline{CE} = \overline{X} = \frac{M_{net}}{\Sigma V} \tag{13.17}$$

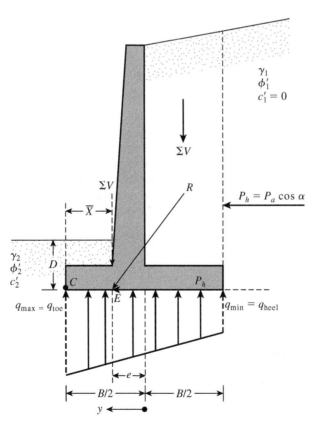

Figure 13.11 Check for bearing capacity failure

Hence, the eccentricity of the resultant R may be expressed as

$$e = \frac{B}{2} - \overline{CE} \tag{13.18}$$

The pressure distribution under the base slab may be determined by using simple principles from the mechanics of materials. First, we have

$$q = \frac{\Sigma V}{A} \pm \frac{M_{net}y}{I} \tag{13.19}$$

where

M_{net} = moment = $(\Sigma V)e$
I = moment of inertia per unit length of the base section
   = $\frac{1}{12}(1)(B^3)$

For maximum and minimum pressures, the value of y in Eq. (13.19) equals $B/2$. Substituting into Eq. (13.19) gives

$$q_{max} = q_{toe} = \frac{\Sigma V}{(B)(1)} + \frac{e(\Sigma V)\dfrac{B}{2}}{\left(\dfrac{1}{12}\right)(B^3)} = \frac{\Sigma V}{B}\left(1 + \frac{6e}{B}\right) \tag{13.20}$$

Similarly,

$$q_{min} = q_{heel} = \frac{\Sigma V}{B}\left(1 - \frac{6e}{B}\right) \tag{13.21}$$

Note that ΣV includes the weight of the soil, as shown in Table 13.1, and that when the value of the eccentricity e becomes greater than $B/6$, q_{min} [Eq. (13.21)] becomes negative. Thus, there will be some tensile stress at the end of the heel section. This stress is not desirable, because the tensile strength of soil is very small. If the analysis of a design shows that $e > B/6$, the design should be reproportioned and calculations redone.

The relationships pertaining to the ultimate bearing capacity of a shallow foundation were discussed in Chapter 4. Recall that [Eq. (4.51)].

$$q_u = c_2' N_c F_{cd} F_{ci} + q N_q F_{qd} F_{qi} + \tfrac{1}{2}\gamma_2 B' N_\gamma F_{\gamma d} F_{\gamma i} \tag{13.22}$$

where

$q = \gamma_2 D$
$B' = B - 2e$
$F_{cd} = F_{qd} - \dfrac{1 - F_{qd}}{N_c \tan \phi_2'}$
$F_{qd} = 1 + 2 \tan \phi_2'(1 - \sin \phi_2')^2 \dfrac{D}{B'}$
$F_{\gamma d} = 1$

$$F_{ci} = F_{qi} = \left(1 - \frac{\psi^{\circ}}{90^{\circ}}\right)^2$$

$$F_{\gamma i} = \left(1 - \frac{\psi^{\circ}}{\phi_2'^{\circ}}\right)^2$$

$$\psi^{\circ} = \tan^{-1}\left(\frac{P_a \cos \alpha}{\Sigma V}\right)$$

Note that the shape factors F_{cs}, F_{qs}, and $F_{\gamma s}$ given in Chapter 4 are all equal to unity, because they can be treated as a continuous foundation. For this reason, the shape factors are not shown in Eq. (13.22).

Once the ultimate bearing capacity of the soil has been calculated by using Eq. (13.22), the factor of safety against bearing capacity failure can be determined:

$$\text{FS}_{(\text{bearing capacity})} = \frac{q_u}{q_{\max}} \tag{13.23}$$

Generally, a factor of safety of 3 is required. In Chapter 4, we noted that the ultimate bearing capacity of shallow foundations occurs at a settlement of about 10% of the foundation width. In the case of retaining walls, the width B is large. Hence, the ultimate load q_u will occur at a fairly large foundation settlement. A factor of safety of 3 against bearing capacity failure may not ensure that settlement of the structure will be within the tolerable limit in all cases. Thus, this situation needs further investigation.

Example 13.1

The cross section of a cantilever retaining wall is shown in Figure 13.12. Calculate the factors of safety with respect to overturning, sliding, and bearing capacity.

Solution
From the figure,

$$H' = H_1 + H_2 + H_3 = 2.6 \tan 10^{\circ} + 6 + 0.7$$
$$= 0.458 + 6 + 0.7 = 7.158 \text{ m}$$

The Rankine active force per unit length of wall $= P_p = \frac{1}{2}\gamma_1 H'^2 K_a$. For $\phi_1' = 30^{\circ}$ and $\alpha = 10^{\circ}$, K_a is equal to 0.3495. (See Table 12.1.) Thus,

$$P_a = \frac{1}{2}(18)(7.158)^2(0.3495) = 161.2 \text{ kN/m}$$

$$P_v = P_a \sin 10^{\circ} = 161.2 (\sin 10^{\circ}) = 28.0 \text{ kN/m}$$

and

$$P_h = P_a \cos 10^{\circ} = 161.2 (\cos 10^{\circ}) = 158.75 \text{ kN/m}$$

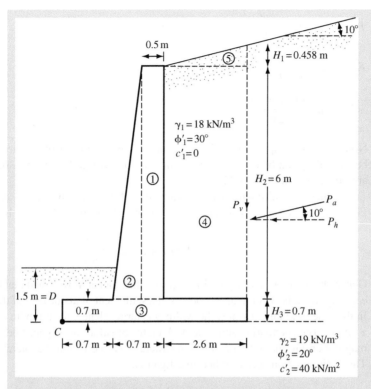

Figure 13.12 Calculation of stability of a retaining wall

Factor of Safety against Overturning

The following table can now be prepared for determining the resisting moment:

Section no.[a]	Area (m²)	Weight/unit length (kN/m)	Moment arm from point C (m)	Moment (kN-m/m)
1	$6 \times 0.5 = 3$	70.74	1.15	81.35
2	$\frac{1}{2}(0.2)6 = 0.6$	14.15	0.833	11.79
3	$4 \times 0.7 = 2.8$	66.02	2.0	132.04
4	$6 \times 2.6 = 15.6$	280.80	2.7	758.16
5	$\frac{1}{2}(2.6)(0.458) = 0.595$	10.71	3.13	33.52
		$P_v = 28.0$	4.0	112.0
		$\Sigma V = 470.42$		$1128.86 = \Sigma M_R$

[a]For section numbers, refer to Figure 13.12
$\gamma_{concrete} = 23.58$ kN/m³

The overturning moment

$$M_o = P_h\left(\frac{H'}{3}\right) = 158.75\left(\frac{7.158}{3}\right) = 378.78 \text{ kN-m/m}$$

and

$$\text{FS}_{\text{(overturning)}} = \frac{\Sigma M_R}{M_o} = \frac{1128.86}{378.78} = \mathbf{2.98 > 2, OK}$$

Factor of Safety against Sliding
From Eq. (12.11),

$$\text{FS}_{\text{(sliding)}} = \frac{(\Sigma V)\tan(k_1\phi_2') + Bk_2c_2' + P_p}{P_a\cos\alpha}$$

Let $k_1 = k_2 = \frac{2}{3}$. Also,

$$P_p = \tfrac{1}{2}K_p\gamma_2 D^2 + 2c_2'\sqrt{K_p}D$$

$$K_p = \tan^2\left(45 + \frac{\phi_2'}{2}\right) = \tan^2(45 + 10) = 2.04$$

and

$$D = 1.5\text{ m}$$

So

$$P_p = \tfrac{1}{2}(2.04)(19)(1.5)^2 + 2(40)(\sqrt{2.04})(1.5)$$
$$= 43.61 + 171.39 = 215\text{ kN/m}$$

Hence,

$$\text{FS}_{\text{(sliding)}} = \frac{(470.42)\tan\left(\dfrac{2 \times 20}{3}\right) + (4)\left(\dfrac{2}{3}\right)(40) + 215}{158.75}$$

$$= \frac{111.49 + 106.67 + 215}{158.75} = \mathbf{2.73 > 1.5, OK}$$

Note: For some designs, the depth D in a passive pressure calculation may be taken to be *equal to the thickness of the base slab.*

Factor of Safety against Bearing Capacity Failure
Combining Eqs. (13.16), (13.17), and (13.18) yields

$$e = \frac{B}{2} - \frac{\Sigma M_R - \Sigma M_o}{\Sigma V} = \frac{4}{2} - \frac{1128.86 - 378.78}{470.42}$$

$$= 0.406\text{ m} < \frac{B}{6} = \frac{4}{6} = 0.666\text{ m}$$

Again, from Eqs. (13.20) and (13.21)

$$q^{\text{toe}}_{\text{heel}} = \frac{\Sigma V}{B}\left(1 \pm \frac{6e}{B}\right) = \frac{470.42}{4}\left(1 \pm \frac{6 \times 0.406}{4}\right) = 189.2\text{ kN/m}^2\text{ (toe)}$$

$$= 45.98\text{ kN/m}^2\text{ (heel)}$$

The ultimate bearing capacity of the soil can be determined from Eq. (13.22)

$$q_u = c_2' N_c F_{cd} F_{ci} + q N_q F_{qd} F_{qi} + \frac{1}{2} \gamma_2 B' N_\gamma F_{\gamma d} F_{\gamma i}$$

For $\phi_2' = 20°$ (see Table 4.2), $N_c = 14.83$, $N_q = 6.4$, and $N_\gamma = 5.39$. Also,

$$q = \gamma_2 D = (19)(1.5) = 28.5 \text{ kN/m}^2$$

$$B' = B - 2e = 4 - 2(0.406) = 3.188 \text{ m}$$

$$F_{cd} = F_{qd} - \frac{1 - F_{qd}}{N_c \tan\phi_2'} = 1.148 - \frac{1 - 1.148}{(14.83)(\tan 20)} = 1.175$$

$$F_{qd} = 1 + 2\tan\phi_2'(1 - \sin\phi_2')^2\left(\frac{D}{B'}\right) = 1 + 0.315\left(\frac{1.5}{3.188}\right) = 1.148$$

$$F_{\gamma d} = 1$$

$$F_{ci} = F_{qi} = \left(1 - \frac{\psi°}{90°}\right)^2$$

and

$$\psi = \tan^{-1}\left(\frac{P_a \cos\alpha}{\Sigma V}\right) = \tan^{-1}\left(\frac{158.75}{470.42}\right) = 18.65°$$

So

$$F_{ci} = F_{qi} = \left(1 - \frac{18.65}{90}\right)^2 = 0.628$$

and

$$F_{\gamma i} = \left(1 - \frac{\psi}{\phi_2'}\right)^2 = \left(1 - \frac{18.65}{20}\right)^2 \approx 0$$

Hence,

$$q_u = (40)(14.83)(1.175)(0.628) + (28.5)(6.4)(1.148)(0.628)$$

$$+ \tfrac{1}{2}(19)(5.93)(3.188)(1)(0)$$

$$= 437.72 + 131.5 + 0 = 569.22 \text{ kN/m}^2$$

and

$$\text{FS}_{\text{(bearing capacity)}} = \frac{q_u}{q_{\text{toe}}} = \frac{569.22}{189.2} = \textbf{3.0 OK}$$

Example 13.2

A gravity retaining wall is shown in Figure 13.13. Use $\delta' = 2/3\phi'_1$ and Coulomb's active earth pressure theory. Determine

    a. The factor of safety against overturning
    b. The factor of safety against sliding
    c. The pressure on the soil at the toe and heel

Solution
The height

$$H' = 5 + 1.5 = 6.5 \text{ m}$$

Coulomb's active force is

$$P_a = \tfrac{1}{2}\gamma_1 H'^2 K_a$$

With $\alpha = 0°$, $\beta = 75°$, $\delta' = 2/3\phi'_1$, and $\phi'_1 = 32°$, $K_a = 0.4023$. (See Table 12.6.) So,

$$P_a = \tfrac{1}{2}(18.5)(6.5)^2(0.4023) = 157.22 \text{ kN/m}$$

$$P_h = P_a \cos\left(15 + \tfrac{2}{3}\phi'_1\right) = 157.22 \cos 36.33 = 126.65 \text{ kN/m}$$

and

$$P_v = P_a \sin\left(15 + \tfrac{2}{3}\phi'_1\right) = 157.22 \sin 36.33 = 93.14 \text{ kN/m}$$

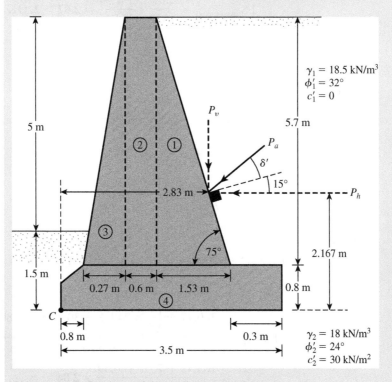

Figure 13.13 Gravity retaining wall (not to scale)

Part a: Factor of Safety against Overturning
From Figure 13.13, one can prepare the following table:

Area no.	Area (m²)	Weight* (kN/m)	Moment arm from C (m)	Moment (kN-m/m)
1	$\frac{1}{2}(5.7)(1.53) = 4.36$	102.81	2.18	224.13
2	$(0.6)(5.7) = 3.42$	80.64	1.37	110.48
3	$\frac{1}{2}(0.27)(5.7) = 0.77$	18.16	0.98	17.80
4	$\approx (3.5)(0.8) = 2.8$	66.02	1.75	115.54
		$P_v = 93.14$	2.83	263.59
		$\Sigma V = 360.77$ kN/m		$\Sigma M_R = 731.54$ kN-m/m

*$\gamma_{concrete} = 23.58$ kN/m³

Note that the weight of the soil above the back face of the wall is not taken into account in the preceding table. We have

$$\text{Overturning moment} = M_o = P_h\left(\frac{H'}{3}\right) = 126.65(2.167) = 274.45 \text{ kN-m/m}$$

Hence,

$$FS_{(overturning)} = \frac{\Sigma M_R}{\Sigma M_o} = \frac{731.54}{274.45} = \mathbf{2.67 > 2, OK}$$

Part b: Factor of Safety against Sliding
We have

$$FS_{(sliding)} = \frac{(\Sigma V)\tan\left(\frac{2}{3}\phi_2'\right) + \frac{2}{3}c_2'B + P_p}{P_h}$$

$$P_p = \frac{1}{2}K_p\gamma_2 D^2 + 2c_2'\sqrt{K_p}D$$

and

$$K_p = \tan^2\left(45 + \frac{24}{2}\right) = 2.37$$

Hence,

$$P_p = \frac{1}{2}(2.37)(18)(1.5)^2 + 2(30)(1.54)(1.5) = 186.59 \text{ kN/m}$$

So

$$FS_{(sliding)} = \frac{360.77\tan\left(\frac{2}{3} \times 24\right) + \frac{2}{3}(30)(3.5) + 186.59}{126.65}$$

$$= \frac{103.45 + 70 + 186.59}{126.65} = \mathbf{2.84}$$

If P_p is ignored, the factor of safety is **1.37**.

Part c: Pressure on Soil at Toe and Heel
From Eqs. (13.16), (13.17), and (13.18),

$$e = \frac{B}{2} - \frac{\Sigma M_R - \Sigma M_o}{\Sigma V} = \frac{3.5}{2} - \frac{731.54 - 274.45}{360.77} = 0.483 < \frac{B}{6} = 0.583$$

$$q_{\text{toe}} = \frac{\Sigma V}{B}\left[1 + \frac{6e}{B}\right] = \frac{360.77}{3.5}\left[1 + \frac{(6)(0.483)}{3.5}\right] = \mathbf{188.43 \ kN/m^2}$$

and

$$q_{\text{heel}} = \frac{V}{B}\left[1 - \frac{6e}{B}\right] = \frac{360.77}{3.5}\left[1 - \frac{(6)(0.483)}{3.5}\right] = \mathbf{17.73 \ kN/m^2} \qquad \blacksquare$$

13.8 Construction Joints and Drainage from Backfill

Construction Joints

A retaining wall may be constructed with one or more of the following joints:

1. *Construction joints* (see Figure 13.14a) are vertical and horizontal joints that are placed between two successive pours of concrete. To increase the shear at the joints, keys may be used. If keys are not used, the surface of the first pour is cleaned and roughened before the next pour of concrete.
2. *Contraction joints* (Figure 13.14b) are vertical joints (grooves) placed in the face of a wall (from the top of the base slab to the top of the wall) that allow the concrete to shrink without noticeable harm. The grooves may be about 6 to 8 mm ($\approx$0.25 to 0.3 in.) wide and 12 to 16 mm ($\approx$0.5 to 0.6 in.) deep.
3. *Expansion joints* (Figure 13.14c) allow for the expansion of concrete caused by temperature changes; vertical expansion joints from the base to the top of the wall may also be used. These joints may be filled with flexible joint fillers. In most cases, horizontal reinforcing steel bars running across the stem are continuous through all joints. The steel is greased to allow the concrete to expand.

Drainage from the Backfill

As the result of rainfall or other wet conditions, the backfill material for a retaining wall may become saturated, thereby increasing the pressure on the wall and perhaps creating an unstable condition. For this reason, adequate drainage must be provided by means of *weep holes* or *perforated drainage pipes*. (See Figure 13.15.)

When provided, weep holes should have a minimum diameter of about 0.1 m (4 in.) and be adequately spaced. Note that there is always a possibility that backfill material may be washed into weep holes or drainage pipes and ultimately clog them. Thus, a filter

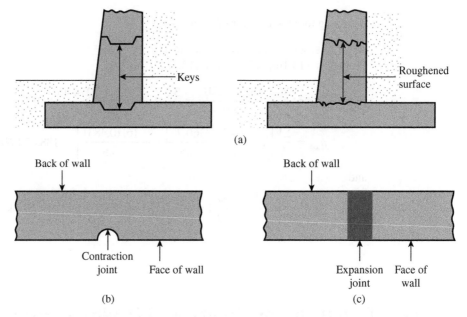

Figure 13.14 (a) Construction joints; (b) contraction joint; (c) expansion joint

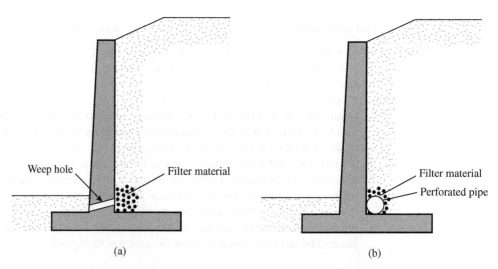

Figure 13.15 Drainage provisions for the backfill of a retaining wall: (a) by weep holes; (b) by a perforated drainage pipe

material needs to be placed behind the weep holes or around the drainage pipes, as the case may be; geotextiles now serve that purpose.

Two main factors influence the choice of filter material: The grain-size distribution of the materials should be such that (a) the soil to be protected is not washed into the filter and (b) excessive hydrostatic pressure head is not created in the soil with a lower hydraulic conductivity (in this case, the backfill material). The preceding conditions can be satisfied if the following requirements are met (Terzaghi and Peck, 1967):

$$\frac{D_{15(F)}}{D_{85(B)}} < 5 \qquad \text{[to satisfy condition(a)]} \qquad (13.24)$$

$$\frac{D_{15(F)}}{D_{15(B)}} > 4 \qquad \text{[to satisfy condition(b)]} \qquad (13.25)$$

In these relations, the subscripts F and B refer to the *filter* and the *base* material (i.e., the backfill soil), respectively. Also, D_{15} and D_{85} refer to the diameters through which 15% and 85% of the soil (filter or base, as the case may be) will pass. Example 13.3 gives the procedure for designing a filter.

Example 13.3

Figure 13.16 shows the grain-size distribution of a backfill material. Using the conditions outlined in Section 13.8, determine the range of the grain-size distribution for the filter material.

Solution
From the grain-size distribution curve given in the figure, the following values can be determined:

$$D_{15(B)} = 0.04 \text{ mm}$$
$$D_{85(B)} = 0.25 \text{ mm}$$
$$D_{50(B)} = 0.13 \text{ mm}$$

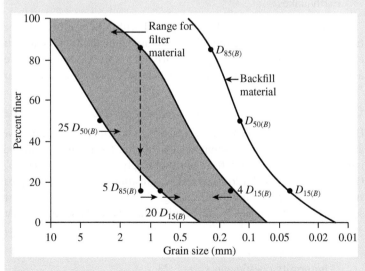

Figure 13.16 Determination of grain-size distribution of filter material

> **Conditions of Filter**
>
> **1.** $D_{15(F)}$ should be less than $5D_{85(B)}$; that is, $5 \times 0.25 = 1.25$ mm.
> **2.** $D_{15(F)}$ should be greater than $4D_{15(B)}$; that is, $4 \times 0.04 = 0.16$ mm.
> **3.** $D_{50(F)}$ should be less than $25D_{50(B)}$; that is, $25 \times 0.13 = 3.25$ mm.
> **4.** $D_{15(F)}$ should be less than $20D_{15(B)}$; that is, $20 \times 0.04 = 0.8$ mm.
>
> These limiting points are plotted in Figure 13.16. Through them, two curves can be drawn that are similar in nature to the grain-size distribution curve of the backfill material. These curves define the range of the filter material to be used. ∎

13.9 Comments on Design of Retaining Walls and a Case Study

In Section 13.3, it was suggested that the *active earth pressure coefficient* be used to estimate the lateral force on a retaining wall due to the backfill. It is important to recognize the fact that the active state of the backfill can be established only if the wall yields sufficiently, which does not happen in all cases. The degree to which the wall yields depends on its *height* and the *section modulus*. Furthermore, the lateral force of the backfill depends on several factors identified by Casagrande (1973):

1. Effect of temperature
2. Groundwater fluctuation
3. Readjustment of the soil particles due to creep and prolonged rainfall
4. Tidal changes
5. Heavy wave action
6. Traffic vibration
7. Earthquakes

Insufficient wall yielding combined with other unforeseen factors may generate a larger lateral force on the retaining structure, compared with that obtained from the active earth-pressure theory. This is particularly true in the case of gravity retaining walls, bridge abutments, and other heavy structures that have a large section modulus.

Case Study for the Performance of a Cantilever Retaining Wall

Bentler and Labuz (2006) have reported the performance of a cantilever retaining wall built along Interstate 494 in Bloomington, Minnesota. The retaining wall had 83 panels, each having a length of 9.3 m (30.5 ft). The panel height ranged from 4.0 m to 7.9 m (13 ft to 26 ft). One of the 7.9-m-high (26 ft) panels was instrumented with earth pressure cells, tiltmeters, strain gauges, and inclinometer casings. Figure 13.17 shows a schematic diagram (cross section) of the wall panel. Some details on the backfill and the foundation material are:

- Granular Backfill
    Effective size, $D_{10} = 0.13$ mm
    Uniformity coefficient, $C_u = 3.23$
    Coefficient of gradation, $C_c = 1.4$

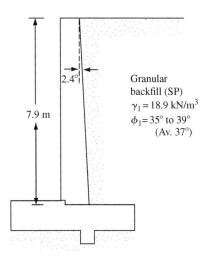

7.9 m

2.4°

Granular
backfill (SP)
$\gamma_1 = 18.9 \text{ kN/m}^3$
$\phi_1 = 35°$ to $39°$
(Av. 37°)

Poorly graded
sand, and sand and
gravel

Figure 13.17 Schematic
diagram of the retaining wall
(drawn to scale)

Unified soil classification − SP
Compacted unit weight, $\gamma_1 = 18.9 \text{ kN/m}^3$ (120 lb/ft³)
Triaxial friction angle, $\phi_1' - 35°$ to $39°$ (average 37°)

• Foundation Material
   Poorly graded sand and sand with gravel (medium dense to dense)

The backfill and compaction of the granular material started on October 28, 2001 in stages and reached a height of 7.6 m (25 ft) on November 21, 2001. The final 0.3 m (1 ft) of soil was placed the following spring. During backfilling, the wall was continuously going through translation. Table 13.3 is a summary of the backfill height and horizontal translation of the wall.

Figure 13.18 shows a typical plot of the variation of lateral earth pressure *after compaction*, σ_a', when the backfill height was 6.1 m (October 31, 2001) along with the

Table 13.3 Horizontal Translation with Backfill Height

Day	Backfill height (m)	Horizontal translation (mm)
1	0.0	0
2	1.1	0
2	2.8	0
3	5.2	2
4	6.1	4
5	6.4	6
11	6.7	9
24	7.3	12
54	7.6	11

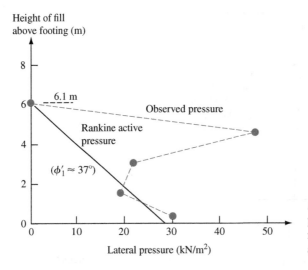

Figure 13.18 Observed lateral pressure distribution after fill height reached 6.1 m (Based on Bentler and Labuz, 2006)

plot of Rankine active earth pressure ($\phi'_1 = 37°$). Note that the measured lateral (horizontal) pressure is higher at most heights than that predicted by the Rankine active pressure theory, which may be due to residual lateral stresses caused by compaction. The measured lateral stress gradually reduced with time. This is demonstrated in Figure 13.19 which shows a plot of the variation of σ'_a with depth (November 27, 2001) when the height of the backfill was 7.6 m. The lateral pressure was lower at practically all depths compared to the Rankine active earth pressure.

Another point of interest is the nature of variation of q_{max} and q_{min} (see Figure 13.11). As shown in Figure 13.11, if the wall rotates about C, q_{max} will be at the toe and q_{min} will be at the heel. However, for the case of the retaining wall under consideration (undergoing horizontal translation), q_{max} was at the heel of the wall with q_{min} at the toe. On November 27, 2001, when the height of the fill was 7.6 m (25 ft), q_{max} at the heel was

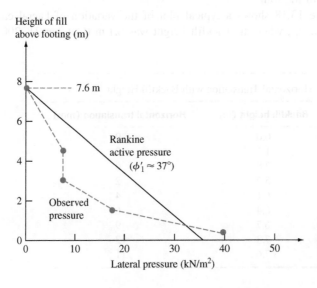

Figure 13.19 Observed pressure distribution on November 27, 2001 (Based on Bentler and Labuz, 2006)

about $140 \, kN/m^2$ (2922 lb/ft^2), which was approximately equal to (γ_1)(height of fill) $=$ $(18.9)(7.6) = 143.6 \, kN/m^2$. Also, at the toe, q_{min} was about $40 \, kN/m^2$ (835 lb/ft^2), which suggests that the moment from lateral force had little effect on the vertical effective stress below the heel.

The lessons learned from this case study are the following:

a. Retaining walls may undergo lateral translation which will affect the variation of q_{max} and q_{min} along the base slab.

b. Initial lateral stress caused by compaction gradually decreases with time and lateral movement of the wall.

Mechanically Stabilized Retaining Walls

More recently, soil reinforcement has been used in the construction and design of foundations, retaining walls, embankment slopes, and other structures. Depending on the type of construction, the reinforcements may be galvanized metal strips, geotextiles, geogrids, or geocomposites. Sections 13.10 and 13.11 provide a general overview of soil reinforcement and various reinforcement materials.

Reinforcement materials such as metallic strips, geotextiles, and geogrids are now being used to reinforce the backfill of retaining walls, which are generally referred to as *mechanically stabilized retaining walls*. The general principles for designing these walls are given in the following sections.

13.10 Soil Reinforcement

The use of reinforced earth is a recent development in the design and construction of foundations and earth-retaining structures. *Reinforced earth* is a construction material made from soil that has been strengthened by tensile elements such as metal rods or strips, nonbiodegradable fabrics (geotextiles), geogrids, and the like. The fundamental idea of reinforcing soil is not new; in fact, it goes back several centuries. However, the present concept of systematic analysis and design was developed by a French engineer, H. Vidal (1966). The French Road Research Laboratory has done extensive research on the applicability and the beneficial effects of the use of reinforced earth as a construction material. This research has been documented in detail by Darbin (1970), Schlosser and Long (1974), and Schlosser and Vidal (1969). The tests that were conducted involved the use of metallic strips as reinforcing material.

Retaining walls with reinforced earth have been constructed around the world since Vidal began his work. The first reinforced-earth retaining wall with metal strips as reinforcement in the United States was constructed in 1972 in southern California.

The beneficial effects of soil reinforcement derive from (a) the soil's increased tensile strength and (b) the shear resistance developed from the friction at the soil-reinforcement interfaces. Such reinforcement is comparable to that of concrete structures. Currently, most reinforced-earth design is done with *free-draining granular soil only*. Thus, the effect of pore water development in cohesive soils, which, in turn, reduces the shear strength of the soil, is avoided.

13.11 Considerations in Soil Reinforcement

Metal Strips

In most instances, galvanized steel strips are used as reinforcement in soil. However, galvanized steel is subject to corrosion. The rate of corrosion depends on several environmental factors. Binquet and Lee (1975) suggested that the average rate of corrosion of galvanized steel strips varies between 0.025 and 0.050 mm/yr. So, in the actual design of reinforcement, allowance must be made for the rate of corrosion. Thus,

$$t_c = t_{\text{design}} + r \, (\text{life span of structure})$$

where

t_c = actual thickness of reinforcing strips to be used in construction
t_{design} = thickness of strips determined from design calculations
r = rate of corrosion

Nonbiodegradable Fabrics

Nonbiodegradable fabrics are generally referred to as *geotextiles*. Since 1970, the use of geotextiles in construction has increased greatly around the world. The fabrics are usually made from petroleum products—polyester, polyethylene, and polypropylene. They may also be made from fiberglass. Geotextiles are not prepared from natural fabrics, because they decay too quickly. Geotextiles may be woven, knitted, or nonwoven.

Woven geotextiles are made of two sets of parallel filaments or strands of yarn systematically interlaced to form a planar structure. *Knitted geotextiles* are formed by interlocking a series of loops of one or more filaments or strands of yarn to form a planar structure. *Nonwoven geotextiles* are formed from filaments or short fibers arranged in an oriented or random pattern in a planar structure. These filaments or short fibers are arranged into a loose web in the beginning and then are bonded by one or a combination of the following processes:

1. *Chemical bonding*—by glue, rubber, latex, a cellulose derivative, or the like
2. *Thermal bonding*—by heat for partial melting of filaments
3. *Mechanical bonding*—by needle punching

Needle-punched nonwoven geotextiles are thick and have high in-plane permeability.
Geotextiles have four primary uses in foundation engineering:

1. *Drainage:* The fabrics can rapidly channel water from soil to various outlets, thereby providing a higher soil shear strength and hence stability.
2. *Filtration:* When placed between two soil layers, one coarse grained and the other fine grained, the fabric allows free seepage of water from one layer to the other. However, it protects the fine-grained soil from being washed into the coarse-grained soil.
3. *Separation:* Geotextiles help keep various soil layers separate after construction and during the projected service period of the structure. For example, in the construction of highways, a clayey subgrade can be kept separate from a granular base course.
4. *Reinforcement:* The tensile strength of geofabrics increases the load-bearing capacity of the soil.

Geogrids

Geogrids are high-modulus polymer materials, such as polypropylene and polyethylene, and are prepared by tensile drawing. Netlon, Ltd., of the United Kingdom was the first producer of geogrids. In 1982, the Tensar Corporation, presently Tensar International Corporation, introduced geogrids into the United States.

Geogrids generally are of two types: (a) uniaxial and (b) biaxial. Figures 13.20a and b show these two types of geogrids, which are produced by Tensar International Corporation.

Commercially available geogrids may be categorized by manufacturing process, principally: extruded, woven, and welded. Extruded geogrids are formed using a thick sheet of polyethylene or polypropylene that is punched and drawn to create apertures and to enhance engineering properties of the resulting ribs and nodes. Woven geogrids are manufactured by grouping polymeric—usually polyester and polypropylene—and weaving them into a mesh pattern that is then coated with a polymeric lacquer. Welded geogrids are manufactured by fusing junctions of polymeric strips. Extruded geogrids have shown good performance when compared to other types for pavement reinforcement applications.

The commercial geogrids currently available for soil reinforcement have nominal rib thicknesses of about 0.5 to 1.5 mm (0.02 to 0.06 in.) and junctions of about 2.5 to 5 mm (0.1 to 0.2 in.). The grids used for soil reinforcement usually have openings or apertures that are rectangular or elliptical. The dimensions of the apertures vary from about 25 to 150 mm (1 to 6 in.). Geogrids are manufactured so that the open areas of the grids are greater than 50% of the total area. They develop reinforcing strength at low strain levels, such as 2% (Carroll, 1988).

The major function of geogrids is *reinforcement*. They are relatively stiff. The apertures are large enough to allow interlocking with surrounding soil or rock (Figure 13.21) to perform the function of reinforcement or segregation (or both). Sarsby (1985) investigated the influence of aperture size on the size of soil particles for maximum frictional

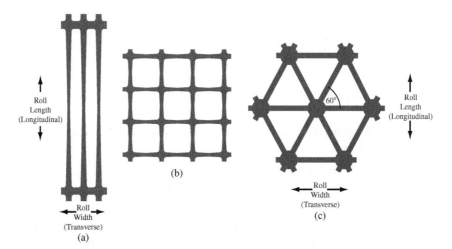

Figure 13.20 Geogrid: (a) uniaxial; (b) biaxial; (c) with triangular apertures (Based on of Tensar International Corporation)

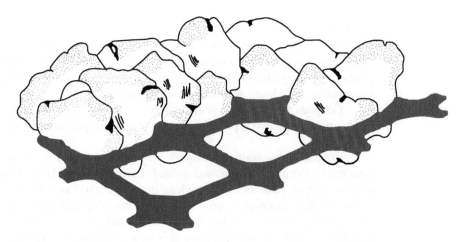

Figure 13.21 Geogrid apertures allowing interlocking with surrounding soil

efficiency (or efficiency against pullout). According to this study, the highest efficiency occurs when

$$B_{GG} > 3.5D_{50} \tag{13.26}$$

where

B_{GG} = minimum width of the geogrid aperture
D_{50} = the particle size through which 50% of the backfill soil passes (i.e., the average particle size)

More recently, geogrids with triangular apertures (Figure 13.20c) have been introduced for construction purposes. Geogrids with triangular apertures are manufactured from a punched polypropylene sheet, which is then oriented in three substantially equilateral directions so that the resulting ribs shall have a high degree of molecular orientation.

13.12 General Design Considerations

The general design procedure of any mechanically stabilized retaining wall can be divided into two parts:

1. Satisfying *internal stability* requirements
2. Checking the *external stability* of the wall

The internal stability checks involve determining tension and pullout resistance in the reinforcing elements and ascertaining the integrity of facing elements. The external stability checks include checks for overturning, sliding, and bearing capacity failure (Figure 13.22). The sections that follow will discuss the retaining-wall design procedures for use with metallic strips, geotextiles, and geogrids.

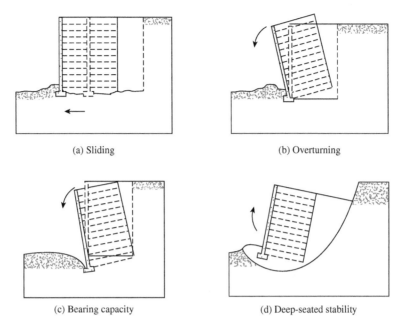

(a) Sliding   (b) Overturning

(c) Bearing capacity   (d) Deep-seated stability

Figure 13.22 External stability checks (After Transportation Research Board, 1995) (Based on *Transportation Research Circular 444: Mechanically Stabilized Earth Walls.* Transportation Research Board, National Research Council, Washington, D.C., 1995.)

13.13 Retaining Walls with Metallic Strip Reinforcement

Reinforced-earth walls are flexible walls. Their main components are

1. *Backfill,* which is granular soil
2. *Reinforcing strips,* which are thin, wide strips placed at regular intervals, and
3. *A cover* or *skin,* on the front face of the wall

Figure 13.23 is a diagram of a reinforced-earth retaining wall. Note that, at any depth, the reinforcing strips or ties are placed with a horizontal spacing of S_H center to center; the vertical spacing of the strips or ties is S_V center to center. The skin can be constructed with sections of relatively flexible thin material. Lee et al. (1973) showed that, with a conservative design, a 5 mm-thick ($\approx$0.2 in.) galvanized steel skin would be enough to hold a wall about 14 to 15 m (45 to 50 ft) high. In most cases, precast concrete slabs also can be used as skin. The slabs are grooved to fit into each other so that soil cannot flow out between the joints. When metal skins are used, they are bolted together, and reinforcing strips are placed between the skins.

Figures 13.24 and 13.25 show a reinforced-earth retaining wall under construction; its skin (facing) is a precast concrete slab. Figure 13.26 shows a metallic reinforcement tie attached to the concrete slab.

The simplest and most common method for the design of ties is the *Rankine method.* We discuss this procedure next.

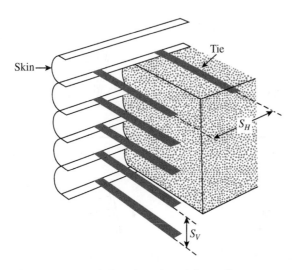

Figure 13.23 Reinforced-earth retaining wall

Figure 13.24 Reinforced-earth retaining wall (with metallic strip) under construction (*Courtesy of Braja M. Das, Henderson, Nevada*)

Figure 13.25 Another view of the retaining wall shown in Figure 13.24 (*Courtesy of Braja M. Das, Henderson, Nevada*)

Figure 13.26 Metallic strip attachment to the precast concrete slab used as the skin (*Courtesy of Braja M. Das, Henderson, Nevada*)

Calculation of Active Horizontal and Vertical Pressure

Figure 13.27 shows a retaining wall with a granular backfill having a unit weight of γ_1 and a friction angle of ϕ_1'. Below the base of the retaining wall, the *in situ* soil has been excavated and recompacted, with granular soil used as backfill. Below the backfill, the *in situ* soil has a unit weight of γ_2, friction angle of ϕ_2', and cohesion of c_2'. A surcharge having an intensity of q per unit area lies atop the retaining wall, which has reinforcement ties at depths $z = 0, S_V, 2S_V, \ldots, NS_V$. The height of the wall is $NS_V = H$.

According to the Rankine active pressure theory (Section 12.3)

$$\sigma_a' = \sigma_o' K_a - 2c'\sqrt{K_a}$$

where σ_a' = Rankine active pressure at any depth z.

For dry granular soils with no surcharge at the top, $c' = 0$, $\sigma_o' = \gamma_1 z$, and $K_a = \tan^2(45 - \phi_1'/2)$. Thus,

$$\sigma_{a(1)}' = \gamma_1 z K_a \tag{13.27}$$

When a surcharge is added at the top, as shown in Figure 13.27,

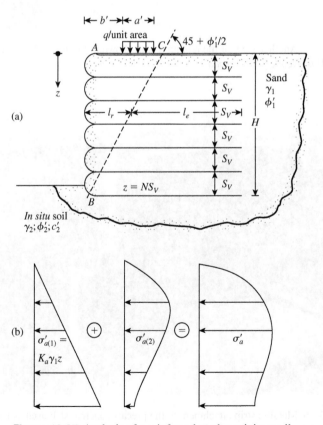

(a)

(b)

Figure 13.27 Analysis of a reinforced-earth retaining wall

$$\sigma'_o = \sigma'_{o(1)} + \sigma'_{o(2)} \tag{13.28}$$

$$\uparrow \qquad \uparrow$$

$$= \gamma_1 z \qquad \text{Due to the}$$

$$\text{Due to} \qquad \text{surcharge}$$

$$\text{soil only}$$

The magnitude of $\sigma'_{o(2)}$ can be calculated by using the 2:1 method of stress distribution described in Eq. (6.18) and Figure 6.7. The 2:1 method of stress distribution is shown in Figure 13.28a. According to Laba and Kennedy (1986),

$$\sigma'_{o(2)} = \frac{qa'}{a' + z} \qquad (\text{for } z \leq 2b') \tag{13.29}$$

and

$$\sigma'_{o(2)} = \frac{qa'}{a' + \dfrac{z}{2} + b'} \qquad (\text{for } z > 2b') \tag{13.30}$$

Also, when a surcharge is added at the top, the lateral pressure at any depth is

$$\sigma'_a = \sigma'_{a(1)} + \sigma'_{a(2)}$$

$$\uparrow \qquad \uparrow$$

$$= K_a \gamma_1 z \qquad \text{Due to the} \tag{13.31}$$

$$\text{Due to} \qquad \text{surcharge}$$

$$\text{soil only}$$

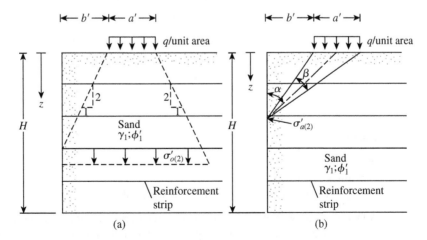

Figure 13.28 (a) Notation for the relationship of $\sigma'_{o(2)}$ in Eqs. (13.29) and (13.30); (b) notation for the relationship of $\sigma'_{a(2)}$ in Eqs. (13.32) and (13.33)

According to Laba and Kennedy (1986), $\sigma'_{a(2)}$ may be expressed (see Figure 13.28b) as

$$\sigma'_{a(2)} = M\left[\frac{2q}{\pi}(\beta - \sin\beta\cos 2\alpha)\right]$$
$$\uparrow$$
$$\text{(in radians)}$$

(13.32)

where

$$M = 1.4 - \frac{0.4b'}{0.14H} \geq 1$$

(13.33)

The net active (lateral) pressure distribution on the retaining wall calculated by using Eqs. (13.31), (13.32), and (13.33) is shown in Figure 13.27b.

Tie Force

The tie force *per unit length of the wall* developed at any depth z (see Figure 13.27) is

$$T = \text{active earth pressure at depth } z$$
$$\times \text{ area of the wall to be supported by the tie}$$
$$= (\sigma'_a)(S_V S_H)$$

(13.34)

Factor of Safety against Tie Failure

The reinforcement ties at each level, and thus the walls, could fail by either (a) tie breaking or (b) tie pullout.

The factor of safety against *tie breaking* may be determined as

$$FS_{(B)} = \frac{\text{yield or breaking strength of each tie}}{\text{maximum force in any tie}}$$
$$= \frac{wtf_y}{\sigma'_a S_V S_H}$$

(13.35)

where

w = width of each tie
t = thickness of each tie
f_y = yield or breaking strength of the tie material

A factor of safety of about 2.5 to 3 is generally recommended for ties at all levels.

Reinforcing ties at any depth z will fail by pullout if the frictional resistance developed along the surfaces of the ties is less than the force to which the ties are being subjected. The *effective length* of the ties along which frictional resistance is developed may be conservatively taken as the length that extends *beyond the limits of the Rankine active failure zone,* which is the zone *ABC* in Figure 13.27. Line *BC* makes an angle of $45 + \phi'_1/2$ with the horizontal. Now, the maximum friction force that can be realized for a tie at depth z is

$$F_R = 2l_e w \sigma'_o \tan \phi'_\mu \tag{13.36}$$

where

l_e = effective length
σ'_o = effective vertical pressure at a depth z
ϕ'_μ = soil–tie friction angle

Thus, the factor of safety against *tie pullout* at any depth z is

$$FS_{(P)} = \frac{F_R}{T} \tag{13.37}$$

Substituting Eqs. (13.34) and (13.36) into Eq. (13.37) yields

$$FS_{(P)} = \frac{2l_e w \sigma'_o \tan \phi'_\mu}{\sigma'_a S_V S_H} \tag{13.38}$$

Total Length of Tie

The total length of ties at any depth is

$$L = l_r + l_e \tag{13.39}$$

where

l_r = length within the Rankine failure zone
l_e = effective length

For a given $FS_{(P)}$ from Eq. (13.38),

$$l_e = \frac{FS_{(P)} \sigma'_a S_V S_H}{2w \sigma'_o \tan \phi'_\mu} \tag{13.40}$$

Again, at any depth z,

$$l_r = \frac{(H - z)}{\tan\left(45 + \dfrac{\phi'_1}{2}\right)} \tag{13.41}$$

So, combining Eqs. (13.39), (13.40), and (13.41) gives

$$L = \frac{(H - z)}{\tan\left(45 + \dfrac{\phi'_1}{2}\right)} + \frac{FS_{(P)} \sigma'_a S_V S_H}{2w \sigma'_o \tan \phi'_\mu} \tag{13.42}$$

13.14 Step-by-Step-Design Procedure Using Metallic Strip Reinforcement

Following is a step-by-step procedure for the design of reinforced-earth retaining walls.

General

Step 1. Determine the height of the wall, H, and the properties of the granular backfill material, such as the unit weight (γ_1) and the angle of friction (ϕ_1').

Step 2. Obtain the soil–tie friction angle, ϕ_μ', and the required value of $FS_{(B)}$ and $FS_{(P)}$.

Internal Stability

Step 3. Assume values for horizontal and vertical tie spacing. Also, assume the width of reinforcing strip, w, to be used.

Step 4. Calculate σ_a' from Eqs. (13.31), (13.32), and (13.33).

Step 5. Calculate the tie forces at various levels from Eq. (12.34).

Step 6. For the known values of $FS_{(B)}$, calculate the thickness of ties, t, required to resist the tie breakout:

$$T = \sigma_a' S_V S_H = \frac{wtf_y}{FS_{(B)}}$$

or

$$t = \frac{(\sigma_a' S_V S_H)[FS_{(B)}]}{wf_y} \tag{13.43}$$

The convention is to keep the magnitude of t the same at all levels, so σ_a' in Eq. (13.43) should equal $\sigma_{a(max)}'$.

Step 7. For the known values of ϕ_μ' and $FS_{(P)}$, determine the length L of the ties at various levels from Eq. (13.42).

Step 8. The magnitudes of S_V, S_H, t, w, and L may be changed to obtain the most economical design.

External Stability

Step 9. Check for *overturning*, using Figure 13.29 as a guide. Taking the moment about B yields the overturning moment for the unit length of the wall:

$$M_o = P_a z' \tag{13.44}$$

Here,

$$P_a = \text{active force} = \int_0^H \sigma_a' dz$$

The resisting moment per unit length of the wall is

$$M_R = W_1 x_1 + W_2 x_2 + \cdots + qa'\left(b' + \frac{a'}{2}\right) \tag{13.45}$$

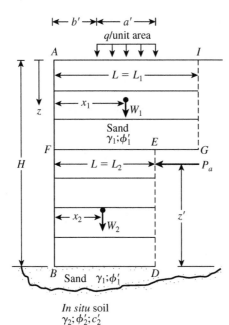

In situ soil
$\gamma_2; \phi_2'; c_2'$

Figure 13.29 Stability check for the retaining wall

where

$W_1 = (\text{area } AFEGI)(1)(\gamma_1)$

$W_2 = (\text{area } FBDE)(1)(\gamma_1)$

$\vdots$

So,

$$FS_{(\text{overturning})} = \frac{M_R}{M_o}$$

$$= \frac{W_1 x_1 + W_2 x_2 + \cdots + qa'\left(b' + \dfrac{a'}{2}\right)}{\left(\displaystyle\int_0^H \sigma_a' \, dz\right) z'} \tag{13.46}$$

Step 10. The check for *sliding* can be done by using Eq. (13.11), or

$$FS_{(\text{sliding})} = \frac{(W_1 + W_2 + \cdots + qa')[\tan(k\phi_1')]}{P_a} \tag{13.47}$$

where $k \approx \frac{2}{3}$.

Step 11. Check for ultimate bearing capacity failure, which can be given as

$$q_u = c_2' N_c + \tfrac{1}{2}\gamma_2 L_2 N_\gamma \tag{13.48}$$

The bearing capacity factors N_c and N_γ correspond to the soil friction angle ϕ'_2. (See Table 4.2.)

From Eq. 13.28, the vertical stress at $z = H$ is

$$\sigma'_{o(H)} = \gamma_1 H + \sigma'_{o(2)} \tag{13.49}$$

So the factor of safety against bearing capacity failure is

$$\text{FS}_{\text{(bearing capacity)}} = \frac{q_{\text{ult}}}{\sigma'_{o(H)}} \tag{13.50}$$

Generally, minimum values of $\text{FS}_{\text{(overturning)}} = 3$, $\text{FS}_{\text{(sliding)}} = 3$, and $\text{FS}_{\text{(bearing capacity failure)}} = 3$ to 5 are recommended.

Example 13.4

A 10-m-high retaining wall with galvanized steel-strip reinforcement in a granular backfill has to be constructed. Referring to Figure 13.27, given:

Granular backfill: $\phi'_1 = 36°$
                             $\gamma_1 = 16.5 \text{ kN/m}^3$

Foundation soil: $\phi'_2 = 28°$
                           $\gamma_2 = 17.3 \text{ kN/m}^3$
                           $c'_2 = 50 \text{ kN/m}^2$

Galvanized steel reinforcement:

     Width of strip, $w = 75 \text{ mm}$
                         $S_V = 0.6 \text{ m center-to-center}$
                         $S_H = 1 \text{ m center-to-center}$
                         $f_y = 240,00 \text{ kN/m}^2$
                         $\phi'_\mu = 20°$

         Required $\text{FS}_{(B)} = 3$

         Required $\text{FS}_{(P)} = 3$

Check for the external and internal stability. Assume the corrosion rate of the galvanized steel to be 0.025 mm/year and the life span of the structure to be 50 years.

Solution
Internal Stability Check

Tie thickness: Maximum tie force, $T_{\max} = \sigma'_{a(\max)} S_V S_H$

$$\sigma_{a(\max)} = \gamma_1 H K_a = \gamma_1 H \tan^2\left(45 - \frac{\phi'_1}{2}\right)$$

so

$$T_{max} = \gamma_1 H \tan^2\left(45 - \frac{\phi_1'}{2}\right) S_V S_H$$

From Eq. (13.43), for *tie break,*

$$t = \frac{(\sigma_a' S_V S_H)[\text{FS}_{(B)}]}{w f_y} = \frac{\left[\gamma_1 H \tan^2\left(45 - \frac{\phi_1'}{2}\right) S_V S_H\right] \text{FS}_{(B)}}{w f_y}$$

or

$$t = \frac{\left[(16.5)(10) \tan^2\left(45 - \frac{36}{2}\right)(0.6)(1)\right](3)}{(0.075 \text{ m})(240{,}000 \text{ kN/m}^2)} = 0.00428 \text{ m} = 4.28 \text{ mm}$$

If the rate of corrosion is 0.025 mm/yr and the life span of the structure is 50 yr, then the actual thickness, *t*, of the ties will be

$$t = 4.28 + (0.025)(50) = 5.53 \text{ mm}$$

So a **tie thickness of 6 m** would be enough.

Tie length: Refer to Eq. (13.42). For this case, $\sigma_a' = \gamma_1 z K_a$ and $\sigma_o' = \gamma_1 z$, so

$$L = \frac{(H - z)}{\tan\left(45 + \frac{\phi_1'}{2}\right)} + \frac{\text{FS}_{(P)} \gamma_1 z K_a S_V S_H}{2 w \gamma_1 z \tan \phi_\mu'}$$

Now the following table can be prepared. (*Note:* $\text{FS}_{(P)} = 3$, $H = 10$ m, $w = 75$ mm, and $\phi_\mu' = 20°$.)

z(m)	Tie length L (m) [Eq. (13.42)]
2	12.65
4	11.63
6	10.61
8	9.59
10	8.57

So use a **tie length of L = 13 m.**

External Stability Check
Check for overturning: Refer to Figure 13.30. For this case, using Eq. (13.46)

$$\text{FS}_{(\text{overturning})} = \frac{W_1 x_1}{\left[\int_0^H \sigma_a' \, dz\right] z'}$$

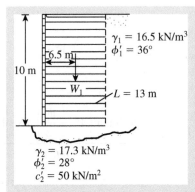

$\gamma_1 = 16.5 \text{ kN/m}^3$
$\phi_1' = 36°$

6.5 m

10 m

W_1

$L = 13 \text{ m}$

$\gamma_2 = 17.3 \text{ kN/m}^3$
$\phi_2' = 28°$
$c_2' = 50 \text{ kN/m}^2$

Figure 13.30 Retaining wall with galvanized steel-strip reinforcement in the backfill

$$W_1 = \gamma_1 HL = (16.5)(10)(13) = 2145 \text{ kN/m}$$

$$x_1 = 6.5 \text{ m}$$

$$P_a = \int_0^H \sigma_a' \, dz = \tfrac{1}{2}\gamma_1 K_a H^2 = (\tfrac{1}{2})(16.5)(0.26)(10)^2 = 214.5 \text{ kN/m}$$

$$z' = \frac{10}{3} = 3.33 \text{ m}$$

$$\text{FS}_{(\text{overturning})} = \frac{(2145)(6.5)}{(214.5)(3.33)} = \mathbf{19.52 > 3\text{---OK}}$$

Check for sliding: From Eq. (13.47)

$$\text{FS}_{(\text{sliding})} = \frac{W_1 \tan(k\phi_1')}{P_a} = \frac{2145 \tan\left[\left(\dfrac{2}{3}\right)(36)\right]}{214.5} = \mathbf{4.45 > 3\text{---OK}}$$

Check for bearing capacity: For $\phi_2' = 28°$, $N_c = 25.8$, $N_\gamma = 16.78$ (Table 4.2). From Eq. (13.48),

$$q_{\text{ult}} = c_2' N_c + \tfrac{1}{2}\gamma_2 L \, N_\gamma$$

$$q_{\text{ult}} = (50)(25.8) + (\tfrac{1}{2})(17.3)(13)(16.72) = 3170.16 \text{ kN/m}^2$$

From Eq. (13.49),

$$\sigma_{o(H)}' = \gamma_1 H = (16.5)(10) = 165 \text{ kN/m}^2$$

$$\text{FS}_{(\text{bearing capacity})} = \frac{q_{\text{ult}}}{\sigma_{o(H)}'} = \frac{3170.16}{165} = \mathbf{19.2 > 5\text{---OK}}$$

13.15 Retaining Walls with Geotextile Reinforcement

Figure 13.31 shows a retaining wall in which layers of geotextile have been used as reinforcement. As in Figure 13.29, the backfill is a granular soil. In this type of retaining wall, the facing of the wall is formed by lapping the sheets as shown with a lap length of l_l. When construction is finished, the exposed face of the wall must be covered; otherwise, the geotextile will deteriorate from exposure to ultraviolet light. *Bitumen emulsion* or *Gunite* is sprayed on the wall face. A wire mesh anchored to the geotextile facing may be necessary to keep the coating on. Figure 13.32 shows the construction of a geotextile-reinforced retaining wall. Figure 13.33 shows a completed geosynthetic-reinforced soil wall. The wall is in

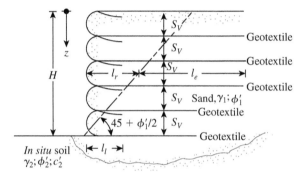

Figure 13.31 Retaining wall with geotextile reinforcement

Figure 13.32 Construction of a geotextile-reinforced retaining wall (*Courtesy of Jonathan T. H. Wu, University of Colorado at Denver, Denver, Colorado*)

Figure 13.33 A completed geotextile-reinforced retaining wall in DeBeque Canyon, Colorado (*Courtesy of Jonathan T. H. Wu, University of Colorado at Denver, Denver, Colorado*)

DeBeque Canyon, Colorado. Note the versatility of the facing type. In this case, single-tier concrete block facing is integrated with a three-tier facing via rock facing.

The design of this type of retaining wall is similar to that presented in Section 13.14. Following is a step-by-step procedure for design based on the recommendations of Bell et al. (1975) and Koerner (2005):

Internal Stability

Step 1. Determine the active pressure distribution on the wall from the formula

$$\sigma'_a = K_a \sigma'_o = K_a \gamma_1 z \qquad (13.51)$$

where

K_a = Rankine active pressure coefficient = $\tan^2(45 - \phi'_1/2)$
γ_1 = unit weight of the granular backfill
ϕ'_1 = friction angle of the granular backfill

Step 2. Select a geotextile fabric with an allowable tensile strength, T_{all} (lb/ft or kN/m).

The allowable tensile strength for retaining wall construction may be expressed as (Koerner, 2005)

$$T_{all} = \frac{T_{ult}}{RF_{id} \times RF_{cr} \times RF_{cbd}} \qquad (13.52)$$

where

T_{ult} = ultimate tensile strength
RF_{id} = reduction factor for installation damage
RF_{cr} = reduction factor for creep
RF_{cbd} = reduction factor for chemical and biological degradation

The recommended values of the reduction factor are as follows (Koerner, 2005)

RF_{id}	1.1–2.0
RF_{cr}	2–4
RF_{cbd}	1–1.5

Step 3. Determine the vertical spacing of the layers at any depth z from the formula

$$S_V = \frac{T_{all}}{\sigma'_a FS_{(B)}} = \frac{T_{all}}{(\gamma_1 z K_a)[FS_{(B)}]} \tag{13.53}$$

Note that Eq. (13.53) is similar to Eq. (13.35). The magnitude of $FS_{(B)}$ is generally 1.3 to 1.5.

Step 4. Determine the length of each layer of geotextile from the formula

$$L = l_r + l_e \tag{13.54}$$

where

$$l_r = \frac{H - z}{\tan\left(45 + \dfrac{\phi'_1}{2}\right)} \tag{13.55}$$

and

$$l_e = \frac{S_V \sigma'_a [FS_{(P)}]}{2\sigma'_o \tan \phi'_F} \tag{13.56}$$

in which

$$\sigma'_a = \gamma_1 z K_a$$

$$\sigma'_o = \gamma_1 z$$

$$FS_{(P)} = 1.3 \text{ to } 1.5$$

ϕ'_F = friction angle at geotextile–soil interface

$$\approx \tfrac{2}{3}\phi'_1$$

Note that Eqs. (13.54), (13.55), and (13.56) are similar to Eqs. (13.39), (13.41), and (13.40), respectively.

Based on the published results, the assumption of $\phi'_F/\phi'_1 \approx \frac{2}{3}$ is reasonable and appears to be conservative. Martin et al. (1984) presented the following laboratory test results for ϕ'_F/ϕ'_1 between various types of geotextiles and sand.

Type	ϕ'_F/ϕ'_1
Woven—monofilament/concrete sand	0.87
Woven—silt film/concrete sand	0.8
Woven—silt film/rounded sand	0.86
Woven—silt film/silty sand	0.92
Nonwoven—melt-bonded/concrete sand	0.87
Nonwoven—needle-punched/concrete sand	1.0
Nonwoven—needle-punched/rounded sand	0.93
Nonwoven—needle-punched/silty sand	0.91

Step 5. Determine the lap length, l_l, from

$$l_l = \frac{S_V \sigma'_a \text{FS}_{(P)}}{4\sigma'_o \tan \phi'_F}$$

(13.57)

The minimum lap length should be 1 m (3 ft).

External Stability

Step 6. Check the factors of safety against overturning, sliding, and bearing capacity failure as described in Section 13.14 (Steps 9, 10, and 11).

Example 13.5

A geotextile-reinforced retaining wall 5 m high is shown in Figure 13.34. For the granular backfill, $\gamma_1 = 15.7$ kN/m³ and $\phi'_1 = 36°$. For the geotextile, $T_{ult} = 52.5$ kN/m. For the design of the wall, determine $S_V, L,$ and l_l. Use $\text{RF}_{id} = 1.2, \text{RF}_{cr} = 2.5,$ and $\text{RF}_{cbd} = 1.25$.

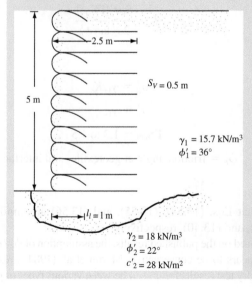

-2.5 m$-$

$S_V = 0.5$ m

5 m

$\gamma_1 = 15.7$ kN/m³
$\phi'_1 = 36°$

$l_l = 1$ m

$\gamma_2 = 18$ kN/m³
$\phi'_2 = 22°$
$c'_2 = 28$ kN/m²

Figure 13.34 Geotextile-reinforced retaining wall

Solution

We have

$$K_a = \tan^2\left(45 - \frac{\phi_1'}{2}\right) = 0.26$$

Determination of S_V

To find S_V, we make a few trials. From Eq. (13.53),

$$S_V = \frac{T_{\text{all}}}{(\gamma_1 z K_a)[\text{FS}_{(B)}]}$$

From Eq. (13.52),

$$T_{\text{all}} = \frac{T_{\text{uef}}}{\text{RF}_{\text{id}} \times \text{RF}_{\text{cr}} \times \text{RF}_{\text{cbd}}} = \frac{52.5}{1.2 \times 2.5 \times 1.25} = 14 \text{ kN/m}$$

With $\text{FS}_{(B)} = 1.5$ at $z = 2$ m,

$$S_V = \frac{14}{(15.7)(2)(0.26)(1.5)} = 1.14 \text{ m}$$

At $z = 4$ m,

$$S_V = \frac{14}{(15.7)(4)(0.26)(1.5)} = 0.57 \text{ m}$$

At $z = 5$ m,

$$S_V = \frac{14}{(15.7)(5)(0.26)(1.5))} = 0.46 \text{ m}$$

So, **use $S_V = 0.5$ m for $z = 0$ to $z = 5$ m** (See Figure 13.34.)

Determination of L

From Eqs. (13.54), (13.55), and (13.56),

$$L = \frac{(H - z)}{\tan\left(45 + \frac{\phi_1'}{2}\right)} + \frac{S_V K_a[\text{FS}_{(P)}]}{2 \tan \phi_F'}$$

For $\text{FS}_{(P)} = 1.5$, $\tan \phi_F' = \tan\left[\left(\frac{2}{3}\right)(36)\right] = 0.445$, and it follows that

$$L = (0.51)(H - z) + 0.438 S_V$$

$H = 5$ m, $S_V = 0.5$ m

At $z = 0.5$ m: $L = (0.51)(5 - 0.5) + (0.438)(0.5) = 2.514$ m
At $z = 2.5$ m: $L = (0.51)(5 - 2.5) + (0.438)(0.5) = 1.494$ m

So, **use $L = 2.5$ m throughout.**

Determination of l_l
From Eq. (13.57),

$$l_l = \frac{S_V \sigma'_a [\text{FS}_{(P)}]}{4\sigma'_o \tan \phi'_F}$$

$\sigma'_a = \gamma_1 z K_a$, $\text{FS}_{(P)} = 1.5$; with $\sigma'_o = \gamma_1 z$, $\phi'_F = \frac{2}{3}\phi'_1$. So

$$l_l = \frac{S_V K_a [\text{FS}_{(P)}]}{4 \tan \phi'_F} = \frac{S_V (0.26)(1.5)}{4 \tan \left[\left(\frac{2}{3}\right)(36)\right]} = 0.219 S_V$$

$$l_l = 0.219 S_V = (0.219)(0.5) = 0.11 \text{ m} \leq 1 \text{ m}$$

So, use $l_l = 1$ **m.** ∎

Example 13.6

Consider the results of the internal stability check given in Example 13.5. For the geotextile-reinforced retaining wall, calculate the factor of safety against overturning, sliding, and bearing capacity failure.

Solution
Refer to Figure 13.35.

Factor of Safety Against Overturning

From Eq. (13.46), $\text{FS}_{(\text{overturning})} = \dfrac{W_1 x_1}{(P_a)\left(\dfrac{H}{3}\right)}$

$$W_1 = (5)(2.5)(15.7) = 196.25 \text{ kN/m}$$

$$x_1 = \frac{2.5}{2} = 1.25 \text{ m}$$

$$P_a = \frac{1}{2}\gamma H^2 K_a = \left(\frac{1}{2}\right)(15.7)(5)^2(0.26) = 51.03 \text{ kN/m}$$

Hence,

$$\text{FS}_{(\text{overturning})} = \frac{(196.25)(1.25)}{51.03(5/3)} = 2.88 < 3$$

(increase length of geotextile layers to 3 m)

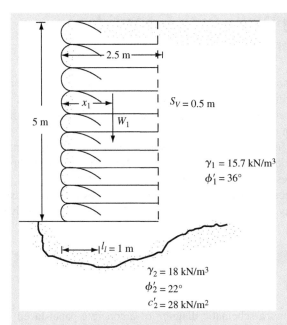

Figure 13.35 Stability check

Factor of Safety Against Sliding

From Eq. (13.47),

$$\text{FS}_{\text{(sliding)}} = \frac{W_1 \tan\left(\frac{2}{3}\phi_1'\right)}{P_a} = \frac{(196.25)\left[\tan\left(\frac{2}{3} \times 36\right)\right]}{51.03} = \mathbf{1.71 > 1.5 - O.K.}$$

Factor of Safety Against Bearing Capacity Failure

From Eq. (13.48), $q_u = c_2' N_c + \frac{1}{2}\gamma_2 L_2 N_\gamma$

Given: $\gamma_2 = 18 \text{ kN/m}^3$, $L_2 = 2.5 \text{ m}$, $c_2' = 28 \text{ kN/m}^2$, and $\phi_2' = 22°$. From Table 4.2, $N_c = 16.88$, and $N_\gamma = 7.13$.

$$q_u = (28)(16.88) + \left(\frac{1}{2}\right)(18)(2.5)(7.13) \approx 633 \text{ kN/m}^2$$

From Eq. (13.50),

$$\text{FS}_{\text{(bearing capacity)}} = \frac{q_u}{\sigma_{o(H)}'} = \frac{633}{\gamma_1 H} = \frac{633}{(15.7)(5)} = \mathbf{8.06 > 3 - O.K.} \quad \blacksquare$$

13.16 Retaining Walls with Geogrid Reinforcement—General

Geogrids can also be used as reinforcement in granular backfill for the construction of retaining walls. Figure 13.36 shows typical schematic diagrams of retaining walls with geogrid reinforcement. Figure 13.37 shows some photographs of geogrid-reinforced retaining walls in the field.

Relatively few field measurements are available for lateral earth pressure on retaining walls constructed with geogrid reinforcement. Figure 13.38 shows a comparison of measured and design lateral pressures (Berg et al. 1986) for two retaining walls constructed with precast panel facing. The figure indicates that the measured earth pressures were substantially smaller than those calculated for the Rankine active case.

13.17 Design Procedure for Geogrid-Reinforced Retaining Wall

Figure 13.39 shows a schematic diagram of a concrete panel-faced wall with a granular backfill reinforced with layers of geogrid. The design process of the wall is essentially similar to that with geotextile reinforcement of the backfill given in Section 13.15. The following is a brief step-by-step procedure.

Internal Stability

Step 1. Determine the active pressure at any depth z as [similar to Eq. (13.51)]:

$$\sigma'_a = K_a \gamma_1 z \tag{13.58}$$

where

$$K_a = \text{Rankine active pressure coefficient} = \tan^2\left(45 - \frac{\phi'_1}{2}\right)$$

Step 2. Select a geogrid with allowable tensile strength, T_{all} [similar to Eq. (13.52)] (Koerner, 2005):

$$T_{\text{all}} = \frac{T_{\text{ult}}}{\text{RF}_{\text{id}} \times \text{RF}_{\text{cr}} \times \text{RF}_{\text{cbd}}} \tag{13.59}$$

where
RF_{id} = reduction factor for installation damage (1.1 to 1.4)
RF_{cr} = reduction factor for creep (2.0 to 3.0)
RF_{cbd} = reduction factor for chemical and biological degradation (1.1 to 1.5).

Step 3. Obtain the vertical spacing of the geogrid layers, S_V, as

$$S_V = \frac{T_{\text{all}} C_r}{\sigma'_a \text{FS}_{(B)}} \tag{13.60}$$

where C_r = coverage ratio for geogrid.

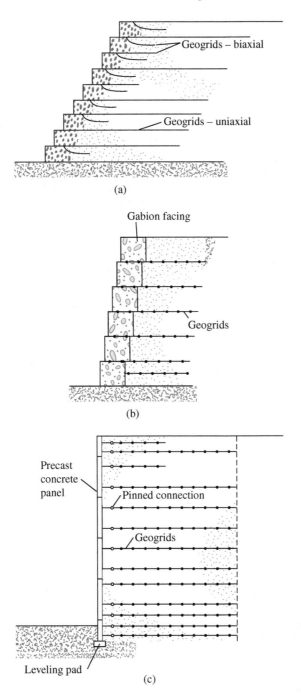

(a)

(b)

(c)

Figure 13.36 Typical schematic diagrams of retaining walls with geogrid reinforcement:
(a) geogrid wraparound wall; (b) wall with gabion facing; (c) concrete panel-faced wall
(Based on Berg et al., 1986)

(a)

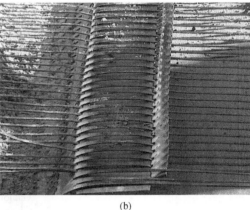

(b)

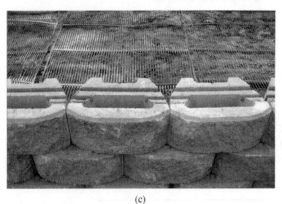

(c)

Figure 13.37 (a) HDPE geogrid-reinforced wall with precast concrete panel facing under construction; (b) mechanical splice between two pieces of geogrid in the working direction; (c) segmented concrete-block faced wall reinforced with uniaxial geogrid (*Courtesy of Tensar International Corporation, Atlanta, Georgia*)

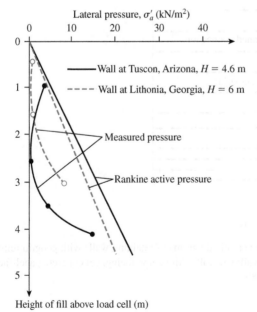

Figure 13.38 Comparison of theoretical and measured lateral pressures in geogrid reinforced retaining walls (Based on Berg et al., 1986)

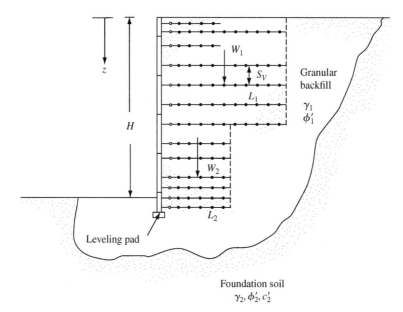

Figure 13.39 Design of geogrid-reinforced retaining wall

The *coverage ratio* is the fractional plan area at any particular elevation that is actually occupied by geogrid. For example, if there is a 0.3-m-wide (1 ft) space between each 1.2-m-wide (4 ft) piece of geogrid, the coverage ratio is

$$C_r = \frac{1.2 \text{ m}}{1.2 \text{ m} + 0.3 \text{ m}} = 0.8$$

Step 4. Calculate the length of each layer of geogrid at a depth z as [Eq. (13.54)]

$$L = l_r + l_e$$

$$l_r = \frac{H - z}{\tan^2\left(45 - \dfrac{\phi_1'}{2}\right)} \tag{13.61}$$

For determination of l_e [similar to Eq. (13.56)],

$$FS_{(P)} = \frac{\text{resistance to pullout at a given normal effective stress}}{\text{pullout force}} \tag{13.62}$$

$$= \frac{(2)(l_e)(C_i\sigma_0' \tan \phi_1')(C_r)}{S_V\sigma_a'}$$

$$= \frac{(2)(l_e)(C_i \tan \phi_1')(C_r)}{S_V K_a}$$

where C_i = interaction coefficient or

$$l_e = \frac{S_V K_a \, FS_{(P)}}{2C_r C_i \tan \phi_1'} \tag{13.63}$$

Thus, at a given depth z, the total length, L, of the geogrid layer is

$$L = l_r + l_e = \frac{H - z}{\tan\left(45 + \dfrac{\phi_1'}{2}\right)} + \frac{S_V K_a \, FS_{(P)}}{2C_r C_i \tan \phi_1'} \tag{13.64}$$

The interaction coefficient, C_i, can be determined experimentally in the laboratory. The following is an approximate range for C_i for various types of backfill.

Gravel, sandy gravel	0.75–0.8
Well graded sand, gravelly sand	0.7–0.75
Fine sand, silty sand	0.55–0.6

External Stability

Check the factors of safety against overturning, sliding, and bearing capacity failure as described in Section 13.14 (Steps 9, 10, and 11).

Example 13.7

Consider a geogrid-reinforced retaining wall. Referring to Figure 13.39, given: $H = 6$ m, $\gamma_1 = 16.5$ kN/m³, $\phi_1' = 35°$, $T_{all} = 45$ kN/m, $FS_{(B)} = 1.5$, $FS_{(P)} = 1.5$, $C_r = 0.8$, and $C_i = 0.75$. For the design of the wall, determine S_V and L.

Solution

$$K_a = \tan^2\left(45 - \frac{\phi_1'}{2}\right) = \tan^2\left(45 - \frac{35}{2}\right) = 0.27$$

Determination of S_V
From Eq. (13.60),

$$S_v = \frac{T_{all} C_r}{\sigma_a' \, FS_{(B)}} = \frac{T_{all} C_r}{\gamma z K_a \, FS_{(B)}} = \frac{(45)(0.8)}{(16.5)(z)(0.27)(1.5)} = \frac{5.39}{z}$$

At $z = 2$ m: $S_v = \dfrac{5.39}{2} = 2.7$ m

At $z = 4$ m: $S_v = \dfrac{5.39}{4} = 1.35$ m

At $z = 5$ m: $S_v = \dfrac{5.39}{5} = 1.08$ m

Use $S_V \approx 1$ m

Determination of L
From Eq. (13.64),

$$L = \frac{H - z}{\tan\left(45 + \dfrac{\phi_1'}{2}\right)} + \frac{S_V K_a \, \text{FS}_{(P)}}{2C_r C_i \tan\phi_1'} = \frac{6 - z}{\tan\left(45 + \dfrac{35}{2}\right)} + \frac{(1\ \text{m})(0.27)(1.5)}{(2)(0.8)(0.75)(\tan 35°)}$$

At $z = 1$ m: $L = 0.52(6 - 1) + 0.482 = 3.08$ m ≈ 3.1 m
At $z = 3$ m: $L = 0.52(6 - 3) + 0.482 = 2.04$ m ≈ 2.1 m
At $z = 5$ m: $L = 0.52(6 - 5) + 0.482 = 1.0$ m

So, use $L = 3$ m for $z = 0$ to 6 m. ∎

Problems

In Problems 13.1 through 13.4, use $\gamma_{\text{concrete}} = 23.58$ kN/m³. Also, in Eq. (13.11), use $k_1 = k_2 = 2/3$ and $P_p = 0$.

13.1 For the cantilever retaining wall shown in Figure P13.1, let the following data be given:

Wall dimensions: $H = 8$ m, $x_1 = 0.4$ m, $x_2 = 0.6$ m, $x_3 = 1.5$ m, $x_4 = 3.5$ m, $x_5 = 0.96$ m, $D = 1.75$ m, $\alpha = 10°$

Soil properties: $\gamma_1 = 16.8$ kN/m³, $\phi_1' = 32°$, $\gamma_2 = 17.6$ kN/m³, $\phi_2' = 28°$, $c_2' = 30$ kN/m²

Calculate the factor of safety with respect to overturning, sliding, and bearing capacity.

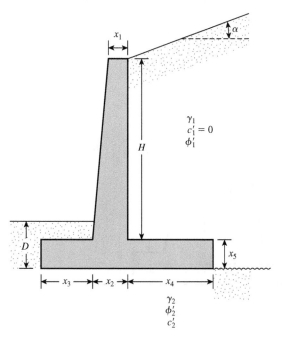

Figure P13.1

13.2 Repeat Problem 13.1 with the following:

Wall dimensions: $H = 6.5$ m, $x_1 = 0.3$ m, $x_2 = 0.6$ m, $x_3 = 0.8$ m, $x_4 = 2$ m, $x_5 = 0.8$ m, $D = 1.5$ m, $\alpha = 0°$

Soil properties: $\gamma_1 = 18.08$ kN/m³, $\phi_1' = 36°$, $\gamma_2 = 19.65$ kN/m³, $\phi_2' = 15°$, $c_2' = 30$ kN/m²

13.3 A gravity retaining wall is shown in Figure P13.3. Calculate the factor of safety with respect to overturning and sliding, given the following data:

Wall dimensions: $H = 6$ m, $x_1 = 0.6$ m, $x_2 = 2$ m, $x_3 = 2$ m, $x_4 = 0.5$ m, $x_5 = 0.75$ m, $x_6 = 0.8$ m, $D = 1.5$ m

Soil properties: $\gamma_1 = 16.5$ kN/m³, $\phi_1' = 32°$, $\gamma_2 = 18$ kN/m³, $\phi_2' = 22°$, $c_2' = 40$ kN/m²

Use the Rankine active earth pressure in your calculation.

13.4 Repeat Problem 13.3 using Coulomb's active earth pressure in your calculation and letting $\delta' = 2/3\ \phi_1'$.

13.5 In Figure 13.27a, use the following parameters:

Wall: $H = 8$ m

Soil: $\gamma_1 = 17$ kN/m³, $\phi_1' = 35°$

Reinforcement: $S_V = 1$ m and $S_H = 1.5$ m

Surcharge: $q = 70$ kN/m², $a' = 1.5$ m, and $b' = 2$ m

Calculate the vertical stress σ_o' [Eqs. (13.28), (13.29) and (13.30)] at $z = 2$ m, 4 m, 6 m, and 8 m.

13.6 For the data given in Problem 13.5, calculate the lateral pressure σ_a' at $z = 2$ m, 4 m, 6 m and 8 m. Use Eqs. (13.31), (13.32), and (13.33).

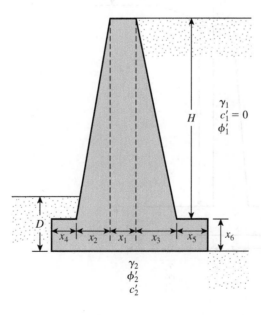

Figure P13.3

13.7 A reinforced earth retaining wall (Figure 13.27) is to be 10-m-high. Here,

Backfill: unit weight, $\gamma_1 = 16$ kN/m³ and soil friction angle, $\phi_1' = 34°$

Reinforcement: vertical spacing, $S_V = 1$ m; horizontal spacing, $S_H = 1.25$ m; width of reinforcement $= 120$ mm., $f_y = 260$ MN/m²; $\phi_\mu = 25°$; factor of safety against tie pullout $= 3$; and factor of safety against tie breaking $= 3$

Determine:
  a. The required thickness of ties
  b. The required maximum length of ties

13.8 In Problem 13.7 assume that the ties at all depths are the length determined in Part b. For the *in situ* soil, $\phi_2' = 25°$, $\gamma_2 = 15.5$ kN/m³, $c_2' = 30$ kN/m². Calculate the factor of safety against (a) overturning, (b) sliding, and (c) bearing capacity failure.

13.9 A retaining wall with geotextile reinforcement is 6-m high. For the granular backfill, $\gamma_1 = 15.9$ kN/m³ and $\phi_1' = 30°$. For the geotextile, $T_{all} = 16$ kN/m. For the design of the wall, determine S_V, L, and l_l. Use $FS_{(B)} = FS_{(P)} = 1.5$.

13.10 With the S_V, L, and l_l determined in Problem 13.9, check the overall stability (i.e., factor of safety against overturning, sliding, and bearing capacity failure) of the wall. For the *in situ* soil, $\gamma_2 = 16.8$ kN/m³, $\phi_2' = 20°$, and $c_2' = 55$ kN/m².

References

BELL, J. R., STILLEY, A. N., and VANDRE, B. (1975). "Fabric Retaining Earth Walls," *Proceedings, Thirteenth Engineering Geology and Soils Engineering Symposium,* Moscow, ID.

BENTLER, J. G. and LABUZ, J. F. (2006). "Performance of a Cantilever Retaining Wall," *Journal of Geotechnical and Geoenvironmental Engineering*, American Society of Civil Engineers, Vol. 132, No. 8, pp. 1062–1070.

BERG, R. R., BONAPARTE, R., ANDERSON, R. P., and Chouery, V. E. (1986). "Design Construction and Performance of Two Tensar Geogrid Reinforced Walls," *Proceedings, Third International Conference on Geotextiles,* Vienna, pp. 401–406.

BINQUET, J. and LEE, K. L. (1975). "Bearing Capacity Analysis of Reinforced Earth Slabs," *Journal of the Geotechnical Engineering Division,* American Society of Civil Engineers, Vol. 101, No. GT12, pp. 1257–1276.

CARROLL, R., Jr. (1988). "Specifying Geogrids," *Geotechnical Fabric Report,* Industrial Fabric Association International, St. Paul, March/April.

CASAGRANDE, L. (1973). "Comments on Conventional Design of Retaining Structure," *Journal of the Soil Mechanics and Foundations Division,* ASCE, Vol. 99, No. SM2, pp. 181–198.

DARBIN, M. (1970). "Reinforced Earth for Construction of Freeways" (in French), *Revue Générale des Routes et Aerodromes,* No. 457, September.

ELMAN, M. T. and TERRY, C. F. (1988). "Retaining Walls with Sloped Heel," *Journal of Geotechnical Engineering*, American Society of Civil Engineers, Vol. 114, No. GT10, pp. 1194–1199.

KOERNER, R. B. (2005). *Design with Geosynthetics,* 5th. ed., Prentice Hall, Englewood Cliffs, NJ.

LABA, J. T. and KENNEDY, J. B. (1986). "Reinforced Earth Retaining Wall Analysis and Design," *Canadian Geotechnical Journal,* Vol. 23, No. 3, pp. 317–326.

LEE, K. L., ADAMS, B. D., and VAGNERON, J. J. (1973). "Reinforced Earth Retaining Walls," *Journal of the Soil Mechanics and Foundations Division,* American Society of Civil Engineers, Vol. 99, No. SM10, pp. 745–763.

MARTIN, J. P., KOERNER, R. M., and WHITTY, J. E. (1984). "Experimental Friction Evaluation of Slippage Between Geomembranes, Geotextiles, and Soils," *Proceedings,* International Conference on Geomembranes, Denver, pp. 191–196.

SARSBY, R. W. (1985). "The Influence of Aperture Size/Particle Size on the Efficiency of Grid Reinforcement," *Proceedings, 2nd Canadian Symposium on Geotextiles and Geomembranes,* Edmonton, pp. 7–12.

SCHLOSSER, F. and LONG, N. (1974). "Recent Results in French Research on Reinforced Earth," *Journal of the Construction Division,* American Society of Civil Engineers, Vol. 100, No. CO3, pp. 113–237.

SCHLOSSER, F. and VIDAL, H. (1969). "Reinforced Earth" (in French), *Bulletin de Liaison des Laboratoires Routier,* Ponts et Chassées, Paris, France, November, pp. 101–144.

TENSAR CORPORATION (1986). Tensar Technical Note. No. TTN:RW1, August.

TERZAGHI, K. and PECK, R. B. (1967). *Soil Mechanics in Engineering Practice,* Wiley, New York.

TRANSPORTATION RESEARCH BOARD (1995). Transportation Research Circular No. 444, National Research Council, Washington, D.C.

VIDAL, H. (1966). "La terre Armée," *Annales de l'Institut Technique du Bâtiment et des Travaux Publiques,* France, July–August, pp. 888–938.

6 Sheet-Pile Walls

14.1 Introduction

Connected or semiconnected sheet piles are often used to build continuous walls for waterfront structures that range from small waterfront pleasure boat launching facilities to large dock facilities. (See Figure 14.1.) In contrast to the construction of other types of retaining wall, the building of sheet-pile walls does not usually require dewatering of the site. Sheet piles are also used for some temporary structures, such as braced cuts. (See Chapter 15.) The principles of sheet-pile wall design are discussed in the current chapter.

Several types of sheet pile are commonly used in construction: (a) wooden sheet piles, (b) precast concrete sheet piles, and (c) steel sheet piles. Aluminum sheet piles are also marketed.

Wooden sheet piles are used only for temporary, light structures that are above the water table. The most common types are ordinary wooden planks and *Wakefield piles*. The wooden planks are about 50 mm × 300 mm (2 in. × 12 in.) in cross section and are driven edge to edge (Figure 14.2a). Wakefield piles are made by nailing three planks together, with the middle plank offset by 50 to 75 mm (2 to 3 in.) (Figure 14.2b). Wooden planks can also be milled to form *tongue-and-groove piles*, as shown in Figure 14.2c. Figure 14.2d shows

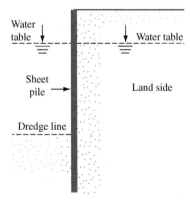

Figure 14.1 Example of waterfront sheet-pile wall

315

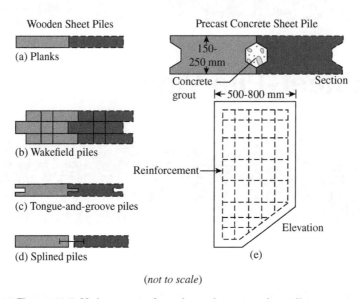

Wooden Sheet Piles

(a) Planks

(b) Wakefield piles

(c) Tongue-and-groove piles

(d) Splined piles

Precast Concrete Sheet Pile

150–250 mm

Concrete grout

Section

|← 500–800 mm →|

Reinforcement →

Elevation

(e)

(*not to scale*)

Figure 14.2 Various types of wooden and concrete sheet pile

another type of wooden sheet pile that has precut grooves. Metal *splines* are driven into the grooves of the adjacent sheetings to hold them together after they are sunk into the ground.

Precast concrete sheet piles are heavy and are designed with reinforcements to withstand the permanent stresses to which the structure will be subjected after construction and also to handle the stresses produced during construction. In cross section, these piles are about 500 to 800 mm (20 to 32 in.) wide and 150 to 250 mm (6 to 10 in.) thick. Figure 14.2e is a schematic diagram of the elevation and the cross section of a reinforced concrete sheet pile.

Steel sheet piles in the United States are about 10 to 13 mm (0.4 to 0.5 in.) thick. European sections may be thinner and wider. Sheet-pile sections may be *Z, deep arch, low arch,* or *straight web* sections. The interlocks of the sheet-pile sections are shaped like a *thumb-and-finger* or *ball-and-socket* joint for watertight connections. Figure 14.3a is a schematic diagram of the thumb-and-finger type of interlocking for straight web sections. The ball-and-socket type of interlocking for *Z* section piles is shown in Figure 14.3b. Figure 14.4 shows some sheet piles at a construction site. Figure 14.5 shows a small enclosure with steel sheet piles for an excavation work. Table 14.1 lists the properties of

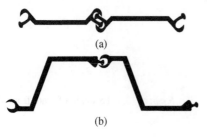

(a)

(b)

Figure 14.3 (a) Thumb-and-finger type sheet-pile connection; (b) ball-and-socket type sheet-pile connection

Figure 14.4 Some steel sheet piles at a construction site (*Courtesy of N. Sivakugan, James Cook University, Australia*)

Figure 14.5 A small enclosure with steel sheet piles for an excavation work (*Courtesy of N. Sivakugan, James Cook University, Australia*)

Table 14.1 Properties of Some Commercially Available Sheet-Pile Sections (Based on Hammer and Steel, Inc., Hazelwood, Missouri, USA)

Section designation	*H* mm (in.)	*L* mm (in.)	*f* mm (in.)	*w* mm (in.)	Section modulus m^3/m of wall (in.^{3}/ft of wall)	Moment of inertia m^4/m of wall (in.^{4}/ft of wall)
PZC-12	318.0 (12.52)	708.2 (27.88)	8.51 (0.335)	8.51 (0.335)	120.42×10^{-5} (22.4)	192.06×10^{-6} (140.6)
PZC-13	319.0 (12.56)	708.2 (27.88)	9.53 (0.375)	9.53 (0.375)	130.1×10^{-5} (24.2)	207.63×10^{-6} (152.0)
PZC-14	320.0 (12.6)	708.2 (27.88)	10.67 (0.420)	10.67 (0.420)	139.78×10^{-5} (26.0)	225.12×10^{-6} (164.8)
PZC-17	386.3 (15.21)	635.0 (25.00)	8.51 (0.335)	8.51 (0.335)	166.67×10^{-5} (31.0)	322.38×10^{-6} (236.6)
PZC-18	387.4 (15.25)	635.0 (25.00)	9.53 (0.375)	9.53 (0.375)	180.1×10^{-5} (33.5)	349.01×10^{-6} (255.5)
PZC-19	388.6 (15.30)	635.0 (25.00)	10.67 (0.420)	10.67 (0.420)	194.07×10^{-5} (36.1)	377.97×10^{-6} (276.7)
PZC-26	449.6 (17.70)	708.2 (27.88)	15.24 (0.60)	13.34 (0.525)	260.2×10^{-5} (48.4)	584.78×10^{-6} (428.1)
PZ-22	235.0 (9.25)	558.8 (22.00)	9.53 (0.375)	9.53 (0.375)	98.92×10^{-5} (18.4)	116.2×10^{-6} (85.1)
PZ-27	307.3 (12.1)	457.2 (18.00)	9.53 (0.375)	9.53 (0.375)	166.66×10^{-5} (31.00)	255.9×10^{-6} (187.3)
PZ-35	383.5 (15.1)	575.1 (22.64)	15.37 (0.605)	12.7 (0.5)	262.9×10^{-5} (48.9)	504.6×10^{-6} (369.4)
PZ-40	416.6 (16.4)	499.1 (19.69)	15.24 (0.6)	12.7 (0.5)	329.5×10^{-5} (61.3)	686.7×10^{-6} (502.7)

(*Continued*)

Table 14.1 (Continued)

Section designation	H mm (in.)	L mm (in.)	f mm (in.)	w mm (in.)	Section modulus m³/m of wall (in.³/ft of wall)	Moment of inertia m⁴/m of wall (in.⁴/ft of wall)
PS-27.5	—	500	—	10.16	10.21×10^{-5}	4.1×10^{-6}
	—	(19.69)	—	(0.4)	(1.9)	(3.0)
PS-31	—	500	—	12.7	10.21×10^{-5}	4.1×10^{-6}
	—	(19.69)	—	(0.5)	(1.9)	(3.0)

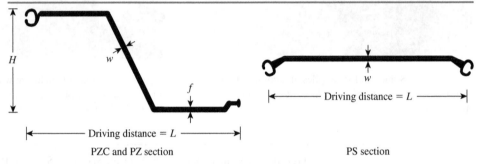

PZC and PZ section   PS section

the steel sheet-pile sections produced by Hammer & Steel, Inc. of Hazelwood, Missouri. The allowable design flexural stress for the steel sheet piles is as follows:

Type of steel	Allowable stress	
ASTM A-328	170 MN/m²	(25,000 lb/in²)
ASTM A-572	210 MN/m²	(30,000 lb/in²)
ASTM A-690	210 MN/m²	(30,000 lb/in²)

Steel sheet piles are convenient to use because of their resistance to the high driving stress that is developed when they are being driven into hard soils. Steel sheet piles are also lightweight and reusable.

14.2 Construction Methods

Sheet-pile walls may be divided into two basic categories: (a) cantilever and (b) anchored.

In the construction of sheet-pile walls, the sheet pile may be driven into the ground and then the backfill placed on the land side, or the sheet pile may first be driven into the ground and the soil in front of the sheet pile dredged. In either case, the soil used for backfill behind the sheet-pile wall is usually granular. The soil below the dredge line may

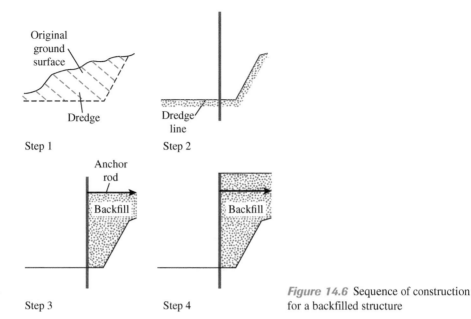

Figure 14.6 Sequence of construction for a backfilled structure

be sandy or clayey. The surface of soil on the water side is referred to as the *mud line* or *dredge line.*

Thus, construction methods generally can be divided into two categories (Tsinker, 1983):

1. Backfilled structure
2. Dredged structure

The sequence of construction for a *backfilled structure* is as follows (see Figure 14.6):

Step 1. Dredge the *in situ* soil in front and back of the proposed structure.
Step 2. Drive the sheet piles.
Step 3. Backfill up to the level of the anchor, and place the anchor system.
Step 4. Backfill up to the top of the wall.

For a cantilever type of wall, only Steps 1, 2, and 4 apply. The sequence of construction for a *dredged structure* is as follows (see Figure 14.7):

Step 1. Drive the sheet piles.
Step 2. Backfill up to the anchor level, and place the anchor system.
Step 3. Backfill up to the top of the wall.
Step 4. Dredge the front side of the wall.

With cantilever sheet-pile walls, Step 2 is not required.

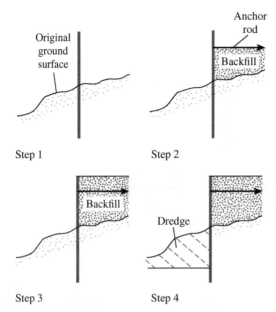

Step 1

Step 2

Step 3

Step 4

Figure 14.7 Sequence of construction for a dredged structure

14.3 Cantilever Sheet-Pile Walls

Cantilever sheet-pile walls are usually recommended for walls of moderate height—about 6 m ($\approx$20 ft) or less, measured above the dredge line. In such walls, the sheet piles act as a wide cantilever beam above the dredge line. The basic principles for estimating net lateral pressure distribution on a cantilever sheet-pile wall can be explained with the aid of Figure 14.8. The figure shows the nature of lateral yielding of a cantilever wall penetrating

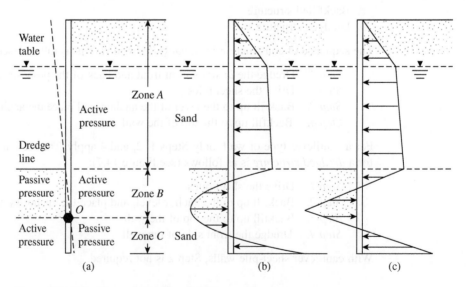

Figure 14.8 Cantilever sheet pile penetrating sand

a sand layer below the dredge line. The wall rotates about point O (Figure 14.8a). Because the hydrostatic pressures at any depth from both sides of the wall will cancel each other, we consider only the effective lateral soil pressures. In zone A, the lateral pressure is just the active pressure from the land side. In zone B, because of the nature of yielding of the wall, there will be active pressure from the land side and passive pressure from the water side. The condition is reversed in zone C—that is, below the point of rotation, O. The net actual pressure distribution on the wall is like that shown in Figure 14.8b. However, for design purposes, Figure 14.8c shows a simplified version.

Sections 14.4 through 14.7 present the mathematical formulation of the analysis of cantilever sheet-pile walls. Note that, in some waterfront structures, the water level may fluctuate as the result of tidal effects. Care should be taken in determining the water level that will affect the net pressure diagram.

14.4 Cantilever Sheet Piling Penetrating Sandy Soils

To develop the relationships for the proper depth of embedment of sheet piles driven into a granular soil, examine Figure 14.9a. The soil retained by the sheet piling above the dredge line also is sand. The water table is at a depth L_1 below the top of the wall. Let the effective angle of friction of the sand be ϕ'. The intensity of the active pressure at a depth $z = L_1$ is

$$\sigma_1' = \gamma L_1 K_a \tag{14.1}$$

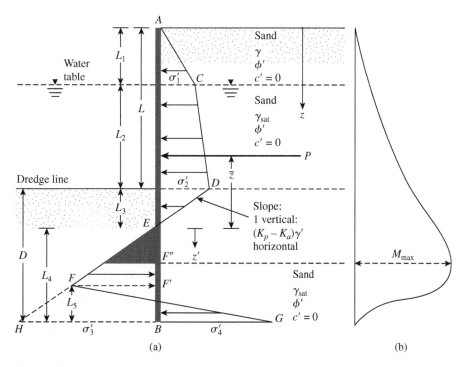

Figure 14.9 Cantilever sheet pile penetrating sand: (a) variation of net pressure diagram; (b) variation of moment

where

K_a = Rankine active pressure coefficient = $\tan^2(45 - \phi'/2)$
γ = unit weight of soil above the water table

Similarly, the active pressure at a depth $z = L_1 + L_2$ (i.e., at the level of the dredge line) is

$$\sigma_2' = (\gamma L_1 + \gamma' L_2)K_a \tag{14.2}$$

where γ' = effective unit weight of soil = $\gamma_{sat} - \gamma_w$.

Note that, at the level of the dredge line, the hydrostatic pressures from both sides of the wall are the same magnitude and cancel each other.

To determine the net lateral pressure below the dredge line up to the point of rotation, O, as shown in Figure 14.8a, an engineer has to consider the passive pressure acting from the left side (the water side) toward the right side (the land side) of the wall and also the active pressure acting from the right side toward the left side of the wall. For such cases, ignoring the hydrostatic pressure from both sides of the wall, the active pressure at depth z is

$$\sigma_a' = [\gamma L_1 + \gamma' L_2 + \gamma'(z - L_1 - L_2)]K_a \tag{14.3}$$

Also, the passive pressure at depth z is

$$\sigma_p' = \gamma'(z - L_1 - L_2)K_p \tag{14.4}$$

where K_p = Rankine passive pressure coefficient = $\tan^2(45 + \phi'/2)$.

Combining Eqs. (14.3) and (14.4) yields the net lateral pressure, namely,

$$\sigma' = \sigma_a' - \sigma_p' = (\gamma L_1 + \gamma' L_2)K_a - \gamma'(z - L_1 - L_2)(K_p - K_a)$$
$$= \sigma_2' - \gamma'(z - L)(K_p - K_a) \tag{14.5}$$

where $L = L_1 + L_2$.

The net pressure, σ' equals zero at a depth L_3 below the dredge line, so

$$\sigma_2' - \gamma'(z - L)(K_p - K_a) = 0$$

or

$$(z - L) = L_3 = \frac{\sigma_2'}{\gamma'(K_p - K_a)} \tag{14.6}$$

Equation (14.6) indicates that the slope of the net pressure distribution line *DEF* is 1 vertical to $(K_p - K_a)\gamma'$ horizontal, so, in the pressure diagram,

$$\overline{HB} = \sigma_3' = L_4(K_p - K_a)\gamma' \tag{14.7}$$

At the bottom of the sheet pile, passive pressure, σ_p', acts from the right toward the left side, and active pressure acts from the left toward the right side of the sheet pile, so, at $z = L + D$,

$$\sigma_p' = (\gamma L_1 + \gamma' L_2 + \gamma' D)K_p \tag{14.8}$$

At the same depth,

$$\sigma'_a = \gamma' D K_a \tag{14.9}$$

Hence, the net lateral pressure at the bottom of the sheet pile is

$$
\begin{aligned}
\sigma'_p - \sigma'_a = \sigma'_4 &= (\gamma L_1 + \gamma' L_2) K_p + \gamma' D (K_p - K_a) \\
&= (\gamma L_1 + \gamma' L_2) K_p + \gamma' L_3 (K_p - K_a) + \gamma' L_4 (K_p - K_a) \\
&= \sigma'_5 + \gamma' L_4 (K_p - K_a)
\end{aligned}
\tag{14.10}
$$

where

$$\sigma'_5 = (\gamma L_1 + \gamma' L_2) K_p + \gamma' L_3 (K_p - K_a) \tag{14.11}$$
$$D = L_3 + L_4 \tag{14.12}$$

For the stability of the wall, the principles of statics can now be applied:

$$\Sigma \text{ horizontal forces per unit length of wall} = 0$$

and

$$\Sigma \text{ moment of the forces per unit length of wall about point } B = 0$$

For the summation of the horizontal forces, we have

Area of the pressure diagram $ACDE$ − area of $EFHB$ + area of $FHBG = 0$

or

$$P - \tfrac{1}{2}\sigma'_3 L_4 + \tfrac{1}{2} L_5 (\sigma'_3 + \sigma'_4) = 0 \tag{14.13}$$

where P = area of the pressure diagram $ACDE$.

Summing the moment of all the forces about point B yields

$$P(L_4 + \bar{z}) - \left(\frac{1}{2} L_4 \sigma'_3\right)\left(\frac{L_4}{3}\right) + \frac{1}{2} L_5 (\sigma'_3 + \sigma'_4)\left(\frac{L_5}{3}\right) = 0 \tag{14.14}$$

From Eq. (14.13),

$$L_5 = \frac{\sigma'_3 L_4 - 2P}{\sigma'_3 + \sigma'_4} \tag{14.15}$$

Combining Eqs. (14.7), (14.10), (14.14), and (14.15) and simplifying them further, we obtain the following fourth-degree equation in terms of L_4:

$$L_4^4 + A_1 L_4^3 - A_2 L_4^2 - A_3 L_4 - A_4 = 0 \tag{14.16}$$

In this equation,

$$A_1 = \frac{\sigma_5'}{\gamma'(K_p - K_a)} \tag{14.17}$$

$$A_2 = \frac{8P}{\gamma'(K_p - K_a)} \tag{14.18}$$

$$A_3 = \frac{6P[2\bar{z}\gamma'(K_p - K_a) + \sigma_5']}{\gamma'^2(K_p - K_a)^2} \tag{14.19}$$

$$A_4 = \frac{P(6\bar{z}\sigma_5' + 4P)}{\gamma'^2(K_p - K_a)^2} \tag{14.20}$$

Step-by-Step Procedure for Obtaining the Pressure Diagram

Based on the preceding theory, a step-by-step procedure for obtaining the pressure diagram for a cantilever sheet-pile wall penetrating a granular soil is as follows:

Step 1. Calculate K_a and K_p.
Step 2. Calculate σ_1' [Eq. (14.1)] and σ_2' [Eq. (14.2)]. (*Note: L_1 and L_2 will be given.*)
Step 3. Calculate L_3 [Eq. (14.6)].
Step 4. Calculate P.
Step 5. Calculate $\bar{z}$ (i.e., the center of pressure for the area $ACDE$) by taking the moment about E.
Step 6. Calculate σ_5' [Eq. (14.11)].
Step 7. Calculate $A_1, A_2, A_3,$ and A_4 [Eqs. (14.17) through (14.20)].
Step 8. Solve Eq. (14.16) by trial and error to determine L_4.
Step 9. Calculate σ_4' [Eq. (14.10)].
Step 10. Calculate σ_3' [Eq. (14.7)].
Step 11. Obtain L_5 from Eq. (14.15).
Step 12. Draw a pressure distribution diagram like the one shown in Figure 14.9a.
Step 13. Obtain the theoretical depth [see Eq. (14.12)] of penetration as $L_3 + L_4$. The actual depth of penetration is increased by about 20 to 30%.

Note that some designers prefer to use a factor of safety on the passive earth pressure coefficient at the beginning. In that case, in Step 1,

$$K_{p(\text{design})} = \frac{K_p}{\text{FS}}$$

where FS = factor of safety (usually between 1.5 and 2).

For this type of analysis, follow Steps 1 through 12 with the value of $K_a = \tan^2(45 - \phi'/2)$ and $K_{p(\text{design})}$ (instead of K_p). The actual depth of penetration can now be determined by adding L_3, obtained from Step 3, and L_4, obtained from Step 8.

Calculation of Maximum Bending Moment

The nature of the variation of the moment diagram for a cantilever sheet-pile wall is shown in Figure 14.9b. The maximum moment will occur between points E and F'. Obtaining the maximum moment (M_{max}) per unit length of the wall requires determining the point of zero shear. For a new axis z' (with origin at point E) for zero shear,

$$P = \tfrac{1}{2}(z')^2(K_p - K_a)\gamma'$$

or

$$z' = \sqrt{\frac{2P}{(K_p - K_a)\gamma'}} \tag{14.21}$$

Once the point of zero shear force is determined (point F'' in Figure 14.9a), the magnitude of the maximum moment can be obtained as

$$M_{max} = P(\bar{z} + z') - [\tfrac{1}{2}\gamma'z'^2(K_p - K_a)](\tfrac{1}{3})z' \tag{14.22}$$

The necessary profile of the sheet piling is then sized according to the allowable flexural stress of the sheet pile material, or

$$S = \frac{M_{max}}{\sigma_{all}} \tag{14.23}$$

where

S = section modulus of the sheet pile required per unit length of the structure
σ_{all} = allowable flexural stress of the sheet pile

Example 14.1

Figure 14.10 shows a cantilever sheet-pile wall penetrating a granular soil. Here, $L_1 = 2$ m, $L_2 = 3$ m, $\gamma = 15.9$ kN/m³, $\gamma_{sat} = 19.33$ kN/m³, and $\phi' = 32°$.

a. What is the theoretical depth of embedment, D?
b. For a 30% increase in D, what should be the total length of the sheet piles?
c. What should be the minimum section modulus of the sheet piles? Use $\sigma_{all} = 172$ MN/m².

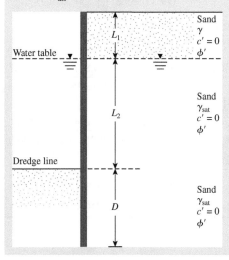

Figure 14.10 Cantilever sheet-pile wall

Solution

Part a

Using Figure 14.9a for the pressure distribution diagram, one can now prepare the following table for a step-by-step calculation.

Quantity required	Eq. no.	Equation and calculation
K_a	—	$\tan^2\!\left(45 - \dfrac{\phi'}{2}\right) = \tan^2\!\left(45 - \dfrac{32}{2}\right) = 0.307$
K_p	—	$\tan^2\!\left(45 + \dfrac{\phi'}{2}\right) = \tan^2\!\left(45 + \dfrac{32}{2}\right) = 3.25$
σ_1'	14.1	$\gamma L_1 K_a = (15.9)(2)(0.307) = 9.763 \text{ kN/m}^2$
σ_2'	14.2	$(\gamma L_1 + \gamma' L_2)K_a = [(15.9)(2) + (19.33 - 9.81)(3)](0.307) = 18.53 \text{ kN/m}^2$
L_3	14.6	$\dfrac{\sigma_2'}{\gamma'(K_p - K_a)} = \dfrac{18.53}{(19.33 - 9.81)(3.25 - 0.307)} = 0.66 \text{ m}$
P	—	$\frac{1}{2}\sigma_1' L_1 + \sigma_1' L_2 + \frac{1}{2}(\sigma_2' - \sigma_1')L_2 + \frac{1}{2}\sigma_2' L_3$
		$= \left(\frac{1}{2}\right)(9.763)(2) + (9.763)(3) + \left(\frac{1}{2}\right)(18.53 - 9.763)(3)$
		$\quad + \left(\frac{1}{2}\right)(18.53)(0.66)$
		$= 9.763 + 29.289 + 13.151 + 6.115 = 58.32 \text{ kN/m}$
$\bar{z}$	—	$\dfrac{\Sigma M_E}{P} = \dfrac{1}{58.32}\left[\begin{array}{l} 9.763(0.66 + 3 + \frac{2}{3}) + 29.289(0.66 + \frac{3}{2}) \\ + 13.151(0.66 + \frac{3}{3}) + 6.115(0.66 \times \frac{2}{3}) \end{array}\right] = 2.23 \text{ m}$
σ_5'	14.11	$(\gamma L_1 + \gamma' L_2)K_p + \gamma' L_3(K_p - K_a) = [(15.9)(2) + (19.33 - 9.81)(3)](3.25)$
		$\qquad\qquad\qquad\qquad\qquad\qquad + (19.33 - 9.81)(0.66)(3.25$
		$\qquad\qquad\qquad\qquad\qquad\qquad - 0.307) = 214.66 \text{ kN/m}^2$
A_1	14.17	$\dfrac{\sigma_5'}{\gamma'(K_p - K_a)} = \dfrac{214.66}{(19.33 - 9.81)(3.25 - 0.307)} = 7.66$
A_2	14.18	$\dfrac{8P}{\gamma'(K_p - K_a)} = \dfrac{(8)(58.32)}{(19.33 - 9.81)(3.25 - 0.307)} = 16.65$
A_3	14.19	$\dfrac{6P[2\bar{z}\gamma'(K_p - K_a) + \sigma_5']}{\gamma'^2(K_p - K_a)^2}$
		$= \dfrac{(6)(58.32)[(2)(2.23)(19.33 - 9.81)(3.25 - 0.307) + 214.66]}{(19.33 - 9.81)^2(3.25 - 0.307)^2}$
		$= 151.93$
A_4	14.20	$\dfrac{P(6\bar{z}\sigma_5' + 4P)}{\gamma'^2(K_p - K_a)^2} = \dfrac{58.32[(6)(2.23)(214.66) + (4)(58.32)]}{(19.33 - 9.81)^2(3.25 - 0.307)^2}$
		$= 230.72$
L_4	14.16	$L_4^4 + A_1 L_4^3 - A_2 L_4^2 - A_3 L_4 - A_4 = 0$
		$L_4^4 + 7.66 L_4^3 - 16.65 L_4^2 - 151.93 L_4 - 230.72 = 0; \; L_4 \approx 4.8 \text{ m}$

Thus,

$$D_{\text{theory}} = L_3 + L_4 = 0.66 + 4.8 = \textbf{5.46 m}$$

Part b

The total length of the sheet piles is

$$L_1 + L_2 + 1.3(L_3 + L_4) = 2 + 3 + 1.3(5.46) = \textbf{12.1 m}$$

Part c

Finally, we have the following table.

Quantity required	Eq. no.	Equation and calculation
z'	14.21	$\sqrt{\dfrac{2P}{(K_p - K_a)\gamma'}} = \sqrt{\dfrac{(2)(58.32)}{(3.25 - 0.307)(19.33 - 9.81)}} = 2.04$ m
$M_{\max}$	14.22	$P(\bar{z} + z') - \left[\dfrac{1}{2}\gamma'z'^2(K_p - K_a)\right]\dfrac{z'}{3} = (58.32)(2.23 + 2.04)$
		$\quad - \left[\left(\dfrac{1}{2}\right)(19.33 - 9.81)(2.04)^2(3.25 - 0.307)\right]\dfrac{2.04}{3}$
		$\quad = 209.39$ kN · m/m
S	14.29	$\dfrac{M_{\max}}{\sigma_{\text{all}}} = \dfrac{209.39 \text{ kN} \cdot \text{m}}{172 \times 10^3 \text{ kN/m}^2} = \textbf{1.217} \times \textbf{10}^{-3} \text{ m}^3\textbf{/m of wall}$ ■

14.5 Special Cases for Cantilever Walls Penetrating a Sandy Soil

Sheet-Pile Wall with the Absence of Water Table

In the absence of the water table, the net pressure diagram on the cantilever sheet-pile wall will be as shown in Figure 14.11, which is a modified version of Figure 14.9. In this case,

$$\sigma_2' = \gamma L K_a \tag{14.24}$$

$$\sigma_3' = L_4(K_p - K_a)\gamma \tag{14.25}$$

$$\sigma_4' = \sigma_5' + \gamma L_4(K_p - K_a) \tag{14.26}$$

$$\sigma_5' = \gamma L K_p + \gamma L_3(K_p - K_a) \tag{14.27}$$

$$L_3 = \frac{\sigma_2'}{\gamma(K_p - K_a)} = \frac{L K_a}{(K_p - K_a)} \tag{14.28}$$

$$P = \tfrac{1}{2}\sigma_2' L + \tfrac{1}{2}\sigma_2' L_3 \tag{14.29}$$

$$\bar{z} = L_3 + \frac{L}{3} = \frac{L K_a}{K_p - K_a} + \frac{L}{3} = \frac{L(2K_a + K_p)}{3(K_p - K_a)} \tag{14.30}$$

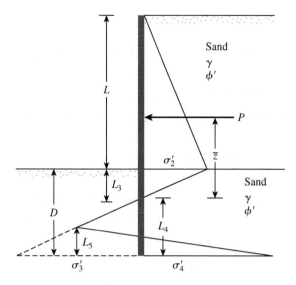

Figure 14.11 Sheet piling penetrating a sandy soil in the absence of the water table

and Eq. (14.16) transforms to

$$L_4^4 + A_1' L_4^3 - A_2' L_4^2 - A_3' L_4 - A_4' = 0 \tag{14.31}$$

where

$$A_1' = \frac{\sigma_5'}{\gamma(K_p - K_a)} \tag{14.32}$$

$$A_2' = \frac{8P}{\gamma(K_p - K_a)} \tag{14.33}$$

$$A_3' = \frac{6P[2\bar{z}\gamma(K_p - K_a) + \sigma_5']}{\gamma^2(K_p - K_a)^2} \tag{14.34}$$

$$A_4' = \frac{P(6\bar{z}\sigma_5' + 4P)}{\gamma^2(K_p - K_a)^2} \tag{14.35}$$

Free Cantilever Sheet Piling

Figure 14.12 shows a free cantilever sheet-pile wall penetrating a sandy soil and subjected to a line load of P per unit length of the wall. For this case,

$$D^4 - \left[\frac{8P}{\gamma(K_p - K_a)}\right]D^2 - \left[\frac{12PL}{\gamma(K_p - K_a)}\right]D - \left[\frac{2P}{\gamma(K_p - K_a)}\right]^2 = 0 \tag{14.36}$$

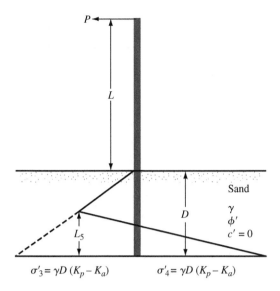

$$\sigma'_3 = \gamma D (K_p - K_a) \qquad \sigma'_4 = \gamma D (K_p - K_a)$$

Figure 14.12 Free cantilever sheet piling penetrating a layer of sand

$$L_5 = \frac{\gamma(K_p - K_a)D^2 - 2P}{2D(K_p - K_a)\gamma} \tag{14.37}$$

$$M_{\max} = P(L + z') - \frac{\gamma z'^3(K_p - K_a)}{6} \tag{14.38}$$

and

$$z' = \sqrt{\frac{2P}{\gamma'(K_p - K_a)}} \tag{14.39}$$

Example 14.2

Redo parts a and b of Example 14.1, assuming the absence of the water table. Use $\gamma = 15.9$ kN/m^3 and $\phi' = 32°$. *Note:* $L = 5$ m.

Solution
Part a

Quantity required	Eq. no.	Equation and calculation
K_a	—	$\tan^2\left(45 - \dfrac{\phi'}{2}\right) = \tan^2\left(45 - \dfrac{32}{2}\right) = 0.307$
K_p	—	$\tan^2\left(45 + \dfrac{\phi'}{2}\right) = \tan^2\left(45 + \dfrac{32}{2}\right) = 3.25$

(Continued)

Quantity required	Eq. no.	Equation and calculation
σ_2'	14.24	$\gamma L K_a = (15.9)(5)(0.307) = 24.41 \text{ kN/m}^2$
L_3	14.28	$\dfrac{L K_a}{K_p - K_a} = \dfrac{(5)(0.307)}{3.25 - 0.307} = 0.521 \text{ m}$
σ_5'	14.27	$\gamma L K_p + \gamma L_3 (K_p - K_a) = (15.9)(5)(3.25) + (15.9)(0.521)(3.25 - 0.307)$ $= 282.76 \text{ kN/m}^2$
P	14.29	$\frac{1}{2}\sigma_2' L + \frac{1}{2}\sigma_2' L_3 = \frac{1}{2}\sigma_2'(L + L_3) = (\frac{1}{2})(24.41)(5 + 0.521) = 67.38 \text{ kN/m}$
$\bar{z}$	14.30	$\dfrac{L(2K_a - K_p)}{3(K_p - K_a)} = \dfrac{5[(2)(0.307) + 3.25]}{3(3.25 - 0.307)} = 2.188 \text{ m}$
A_1'	14.32	$\dfrac{\sigma_5'}{\gamma(K_p - K_a)} = \dfrac{282.76}{(15.9)(3.25 - 0.307)} = 6.04$
A_2'	14.33	$\dfrac{8P}{\gamma(K_p - K_a)} = \dfrac{(8)(67.38)}{(15.9)(3.25 - 0.307)} = 11.52$
A_3'	14.34	$\dfrac{6P[2\bar{z}\gamma(K_p - K_a) + \sigma_5']}{\gamma^2(K_p - K_a)^2}$ $= \dfrac{(6)(67.38)[(2)(2.188)(15.9)(3.25 - 0.307) + 282.76]}{(15.9)^2(3.25 - 0.307)^2} = 90.01$
A_4'	14.35	$\dfrac{P(6\bar{z}\sigma_5' + 4P)}{\gamma^2(K_p - K_a)^2} = \dfrac{(67.38)[(6)(2.188)(282.76) + (4)(67.38)]}{(15.9)^2(3.25 - 0.307)^2} = 122.52$
L_4	14.31	$L_4^4 + A_1' L_4^3 - A_2' L_4^2 - A_3' L_4 - A_4' = 0$ $L_4^4 + 6.04 L_4^3 - 11.52 L_4^2 - 90.01 L_4 - 122.52 = 0; \; L_4 \approx 4.1 \text{ m}$

$$D_{\text{theory}} = L_3 + L_4 = 0.521 + 4.1 \approx \mathbf{4.7 \text{ m}}$$

Part b

Total length, $L + 1.3(D_{\text{theory}}) = 5 + 1.3(4.7) = \mathbf{11.11 \text{ m}}$ ∎

Example 14.3

Refer to Figure 14.12. For $L = 5$ m, $\gamma = 17.3 \text{ kN/m}^3$, $\phi' = 30°$, and $P = 30 \text{ kN/m}$, determine:

a. The theoretical depth of penetration, D
b. The maximum moment, $M_{\max}$ kN-m/m

Solution

$$K_p = \tan^2\left(45 + \frac{\phi'}{2}\right) = \tan^2\left(45 + \frac{30}{2}\right) = 3$$

$$K_a = \tan^2\left(45 - \frac{\phi'}{2}\right) = \tan^2\left(45 - \frac{30}{2}\right) = \frac{1}{3}$$

$$K_p - K_a = 3 - 0.333 = 2.667$$

Part a
From Eq. (14.36),

$$D^4 - \left[\frac{8P}{\gamma(K_p - K_a)}\right]D^2 - \left[\frac{12PL}{\gamma(K_p - K_a)}\right]D - \left[\frac{2P}{\gamma(K_p - K_a)}\right]^2 = 0$$

and

$$\frac{8P}{\gamma(K_p - K_a)} = \frac{(8)(30)}{(17.3)(2.667)} = 5.2$$

$$\frac{12PL}{\gamma(K_p - K_a)} = \frac{(12)(30)(5)}{(17.3)(2.667)} = 39.0$$

$$\frac{2P}{\gamma(K_p - K_a)} = \frac{(2)(30)}{(17.3)(2.667)} = 1.3$$

so

$$D^4 - 5.2\,D^2 - 39D - (1.3)^2 = 0$$

From the preceding equation, $D \approx 4$ m

Part b
From Eq. (14.39),

$$z' = \sqrt{\frac{2P}{\gamma(K_p - K_a)}} = \sqrt{\frac{(2)(30)}{(17.3(2.667)}} = 1.14 \text{ m}$$

From Eq. (14.38),

$$M_{max} = P(L + z') - \frac{\gamma z'^3(K_p - K_a)}{6}$$

$$= (30)(5 + 1.14) - \frac{(17.3)(1.14)^3(2.667)}{6}$$

$$= 184.2 - 11.39 \approx \mathbf{173\,kN\text{-}m/m}$$

■

14.6 Cantilever Sheet Piling Penetrating Clay

At times, cantilever sheet piles must be driven into a clay layer possessing an undrained cohesion $c(\phi = 0)$. The net pressure diagram will be somewhat different from that shown in Figure 14.9a. Figure 14.13 shows a cantilever sheet-pile wall driven into clay with a backfill of granular soil above the level of the dredge line. The water table is at a depth L_1 below the top of the wall. As before, Eqs. (14.1) and (14.2) give the intensity of the net pressures σ_1' and σ_2', and the diagram for pressure distribution above the level of the dredge line can be drawn. The diagram for net pressure distribution below the dredge line can now be determined as follows.

At any depth greater than $L_1 + L_2$, for $\phi = 0$, the Rankine active earth-pressure coefficient $K_a = 1$. Similarly, for $\phi = 0$, the Rankine passive earth-pressure coefficient

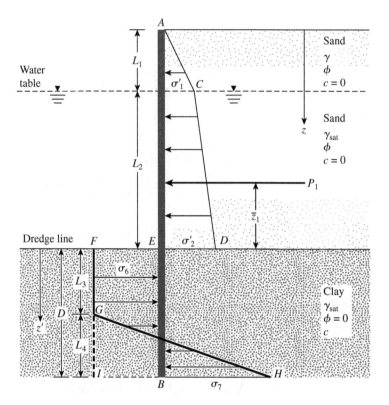

Figure 14.13 Cantilever sheet pile penetrating clay

$K_p = 1$. Thus, above the point of rotation (point O in Figure 14.8a), the active pressure, from right to left is

$$\sigma_a = [\gamma L_1 + \gamma' L_2 + \gamma_{sat}(z - L_1 - L_2)] - 2c \qquad (14.40)$$

Similarly, the passive pressure from left to right may be expressed as

$$\sigma_p = \gamma_{sat}(z - L_1 - L_2) + 2c \qquad (14.41)$$

Thus, the net pressure is

$$\sigma_6 = \sigma_p - \sigma_a = [\gamma_{sat}(z - L_1 - L_2) + 2c]$$
$$- [\gamma L_1 + \gamma' L_2 + \gamma_{sat}(z - L_1 - L_2)] + 2c$$
$$= 4c - (\gamma L_1 + \gamma' L_2) \qquad (14.42)$$

At the bottom of the sheet pile, the passive pressure from right to left is

$$\sigma_p = (\gamma L_1 + \gamma' L_2 + \gamma_{sat}D) + 2c \qquad (14.43)$$

Similarly, the active pressure from left to right is

$$\sigma_a = \gamma_{sat}D - 2c \qquad (14.44)$$

Hence, the net pressure is

$$\sigma_7 = \sigma_p - \sigma_a = 4c + (\gamma L_1 + \gamma' L_2) \tag{14.45}$$

For equilibrium analysis, $\Sigma F_H = 0$; that is, the area of the pressure diagram *ACDE* minus the area of *EFIB* plus the area of *GIH* = 0, or

$$P_1 - [4c - (\gamma L_1 + \gamma' L_2)]D + \tfrac{1}{2}L_4[4c - (\gamma L_1 + \gamma' L_2) + 4c + (\gamma L_1 + \gamma' L_2)] = 0$$

where P_1 = area of the pressure diagram *ACDE*.

Simplifying the preceding equation produces

$$L_4 = \frac{D[4c - (\gamma L_1 + \gamma' L_2)] - P_1}{4c} \tag{14.46}$$

Now, taking the moment about point *B* ($\Sigma M_B = 0$) yields

$$P_1(D + \bar{z}_1) - [4c - (\gamma L_1 + \gamma' L_2)]\frac{D^2}{2} + \frac{1}{2}L_4(8c)\left(\frac{L_4}{3}\right) = 0 \tag{14.47}$$

where $\bar{z}_1$ = distance of the center of pressure of the pressure diagram *ACDE*, measured from the level of the dredge line.

Combining Eqs. (14.46) and (14.47) yields

$$D^2[4c - (\gamma L_1 + \gamma' L_2)] - 2DP_1 - \frac{P_1(P_1 + 12c\bar{z}_1)}{(\gamma L_1 + \gamma' L_2) + 2c} = 0 \tag{14.48}$$

Equation (14.48) may be solved to obtain *D*, the theoretical depth of penetration of the clay layer by the sheet pile.

Step-by-Step Procedure for Obtaining the Pressure Diagram

Step 1. Calculate $K_a = \tan^2(45 - \phi'/2)$ for the granular soil (backfill).
Step 2. Obtain σ'_1 and σ'_2. [See Eqs. (14.1) and (14.2).]
Step 3. Calculate P_1 and $\bar{z}_1$.
Step 4. Use Eq. (14.48) to obtain the theoretical value of *D*.
Step 5. Using Eq. (14.46), calculate L_4.
Step 6. Calculate σ_6 and σ_7. [See Eqs. (14.42) and (14.45).]
Step 7. Draw the pressure distribution diagram as shown in Figure 14.13.
Step 8. The actual depth of penetration is

$$D_{actual} = 1.4 \text{ to } 1.6(D_{theoretical})$$

Maximum Bending Moment

According to Figure 14.13, the maximum moment (zero shear) will be between $L_1 + L_2 < z < L_1 + L_2 + L_3$. Using a new coordinate system z' (with $z' = 0$ at the dredge line) for zero shear gives

$$P_1 - \sigma_6 z' = 0$$

or

$$z' = \frac{P_1}{\sigma_6} \qquad (14.49)$$

The magnitude of the maximum moment may now be obtained:

$$M_{max} = P_1(z' + \bar{z}_1) - \frac{\sigma_6 z'^2}{2} \qquad (14.50)$$

Knowing the maximum bending moment, we determine the section modulus of the sheet-pile section from Eq. (14.23).

Example 14.4

In Figure 14.14, for the sheet-pile wall, determine

a. The theoretical and actual depth of penetration. Use $D_{actual} = 1.5D_{theory}$.
b. The minimum size of sheet-pile section necessary. Use $\sigma_{all} = 172.5 \text{ MN/m}^2$.

Solution

We will follow the step-by-step procedure given in Section 14.6:

Step 1.

$$K_a = \tan^2\left(45 - \frac{\phi'}{2}\right) = \tan^2\left(45 - \frac{32}{2}\right) = 0.307$$

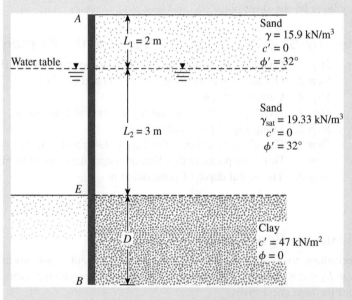

Figure 14.14 Cantilever sheet pile penetrating into saturated clay

Step 2.

$$\sigma'_1 = \gamma L_1 K_a = (15.9)(2)(0.307) = 9.763 \text{ kN/m}^2$$

$$\sigma'_2 = (\gamma L_1 + \gamma' L_2)K_a = [(15.9)(2) + (19.33 - 9.81)3]0.307$$

$$= 18.53 \text{ kN/m}^2$$

Step 3. From the net pressure distribution diagram given in Figure 14.13, we have

$$P_1 = \frac{1}{2}\sigma'_1 L_1 + \sigma'_1 L_2 + \frac{1}{2}(\sigma'_2 - \sigma'_1)L_2$$

$$= 9.763 + 29.289 + 13.151 = 52.2 \text{ kN/m}$$

and

$$\bar{z}_1 = \frac{1}{52.2}\left[9.763\left(3 + \frac{2}{3}\right) + 29.289\left(\frac{3}{2}\right) + 13.151\left(\frac{3}{3}\right)\right]$$

$$= 1.78 \text{ m}$$

Step 4. From Eq. (14.48),

$$D^2[4c - (\gamma L_1 + \gamma' L_2)] - 2DP_1 - \frac{P_1(P_1 + 12c\bar{z}_1)}{(\gamma L_1 + \gamma' L_2) + 2c} = 0$$

Substituting proper values yields

$$D^2\{(4)(47) - [(2)(15.9) + (19.33 - 9.81)3]\} - 2D(52.2)$$

$$- \frac{52.2[52.2 + (12)(47)(1.78)]}{[(15.9)(2) + (19.33 - 9.81)3] + (2)(47)} = 0$$

or

$$127.64D^2 - 104.4D - 357.15 = 0$$

Solving the preceding equation, we obtain $D = 2.13$ m.

Step 5. From Eq. (14.46),

$$L_4 = \frac{D[4c - (\gamma L_1 + \gamma' L_2)] - P_1}{4c}$$

and

$$4c - (\gamma L_1 + \gamma' L_2) = (4)(47) - [(15.9)(2) + (19.33 - 9.81)3]$$

$$= 127.64 \text{ kN/m}^2$$

So,

$$L_4 = \frac{2.13(127.64) - 52.2}{(4)(47)} = 1.17 \text{ m}$$

Step 6.

$$\sigma_6 = 4c - (\gamma L_1 + \gamma' L_2) = 127.64 \text{ kN/m}^2$$

$$\sigma_7 = 4c + (\gamma L_1 + \gamma' L_2) = 248.36 \text{ kN/m}^2$$

Step 7. The net pressure distribution diagram can now be drawn, as shown in Figure 14.13.

Step 8. $D_{actual} \approx 1.5\, D_{theoretical} = 1.5(2.13) \approx \mathbf{3.2\ m}$

Maximum-Moment Calculation

From Eq. (14.49),

$$z' = \frac{P_1}{\sigma_6} = \frac{52.2}{127.64} \approx 0.41\ m$$

Again, from Eq. (14.49),

$$M_{max} = P_1(z' + \bar{z}_1) - \frac{\sigma_6 z'^2}{2}$$

So

$$M_{max} = 52.2(0.41 + 1.78) - \frac{127.64(0.41)^2}{2}$$

$$= 114.32 - 10.73 = 103.59\ kN\text{-}m/m$$

The minimum required section modulus (assuming that $\sigma_{all} = 172.5\ MN/m^2$) is

$$S = \frac{103.59\ kN\text{-}m/m}{172.5 \times 10^3\ kN/m^2} = \mathbf{0.6 \times 10^{-3}\ m^3/m\ of\ the\ wall} \qquad \blacksquare$$

14.7 Special Cases for Cantilever Walls Penetrating Clay

Sheet-Pile Wall in the Absence of Water Table

As in Section 14.5, relationships for special cases for cantilever walls penetrating clay may also be derived. Referring to Figure 14.15, we can write

$$\sigma_2' = \gamma L K_a \qquad (14.51)$$

$$\sigma_6 = 4c - \gamma L \qquad (14.52)$$

$$\sigma_7 = 4c + \gamma L \qquad (14.53)$$

$$P_1 = \tfrac{1}{2}L\sigma_2' = \tfrac{1}{2}\gamma L^2 K_a \qquad (14.54)$$

and

$$L_4 = \frac{D(4c - \gamma L) - \tfrac{1}{2}\gamma L^2 K_a}{4c} \qquad (14.55)$$

The theoretical depth of penetration, D, can be calculated [in a manner similar to the calculation of Eq. (14.48)] as

$$D^2(4c - \gamma L) - 2DP_1 - \frac{P_1(P_1 + 12c\bar{z}_1)}{\gamma L + 2c} = 0 \qquad (14.56)$$

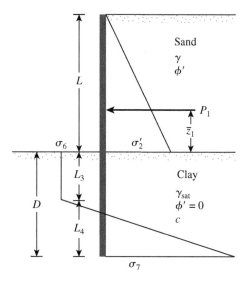

Figure 14.15 Sheet-pile wall penetrating clay

where $\bar{z}_1 = \dfrac{L}{3}$.   (14.57)

The magnitude of the maximum moment in the wall is

$$M_{max} = P_1(z' + \bar{z}_1) - \frac{\sigma_6 z'^2}{2}$$ (14.58)

where $z' = \dfrac{P_1}{\sigma_6} = \dfrac{\frac{1}{2}\gamma L^2 K_a}{4c - \gamma L}$.   (14.59)

Free Cantilever Sheet-Pile Wall Penetrating Clay

Figure 14.16 shows a free cantilever sheet-pile wall penetrating a clay layer. The wall is being subjected to a line load of P per unit length. For this case,

$$\sigma_6 = \sigma_7 = 4c$$ (14.60)

The depth of penetration, D, may be obtained from the relation

$$4D^2 c - 2PD - \frac{P(P + 12cL)}{2c} = 0$$ (14.61)

Also, note that, for a construction of the pressure diagram,

$$L_4 = \frac{4cD - P}{4c}$$ (14.62)

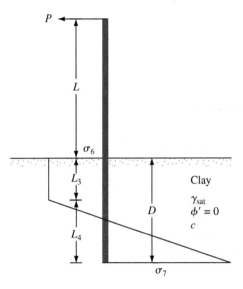

Figure 14.16 Free cantilever sheet piling penetrating clay

The maximum moment in the wall is

$$M_{max} = P(L + z') - \frac{4cz'^2}{2} \tag{14.63}$$

where

$$z' = \frac{P}{4c} \tag{14.64}$$

Example 14.5

Refer to the free cantilever sheet-pile wall shown in Figure 14.16, for which $P = 32$ kN/m, $L = 3.5$ m, and $c = 12$ kN/m^2. Calculate the theoretical depth of penetration.

Solution
From Eq. (14.61),

$$4D^2c - 2PD - \frac{P(P + 12cL)}{2c} = 0$$

$$(4)(D^2)(12) - (2)(32)(D) - \frac{32[32 + (12)(12)(3.5)]}{(2)(12)} = 0$$

$$48D^2 - 64D - 714.7 = 0$$

Hence $D \approx$ **4.6 m**. ∎

Example 14.6

Refer to Figure 14.15 for the sheet-pile wall penetrating clay, Given:

Sand: $L = 6$ m
$\gamma = 16$ kN/m^3
$\phi' = 30°$

Clay: $\gamma_{sat} = 18.9$ kN/m^3
$\phi' = 0$
$c = 95$ kN/m^2

Determine:

a. The theoretical depth of penetration D
b. The magnitude of the maximum moment in the wall

Solution
Part a
From Eq. (14.54),

$$P_1 = \frac{1}{2}\gamma L^2 K_a$$

$$K_a = \tan^2\left(45 - \frac{\phi'}{2}\right) = \tan^2\left(45 - \frac{30}{2}\right) = \frac{1}{3}$$

$$P_1 = \left(\frac{1}{2}\right)(16)(6)^2\left(\frac{1}{3}\right) = 96\,\text{kN/m}$$

From Eq. (14.56),

$$D^2(4c - \gamma L) - 2DP_1 - \frac{P_1(P_1 + 12c\bar{z}_1)}{\gamma L + 2c} = 0$$

From Eq. (14.57),

$$\bar{z}_1 = \frac{L}{3} = \frac{6}{3} = 2\,\text{m}$$

Hence,

$$D^2[(4)(95) - (16)(6)] - (2)(D)(96) - \frac{96[96 + (12)(95)(2)]}{(16)(6) + (2)(95)} = 0$$

or

$$284\,D^2 - 192\,D - 797.5 = 0$$

$$D \approx \mathbf{2.1\,m}$$

Part b
From Eq. (14.58),

$$M_{max} = P_1(z' + \bar{z}_1) - \frac{\sigma_6 z'^2}{2}$$

From Eq. (14.59),

$$z' = \frac{\frac{1}{2}\gamma L^2 K_a}{4c - \gamma L} = \frac{(0.5)(16)(6)^2\left(\frac{1}{3}\right)}{(4 \times 95) - (16 \times 6)} = 0.338\,\text{m}$$

From Eq. (14.52),

$$\sigma_6 = 4c - \gamma L = (4 \times 95) - (16 \times 6) = 284\,\text{kN/m}^2$$

$$M_{\text{max}} = (96)(0.338 + 2) - \frac{(284)(0.338)^2}{2} = 224.45 - 16.22 = \mathbf{208.23\,kN\text{-}m/m}\quad\blacksquare$$

14.8 Anchored Sheet-Pile Walls

When the height of the backfill material behind a cantilever sheet-pile wall exceeds about 6 m ($\approx$20 ft), tying the wall near the top to anchor plates, anchor walls, or anchor piles becomes more economical. This type of construction is referred to as *anchored sheet-pile wall* or an *anchored bulkhead.* Anchors minimize the depth of penetration required by the sheet piles and also reduce the cross-sectional area and weight of the sheet piles needed for construction. However, the tie rods and anchors must be carefully designed.

The two basic methods of designing anchored sheet-pile walls are (a) the *free earth support* method and (b) the *fixed earth support* method. Figure 14.17 shows the assumed nature of deflection of the sheet piles for the two methods.

The free earth support method involves a minimum penetration depth. Below the dredge line, no pivot point exists for the static system. The nature of the variation of the bending moment with depth for both methods is also shown in Figure 14.17. Note that

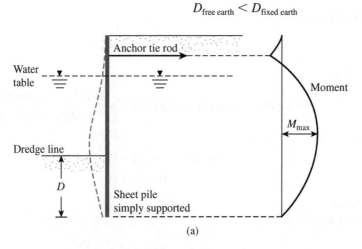

$$D_{\text{free earth}} < D_{\text{fixed earth}}$$

(a)

Figure 14.17 Nature of variation of deflection and moment for anchored sheet piles: (a) free earth support method

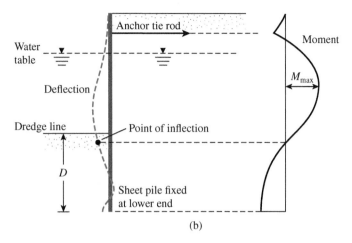

(b)

Figure 14.17 (*Continued*) (b) fixed earth support method

14.9 Free Earth Support Method for Penetration of Sandy Soil

Figure 14.18 shows an anchor sheet-pile wall with a granular soil backfill; the wall has been driven into a granular soil. The tie rod connecting the sheet pile and the anchor is located at a depth l_1 below the top of the sheet-pile wall.

The diagram of the net pressure distribution above the dredge line is similar to that shown in Figure 14.9. At depth $z = L_1$, $\sigma_1' = \gamma L_1 K_a$, and at $z = L_1 + L_2$, $\sigma_2' = (\gamma L_1 + \gamma' L_2)K_a$. Below the dredge line, the net pressure will be zero at $z = L_1 + L_2 + L_3$. The relation for L_3 is given by Eq. (14.6), or

$$L_3 = \frac{\sigma_2'}{\gamma'(K_p - K_a)}$$

At $z = L_1 + L_2 + L_3 + L_4$, the net pressure is given by

$$\sigma_8' = \gamma'(K_p - K_a)L_4 \tag{14.65}$$

Note that the slope of the line *DEF* is 1 vertical to $\gamma'(K_p - K_a)$ horizontal.

For equilibrium of the sheet pile, Σ horizontal forces $= 0$, and Σ moment about $O' = 0$. (*Note:* Point O' is located at the level of the tie rod.)

Summing the forces in the horizontal direction (per unit length of the wall) gives

Area of the pressure diagram *ACDE* − area of *EBF* − $F = 0$

where F = tension in the tie rod/unit length of the wall, or

$$P - \tfrac{1}{2}\sigma_8' L_4 - F = 0$$

or

$$F = P - \tfrac{1}{2}[\gamma'(K_p - K_a)]L_4^2 \tag{14.66}$$

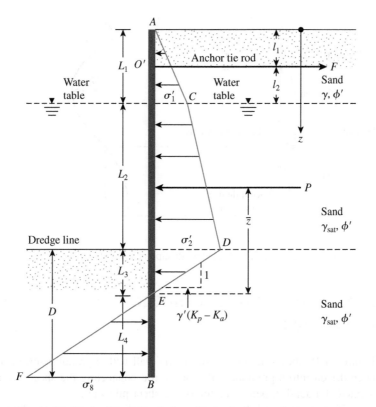

Figure 14.18 Anchored sheet-pile wall penetrating sand

where P = area of the pressure diagram $ACDE$. Now, taking the moment about point O' gives

$$-P[(L_1 + L_2 + L_3) - (\bar{z} + l_1)] + \tfrac{1}{2}[\gamma'(K_p - K_a)]L_4^2(l_2 + L_2 + L_3 + \tfrac{2}{3}L_4) = 0$$

or

$$L_4^3 + 1.5L_4^2(l_2 + L_2 + L_3) - \frac{3P[(L_1 + L_2 + L_3) - (\bar{z} + l_1)]}{\gamma'(K_p - K_a)} = 0 \qquad (14.67)$$

Equation (14.67) may be solved by trial and error to determine the theoretical depth, L_4:

$$D_{\text{theoretical}} = L_3 + L_4$$

The theoretical depth is increased by about 30 to 40% for actual construction, or

$$D_{\text{actual}} = 1.3 \text{ to } 1.4\, D_{\text{theoretical}} \qquad (14.68)$$

The step-by-step procedure in Section 14.4 indicated that a factor of safety can be applied to K_p at the beginning [i.e., $K_{p(\text{design})} = K_p/\text{FS}$]. If that is done, there is no need to increase the theoretical depth by 30 to 40%. This approach is often more conservative.

The maximum theoretical moment to which the sheet pile will be subjected occurs at a depth between $z = L_1$ and $z = L_1 + L_2$. The depth z for zero shear and hence maximum moment may be evaluated from

$$\frac{1}{2}\sigma_1'L_1 - F + \sigma_1'(z - L_1) + \frac{1}{2}K_a\gamma'(z - L_1)^2 = 0 \qquad (14.69)$$

Once the value of z is determined, the magnitude of the maximum moment is easily obtained.

Example 14.7

Let $L_1 = 3.05$ m, $L_2 = 6.1$ m, $l_1 = 1.53$ m, $l_2 = 1.52$ m, $c' = 0$, $\phi' = 30°$, $\gamma = 16$ kN/m³, $\gamma_{sat} = 19.5$ kN/m³, and $E = 207 \times 10^3$ MN/m² in Figure 14.18.

    a. Determine the theoretical and actual depths of penetration. (*Note:* $D_{actual} = 1.3D_{theory}$.)

    b. Find the anchor force per unit length of the wall.

    c. Determine the maximum moment, M_{max}.

Solution

Part a

We use the following table.

Quantity required	Eq. no.	Equation and calculation
K_a	—	$\tan^2\left(45 - \dfrac{\phi'}{2}\right) = \tan^2\left(45 - \dfrac{30}{2}\right) = \dfrac{1}{3}$
K_P	—	$\tan^2\left(45 + \dfrac{\phi'}{2}\right) = \tan^2\left(45 + \dfrac{30}{2}\right) = 3$
$K_p - K_a$	—	$3 - 0.333 = 2.667$
γ'	—	$\gamma_{sat} - \gamma_w = 19.5 - 9.81 = 9.69$ kN/m³
σ_1'	14.1	$\gamma L_1 K_a = (16)(3.05)(\frac{1}{3}) = 16.27$ kN/m²
σ_2'	14.2	$(\gamma L_1 + \gamma' L_2)K_a = [(16)(3.05) + (9.69)(6.1)]\frac{1}{3} = 35.97$ kN/m²
L_3	14.6	$\dfrac{\sigma_2'}{\gamma'(K_p - K_a)} = \dfrac{35.97}{(9.69)(2.667)} = 1.39$ m
P	—	$\frac{1}{2}\sigma_1'L_1 + \sigma_2'L_2 + \frac{1}{2}(\sigma_2' - \sigma_1')L_2 + \frac{1}{2}\sigma_2'L_3 = (\frac{1}{2})(16.27)(3.05)$ $+ (16.27)(6.1) + (\frac{1}{2})(35.97 - 16.27)(6.1) + (\frac{1}{2})(35.97)(1.39)$ $= 24.81 + 99.25 + 60.01 + 25.0 = 209.07$ kN/m
$\bar{z}$	—	$\dfrac{\Sigma M_E}{P} = \left[(24.81)\left(1.39 + 6.1 + \dfrac{3.05}{3}\right) + (99.25)\left(1.39 + \dfrac{6.1}{2}\right) \right.$ $\left. + (60.01)\left(1.39 + \dfrac{6.1}{3}\right) + (25.0)\left(\dfrac{2 \times 1.39}{3}\right)\right]\dfrac{1}{209.07}$ $= 4.21$ m

(Continued)

Quantity required	Eq. no.	Equation and calculation
L_4	14.67	$L_4^3 + 1.5L_4^2(l_2 + L_2 + L_3) - \dfrac{3P[(L_1 + L_2 + L_3) - (\bar{z} + l_1)]}{\gamma'(K_p - K_a)} = 0$
		$L_4^3 + 1.5L_4^2(1.52 + 6.1 + 1.39)$
		$\quad - \dfrac{(3)(209.07)[(3.05 + 6.1 + 1.39) - (4.21 + 1.53)]}{(9.69)(2.667)} = 0$
		$L_4 = 2.7$ m
D_{theory}	—	$L_3 + L_4 = 1.39 + 2.7 = 4.09 \approx \mathbf{4.1 \text{ m}}$
D_{actual}	—	$1.3D_{\text{theory}} = (1.3)(4.1) = \mathbf{5.33 \text{ m}}$

Part b
The anchor force per unit length of the wall is

$$F = P - \tfrac{1}{2}\gamma'(K_p - K_a)L_4^2$$
$$= 209.07 - \left(\tfrac{1}{2}\right)(9.69)(2.667)(2.7)^2 = 114.87 \text{ kN/m} \approx \mathbf{115 \text{ kN/m}}$$

Part c
From Eq. (14.69), for zero shear,

$$\tfrac{1}{2}\sigma_1'L_1 - F + \sigma_1'(z - L_1) + \tfrac{1}{2}K_a\gamma'(z - L_1)^2 = 0$$

Let $z - L_1 = x$, so that

$$\tfrac{1}{2}\sigma_1'L_1 - F + \sigma_1'x + \tfrac{1}{2}K_a\gamma'x^2 = 0$$

or

$$\left(\tfrac{1}{2}\right)(16.27)(3.05) - 115 + (16.27)(x) + \left(\tfrac{1}{2}\right)\left(\tfrac{1}{3}\right)(9.69)x^2 = 0$$

giving $x^2 + 10.07x - 55.84 = 0$

Now, $x = 4$ m and $z = x + L_1 = 4 + 3.05 = 7.05$ m. Taking the moment about the point of zero shear, we obtain

$$M_{\max} = -\frac{1}{2}\sigma_1'L_1\left(x + \frac{3.05}{3}\right) + F(x + 1.52) - \sigma_1'\frac{x^2}{2} - \frac{1}{2}K_a\gamma'x^2\left(\frac{x}{3}\right)$$

or

$$M_{\max} = -\left(\frac{1}{2}\right)(16.27)(3.05)\left(4 + \frac{3.05}{3}\right) + (115)(4 + 1.52) - (16.27)\left(\frac{4^2}{2}\right)$$

$$\quad -\left(\frac{1}{2}\right)\left(\frac{1}{3}\right)(9.69)(4)^2\left(\frac{4}{3}\right) = \mathbf{344.9 \text{ kN-m/m}}$$

14.10 Design Charts for Free Earth Support Method (Penetration into Sandy Soil)

Using the free earth support method, Hagerty and Nofal (1992) provided simplified design charts for quick estimation of the depth of penetration, D, anchor force, F, and maximum moment, M_{max}, for anchored sheet-pile walls penetrating into sandy soil, as shown in Figure 14.18. They made the following assumptions for their analysis.

a. The soil friction angle, ϕ', above and below the dredge line is the same.
b. The angle of friction between the sheet-pile wall and the soil is $\phi'/2$.
c. The passive earth pressure below the dredge line has a logarithmic spiral failure surface.
d. For active earth-pressure calculation, Coulomb's theory is valid.
   The magnitudes of D, F, and M_{max} may be calculated from the following relationships:

$$\frac{D}{L_1 + L_2} = (GD)(CDL_1) \tag{14.70}$$

$$\frac{F}{\gamma_a(L_1 + L_2)^2} = (GF)(CFL_1) \tag{14.71}$$

$$\frac{M_{max}}{\gamma_a(L_1 + L_2)^3} = (GM)(CML_1) \tag{14.72}$$

where

γ_a = average unit weight of soil

$$= \frac{\gamma L_1^2 + (\gamma_{sat} - \gamma_w)L_2^2 + 2\gamma L_1 L_2}{(L_1 + L_2)^2} \tag{14.73}$$

GD = generalized nondimensional embedment

$$= \frac{D}{L_1 + L_2} \qquad (\text{for } L_1 = 0 \text{ and } L_2 = L_1 + L_2)$$

GF = generalized nondimensional anchor force

$$= \frac{F}{\gamma_a(L_1 + L_2)^2} \qquad (\text{for } L_1 = 0 \text{ and } L_2 = L_1 + L_2)$$

GM = generalized nondimensional moment

$$= \frac{M_{\max}}{\gamma_a(L_1 + L_2)^3} \qquad \text{(for } L_1 = 0 \text{ and } L_2 = L_1 + L_2)$$

CDL_1, CFL_1, CML_1 = correction factors for $L_1 \neq 0$

The variations of GD, GF, GM, CDL_1, CFL_1, and CML_1 are shown in Figures 14.19, 14.20, 14.21, 14.22, 14.23, and 14.24, respectively.

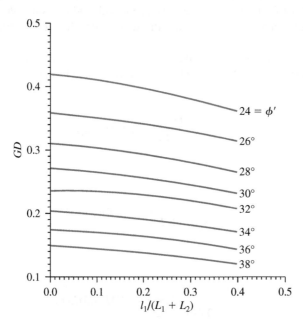

Figure 14.19 Variation of GD with $l_1/(L_1 + L_2)$ and ϕ' [Based on Hagerty, D. J., and Nofal, M. M. (1992). "Design Aids: Anchored Bulkheads in Sand," *Canadian Geotechnical Journal*, Vol. 29, No. 5, pp. 789–795.]

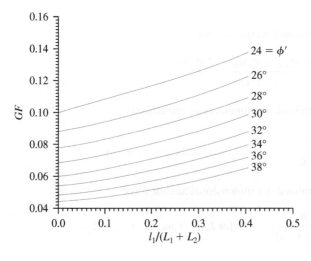

Figure 14.20 Variation of GF with $l_1/(L_1 + L_2)$ and ϕ' [Based on Hagerty, D. J., and Nofal, M. M. (1992). "Design Aids: Anchored Bulkheads in Sand," *Canadian Geotechnical Journal*, Vol. 29, No. 5, pp. 789–795.]

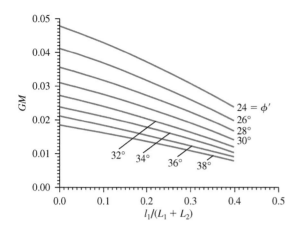

Figure 14.21 Variation of GM with $l_1/(L_1 + L_2)$ and ϕ' [Based on Hagerty, D. J. and Nofal, M. M. (1992), "Design Aids: Anchored Bulkheads in Sand," *Canadian Geotechnical Journal*, Vol. 29, No. 5, pp. 789–795.]

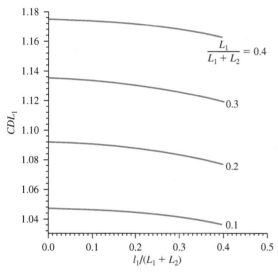

Figure 14.22 Variation of CDL_1 with $L_1/(L_1 + L_2)$ and $l_1/(L_1 + L_2)$ [Based on Hagerty, D. J., and Nofal, M. M. (1992). "Design Aids: Anchored Bulkheads in Sand," *Canadian Geotechnical Journal*, Vol. 29, No. 5, pp. 789–795.]

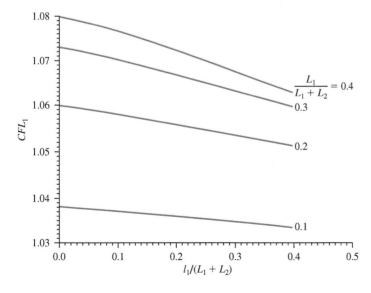

Figure 14.23 Variation of CFL_1 with $L_1/(L_1 + L_2)$ and $l_1/(L_1 + L_2)$ [Based on Hagerty, D. J., and Nofal, M. M. (1992). "Design Aids: Anchored Bulkheads in Sand," *Canadian Geotechnical Journal*, Vol. 29, No. 5, pp. 789–795.]

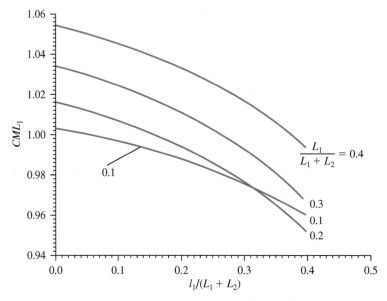

Figure 14.24 Variation of CML_1 with $L_1/(L_1 + L_2)$ and $l_1/(L_1 + L_2)$ [Based on Hagerty, D. J., and Nofal, M. M. (1992). "Design Aids: Anchored Bulkheads in Sand," *Canadian Geotechnical Journal*, Vol. 29, No. 5, pp. 789–795.]

Example 14.8

Refer to Figure 14.18. Given: $L_1 = 2$ m, $L_2 = 3$ m, $l_1 = l_2 = 1$ m, $c = 0$, $\phi' = 32°$ $\gamma = 15.9$ kN/m³, and $\gamma_{sat} = 19.33$ kN/m³. Determine:

a. Theoretical and actual depth of penetration *Note:* $D_{actual} = 1.4 D_{theory}$.
b. Anchor force per unit length of wall
c. Maximum moment, M_{max}

Use the charts presented in Section 14.10.

Solution
Part a
From Eq. (14.70),

$$\frac{D}{L_1 + L_2} = (GD)(CDL_1)$$

$$\frac{l_1}{L_1 + L_2} = \frac{1}{2 + 3} = 0.2$$

From Figure 14.19 for $l_1/(L_1 + L_2) = 0.2$ and $\phi' = 32°$, $GD = 0.22$. From Figure 14.22, for

$$\frac{L_1}{L_1 + L_2} = \frac{2}{2 + 3} = 0.4 \qquad \text{and} \qquad \frac{l_1}{L_1 + L_2} = 0.2$$

$CDL_1 \approx 1.172$. So

$$D_{\text{theory}} = (L_1 + L_2)(GD)(CDL_1) = (5)(0.22)(1.172) \approx 1.3$$

$$D_{\text{actual}} \approx (1.4)(1.3) = 1.82 \approx \mathbf{2\ m}$$

Part b

From Figure 14.20 for $l_1/(L_1 + L_2) = 0.2$ and $\phi' = 32°$, $GF \approx 0.074$. Also, from Figure 14.23, for

$$\frac{L_1}{L_1 + L_2} = \frac{2}{2+3} = 0.4, \quad \frac{l_1}{L_1 + L_2} = 0.2, \quad \text{and} \quad \phi' = 32°$$

$CFL_1 = 1.073$. From Eq. (14.73),

$$\gamma_a = \frac{\gamma L_1^2 + \gamma' L_2^2 + 2\gamma L_1 L_2}{(L_1 + L_2)^2}$$

$$= \frac{(15.9)(2)^2 + (19.33 - 9.81)(3)^2 + (2)(15.9)(2)(3)}{(2+3)^2} = 13.6\ \text{kN/m}^3$$

Using Eq. (14.71) yields

$$F = \gamma_a(L_1 + L_2)^2(GF)(CFL_1) = (13.6)(5)^2(0.074)(1.073) \approx \mathbf{27\ kN/m}$$

Part c

From Figure 14.21, for $l_1/(L_1 + L_2) = 0.2$ and $\phi' = 32°$, $GM = 0.021$. Also, from Figure 14.24, for

$$\frac{L_1}{L_1 + L_2} = \frac{2}{2+3} = 0.4, \quad \frac{l_1}{L_1 + L_2} = 0.2, \quad \text{and} \quad \phi' = 32°$$

$CML_1 = 1.036$. Hence from Eq. (14.72),

$$M_{\text{max}} = \gamma_a(L_1 + L_2)^3(GM)(CML_1) = (13.6)(5)^3(0.021)(1.036) = \mathbf{36.99\ kN \cdot m/m} \quad \blacksquare$$

14.11 Moment Reduction for Anchored Sheet-Pile Walls Penetrating into Sand

Sheet piles are flexible, and hence sheet-pile walls yield (i.e., become displaced laterally), which redistributes the lateral earth pressure. This change tends to reduce the maximum bending moment, M_{max}, as calculated by the procedure outlined in Section 14.9. For that reason, Rowe (1952, 1957) suggested a procedure for reducing the maximum design moment on the sheet-pile walls *obtained from the free earth support method*. This section discusses the procedure of moment reduction for sheet piles *penetrating into sand*.

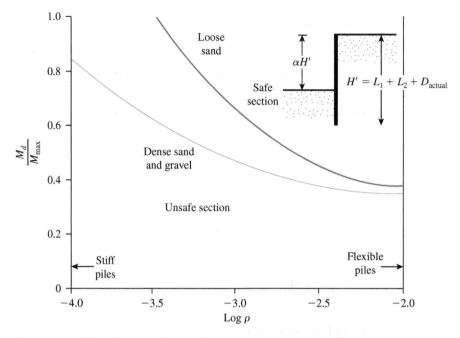

Figure 14.25 Plot of log ρ against M_d/M_{max} for sheet-pile walls penetrating sand (Based on Rowe, P. W. (1952). "Anchored Sheet-Pile Walls," *Proceedings, Institute of Civil Engineers,* Vol. 1, Part 1, pp. 27–70.)

In Figure 14.25, which is valid for the case of a sheet pile penetrating sand, the following notation is used:

1. H' = total height of pile driven (i.e., $L_1 + L_2 + D_{actual}$)

2. Relative flexibility of pile = $\rho = 10.91 \times 10^{-7}\left(\dfrac{H'^4}{EI}\right)$ (14.74a)

where

H' is in meters
E = modulus of elasticity of the pile material (MN/m^2)
I = moment of inertia of the pile section per meter of the wall (m^4/m of wall)

3. M_d = design moment
4. M_{max} = maximum theoretical moment

In English units, Eq. (14.74a) takes the form

$$\rho = \frac{H'^4}{EI}$$ (14.74b)

where H' is in ft, E is in lb/in^2, and I is in in^4/ft of the wall.

The procedure for the use of the moment reduction diagram (see Figure 14.25) is as follows:

Step 1. Choose a sheet-pile section (e.g., from among those given in Table 14.1).

Step 2. Find the modulus S of the selected section (Step 1) per unit length of the wall.

Step 3. Determine the moment of inertia of the section (Step 1) per unit length of the wall.

Step 4. Obtain H' and calculate ρ [see Eq. (14.74a) or Eq. (14.74b)].

Step 5. Find log ρ.

Step 6. Find the moment capacity of the pile section chosen in Step 1 as $M_d = \sigma_{all}S$.

Step 7. Determine M_d/M_{max}. Note that M_{max} is the maximum theoretical moment determined before.

Step 8. Plot log ρ (Step 5) and M_d/M_{max} in Figure 14.25.

Step 9. Repeat Steps 1 through 8 for several sections. The points that fall above the curve (in loose sand or dense sand, as the case may be) are *safe sections*.

The points that fall below the curve are *unsafe sections*. The cheapest section may now be chosen from those points which fall above the proper curve. Note that the section chosen will have an $M_d < M_{max}$.

Example 14.9

Refer to Example 14.7. Use Rowe's moment reduction diagram (Figure 14.25) to find an appropriate sheet-pile section. For the sheet pile, use $E = 207 \times 10^3$ MN/m^2 and $\sigma_{all} = 172{,}500$ kN/m^2.

Solution

$$H' = L_1 + L_2 + D_{actual} = 3.05 + 6.1 + 5.33 = 14.48 \text{ m}$$

$M_{max} = 344.9$ kN $\cdot$ m/m. Now the following table can be prepared.

Section	I(m^4/m)	H'(m)	$\rho = 10.91 \times 10^{-7}\left(\dfrac{H'^4}{EI}\right)$	log ρ	S(m^3/m)	$M_d = S\sigma_{all}$ (kN $\cdot$ m/m)	$\dfrac{M_d}{M_{max}}$
PZ-22	116.2×10^{-6}	14.48	19.94×10^{-4}	-2.7	98.92×10^{-5}	170.64	0.495
PZ-27	255.9×10^{-6}	14.48	9.05×10^{-4}	-3.04	166.66×10^{-5}	287.49	0.834

Figure 14.26 gives a plot of M_d/M_{max} versus ρ. It can be seen that **PZ-27** will be sufficient.

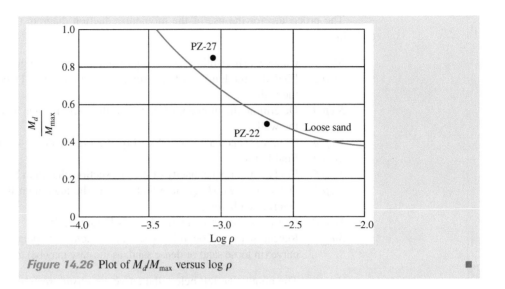

Figure 14.26 Plot of M_d/M_{max} versus log ρ

14.12   Computational Pressure Diagram Method for Penetration into Sandy Soil

The computational pressure diagram (CPD) method for sheet pile penetrating a sandy soil is a simplified method of design and an alternative to the free earth method described in Sections 14.9 and 14.11 (Nataraj and Hoadley, 1984). In this method, the net pressure diagram shown in Figure 14.18 is replaced by rectangular pressure diagrams, as in Figure 14.27. Note that $\overline{\sigma}'_a$ is the width of the net active pressure diagram above the dredge line and $\overline{\sigma}'_p$ is

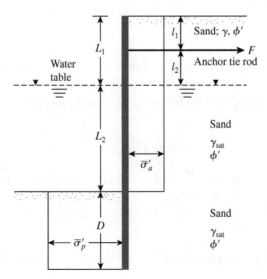

Figure 14.27 Computational pressure diagram method (*Note: $L_1 + L_2 = L$*)

Table 14.2 Range of Values for C and R [from Eqs. (14.75) and (14.76)]

Soil type	C^a	R
Loose sand	0.8–0.85	0.3–0.5
Medium sand	0.7–0.75	0.55–0.65
Dense sand	0.55–0.65	0.60–0.75

[a]Valid for the case in which there is no surcharge above the granular backfill (i.e., on the right side of the wall, as shown in Figure 14.27)

the width of the net passive pressure diagram below the dredge line. The magnitudes of $\overline{\sigma}'_a$ and $\overline{\sigma}'_p$ may respectively be expressed as

$$\overline{\sigma}'_a = CK_a\gamma'_{av}L \tag{14.75}$$

and

$$\overline{\sigma}'_p = RCK_a\gamma'_{av}L = R\overline{\sigma}'_a \tag{14.76}$$

where

γ'_{av} = average effective unit weight of sand

$$\approx \frac{\gamma L_1 + \gamma' L_2}{L_1 + L_2} \tag{14.77}$$

C = coefficient

$$R = \text{coefficient} = \frac{L(L - 2l_1)}{D(2L + D - 2l_1)} \tag{14.78}$$

The range of values for C and R is given in Table 14.2.

The depth of penetration, D, anchor force per unit length of the wall, F, and maximum moment in the wall, M_{max}, are obtained from the following relationships.

Depth of Penetration

For the depth of penetration, we have

$$D^2 + 2DL\left[1 - \left(\frac{l_1}{L}\right)\right] - \left(\frac{L^2}{R}\right)\left[1 - 2\left(\frac{l_1}{L}\right)\right] = 0 \tag{14.79}$$

Anchor Force

The anchor force is

$$F = \overline{\sigma}'_a(L - RD) \tag{14.80}$$

Maximum Moment

The maximum moment is calculated from

$$M_{max} = 0.5\bar{\sigma}'_a L^2 \left[\left(1 - \frac{RD}{L} \right)^2 - \left(\frac{2l_1}{L} \right) \left(1 - \frac{RD}{L} \right) \right] \qquad (14.81)$$

Note the following qualifications:

1. The magnitude of D obtained from Eq. (14.79) is about 1.25 to 1.5 times the value of D_{theory} obtained by the conventional free earth support method (see Section 14.9), so

$$D \approx D_{actual}$$
$$\uparrow \qquad \uparrow$$
Eq. (14.79)　Eq. (14.68)

2. The magnitude of F obtained by using Eq. (14.80) is about 1.2 to 1.6 times the value obtained by using Eq. (14.66). Thus, an additional factor of safety for the actual design of anchors need not be used.
3. The magnitude of M_{max} obtained from Eq. (14.81) is about 0.6 to 0.75 times the value of M_{max} obtained by the conventional free earth support method. Hence, the former value of M_{max} can be used as the actual design value, and Rowe's moment reduction need not be applied.

Example 14.10

For the anchored sheet-pile wall shown in Figure 14.28, determine (a) D, (b) F, and (c) M_{max}. Use the CPD method; assume that $C = 0.68$ and $R = 0.6$.

Solution
Part a

$$\gamma' = \gamma_{sat} - \gamma_w = 19.24 - 9.81 = 9.43 \text{ kN/m}^3$$

From Eq. (14.77)

$$\gamma'_{av} = \frac{\gamma L_1 + \gamma' L_2}{L_1 + L_2} = \frac{(17.3)(3) + (9.43)(6)}{3 + 6} = 12.05 \text{ kN/m}^3$$

$$K_a = \tan^2 \left(45 - \frac{\phi'}{2} \right) = \tan^2 \left(45 - \frac{35}{2} \right) = 0.271$$

$$\bar{\sigma}'_a = C K_a \gamma'_{av} L = (0.68)(0.271)(12.05)(9) = 19.99 \text{ kN/m}^2$$

$$\bar{\sigma}'_p = R\bar{\sigma}'_a = (0.6)(19.99) = 11.99 \text{ kN/m}^2$$

From Eq. (14.80)

$$D^2 + 2DL \left[1 - \left(\frac{l_1}{L} \right) \right] - \frac{L^2}{R} \left[1 - 2 \left(\frac{l_1}{L} \right) \right] = 0$$

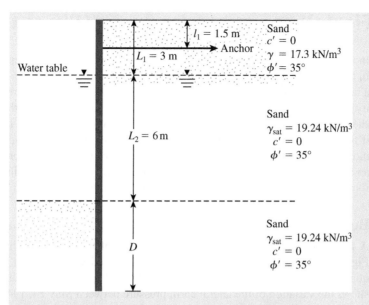

Figure 14.28

or

$$D^2 + 2(D)(9)\left[1 - \left(\frac{1.5}{9}\right)\right] - \frac{(9)^2}{0.6}\left[1 - 2\left(\frac{1.5}{9}\right)\right] = D^2 + 50D - 1000 = 0$$

Hence $D \approx$ **4.6 m.**
Check for the assumption of R:

$$R = \frac{L(L - 2l_1)}{D(2L + D - 2l_1)} = \frac{9[9 - (2)(1.5)]}{4.6[(2)(9) + 4.6 - (2)(1.5)]} \approx \mathbf{0.6 - OK}$$

Part b
From Eq. (14.80)

$$F = \overline{\sigma}'_a(L - RD) = 19.99[9 - (0.6)(4.6)] = \mathbf{124.74\ kN/m}$$

Part c
From Eq. (14.81)

$$M_{max} = 0.5\overline{\sigma}'_a L^2\left[\left(1 - \frac{RD}{L}\right)^2 - \left(\frac{2l_1}{L}\right)\left(1 - \frac{RD}{L}\right)\right]$$

$$1 - \frac{RD}{L} = 1 - \frac{(0.6)(4.6)}{9} = 0.693$$

So,

$$M_{max} = (0.5)(19.99)(9)^2\left[(0.693)^2 - \frac{(2)(1.5)(0.693)}{9}\right] = \mathbf{201.6\ kN\text{-}m/m} \quad \blacksquare$$

14.13 Field Observations for Anchor Sheet-Pile Walls

In the preceding sections, large factors of safety were used for the depth of penetration, D. In most cases, designers use smaller magnitudes of soil friction angle, ϕ', thereby ensuring a built-in factor of safety for the active earth pressure. This procedure is followed primarily because of the uncertainties involved in predicting the actual earth pressure to which a sheet-pile wall in the field will be subjected. In addition, Casagrande (1973) observed that, if the soil behind the sheet-pile wall has grain sizes that are predominantly smaller than those of coarse sand, the active earth pressure after construction sometimes increases to an at-rest earth-pressure condition. Such an increase causes a large increase in the anchor force, F. The following two case histories are given by Casagrande (1973).

Bulkhead of Pier C—Long Beach Harbor, California (1949)

A typical cross section of the Pier C bulkhead of the Long Beach harbor is shown in Figure 14.29. Except for a rockfill dike constructed with 76 mm (3 in.) maximum-size quarry wastes, the backfill of the sheet-pile wall consisted of fine sand. Figure 14.30 shows the variation of the lateral earth pressure between May 24, 1949 (the day construction was completed) and August 6, 1949. On May 24, the lateral earth pressure reached an active state, as shown in Figure 14.30a, due to the wall yielding. Between May 24 and June 3, the anchor resisted further yielding and the lateral earth pressure increased to the at-rest state (Figure 14.30b). However, the flexibility of the sheet piles ultimately resulted in a gradual decrease in the lateral earth-pressure distribution on the sheet piles (see Figure 14.30c).

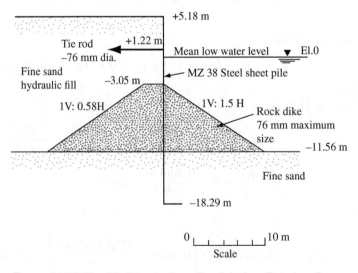

Figure 14.29 Pier C bulkhead—Long Beach harbor (Based on Casagrande, 1973)

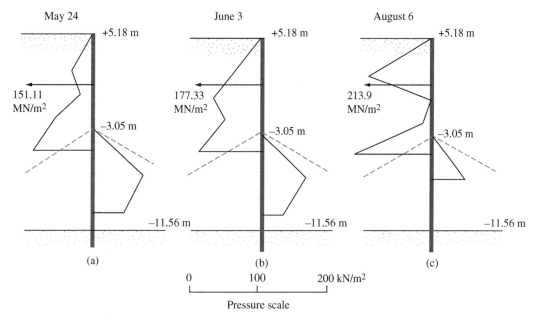

May 24

+5.18 m

151.11
MN/m²

−3.05 m

−11.56 m

(a)

June 3

+5.18 m

177.33
MN/m²

−3.05 m

−11.56 m

(b)

August 6

+5.18 m

213.9
MN/m²

−3.05 m

−11.56 m

(c)

0   100   200 kN/m²

Pressure scale

Figure 14.30 Measured stresses at Station 27 + 30—Pier C bulkhead, Long Beach
(Based on Casagrande, 1973)

With time, the stress on the tie rods for the anchor increased as shown in the following table.

Date	Stress on anchor tie rod (MN/m²)
May 24, 1949	151.11
June 3, 1949	177.33
June 11, 1949	193.2
July 12, 1949	203.55
August 6, 1949	213.9

These observations show that the magnitude of the active earth pressure may vary with time and depend greatly on the flexibility of the sheet piles. Also, the actual variations in the lateral earth-pressure diagram may not be identical to those used for design.

Bulkhead—Toledo, Ohio (1961)

A typical cross section of a Toledo bulkhead completed in 1961 is shown in Figure 14.31. The foundation soil was primarily fine to medium sand, but the dredge line did cut into highly overconsolidated clay. Figure 14.31 also shows the actual measured values of bending moment along the sheet-pile wall. Casagrande (1973) used the Rankine active earth-pressure distribution to calculate the maximum bending moment according to the free earth support method with and without Rowe's moment reduction.

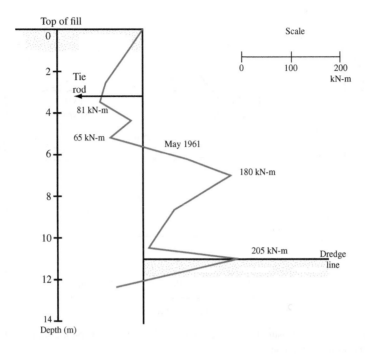

Figure 14.31 Bending moment from strain-gage measurements at test location 3, Toledo bulkhead (Based on Casagrande, 1973)

Design method	Maximum predicted bending moment, M_{max}
Free earth support method	146.5 kN-m
Free earth support method with Rowe's moment reduction	78.6 kN-m

Comparisons of these magnitudes of M_{max} with those actually observed show that the field values are substantially larger. The reason probably is that the backfill was primarily fine sand and the measured active earth-pressure distribution was larger than that predicted theoretically.

14.14   Free Earth Support Method for Penetration of Clay

Figure 14.32 shows an anchored sheet-pile wall penetrating a clay soil and with a granular soil backfill. The diagram of pressure distribution above the dredge line is similar to that shown in Figure 14.13. From Eq. (14.42), the net pressure distribution below the dredge line (from $z = L_1 + L_2$ to $z = L_1 + L_2 + D$) is

$$\sigma_6 = 4c - (\gamma L_1 + \gamma' L_2)$$

For static equilibrium, the sum of the forces in the horizontal direction is

$$P_1 - \sigma_6 D = F \qquad (14.82)$$

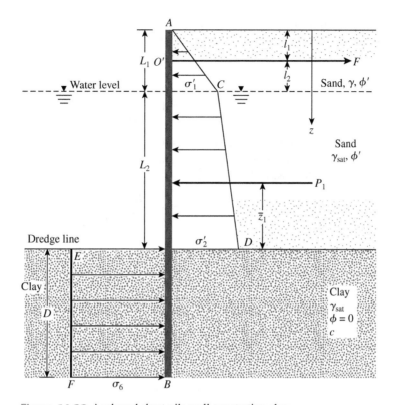

Figure 14.32 Anchored sheet-pile wall penetrating clay

where

P_1 = area of the pressure diagram ACD
F = anchor force per unit length of the sheet-pile wall

Again, taking the moment about O' produces

$$P_1(L_1 + L_2 - l_1 - \bar{z}_1) - \sigma_6 D\left(l_2 + L_2 + \frac{D}{2}\right) = 0$$

Simplification yields

$$\sigma_6 D^2 + 2\sigma_6 D(L_1 + L_2 - l_1) - 2P_1(L_1 + L_2 - l_1 - \bar{z}_1) = 0 \qquad (14.83)$$

Equation (14.83) gives the theoretical depth of penetration, D.

    As in Section 14.9, the maximum moment in this case occurs at a depth $L_1 < z < L_1 + L_2$. The depth of zero shear (and thus the maximum moment) may be determined from Eq. (14.69).

A moment reduction technique similar to that in Section 14.11 for anchored sheet piles penetrating into clay has also been developed by Rowe (1952, 1957). This technique is presented in Figure 14.33, in which the following notation is used:

1. The stability number is

$$S_n = 1.25\frac{c}{(\gamma L_1 + \gamma' L_2)}$$ (14.84)

where c = undrained cohesion ($\phi = 0$).

For the definition of γ, γ', L_1, and L_2, see Figure 14.32.

2. The nondimensional wall height is

$$\alpha = \frac{L_1 + L_2}{L_1 + L_2 + D_{actual}}$$ (14.85)

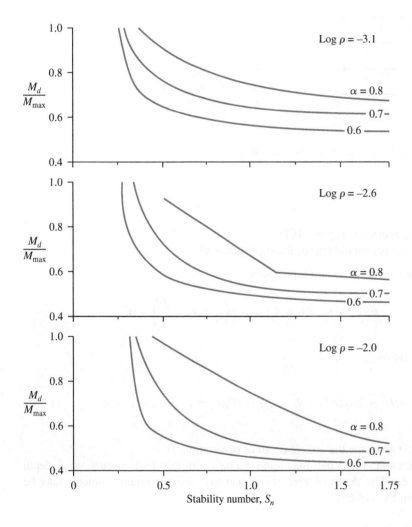

Figure 14.33 Plot of M_d/M_{max} against stability number for sheet-pile wall penetrating clay [Based on Rowe, P. W. (1957). "Sheet-Pile Walls in Clay," *Proceedings, Institute of Civil Engineers,* Vol. 7, pp. 654–692.]

3. The flexibility number is ρ [see Eq. (14.74a) or Eq. (14.74b)]
4. M_d = design moment
   M_{max} = maximum theoretical moment

The procedure for moment reduction, using Figure 14.33, is as follows:

Step 1. Obtain $H' = L_1 + L_2 + D_{actual}$.
Step 2. Determine $\alpha = (L_1 + L_2)/H'$.
Step 3. Determine S_n [from Eq. (14.84)].
Step 4. For the magnitudes of α and S_n obtained in Steps 2 and 3, determine M_d/M_{max} for various values of log ρ from Figure 14.33, and plot M_d/M_{max} against log ρ.
Step 5. Follow Steps 1 through 9 as outlined for the case of moment reduction of sheet-pile walls penetrating granular soil. (See Section 14.11.)

Example 14.11

In Figure 14.32, let $L_1 = 3$ m, $L_2 = 6$ m, and $l_1 = 1.5$ m. Also, let $\gamma = 17$ kN/m^3, $\gamma_{sat} = 20$ kN/m^3, $\phi' = 35°$, and $c = 41$ kN/m^2.

a. Determine the theoretical depth of embedment of the sheet-pile wall.
b. Calculate the anchor force per unit length of the wall.

Solution
Part a
We have

$$K_a = \tan^2\left(45 - \frac{\phi'}{2}\right) = \tan^2\left(45 - \frac{35}{2}\right) = 0.271$$

and

$$K_p = \tan^2\left(45 + \frac{\phi'}{2}\right) = \tan^2\left(45 + \frac{35}{2}\right) = 3.69$$

From the pressure diagram in Figure 14.34,

$$\sigma_1' = \gamma L_1 K_a = (17)(3)(0.271) = 13.82 \text{ kN/m}^2$$

$$\sigma_2' = (\gamma L_1 + \gamma' L_2)K_a = [(17)(3) + (20 - 9.81)(6)](0.271) = 30.39 \text{ kN/m}^2$$

$$P_1 = \text{areas } 1 + 2 + 3 = 1/2(3)(13.82) + (13.82)(6) + 1/2(30.39 - 13.82)(6)$$

$$= 20.73 + 82.92 + 49.71 = 153.36 \text{ kN/m}$$

and

$$\bar{z}_1 = \frac{(20.73)\left(6 + \dfrac{3}{3}\right) + (82.92)\left(\dfrac{6}{2}\right) + (49.71)\left(\dfrac{6}{3}\right)}{153.36} = 3.2 \text{ m}$$

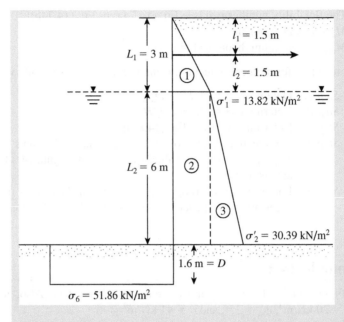

Figure 14.34 Free earth support method, with sheet pile penetrating into clay

From Eq. (14.83),

$$\sigma_6 D^2 + 2\sigma_6 D(L_1 + L_2 - l_1) - 2P_1(L_1 + L_2 - l_1 - \bar{z}_1) = 0$$
$$\sigma_6 = 4c - (\gamma L_1 + \gamma' L_2) = (4)(41) - [(17)(3)$$
$$+ (20 - 9.81)(6)] = 51.86 \text{ kN/m}^2$$

So,

$$(51.86)D^2 + (2)(51.86)(D)(3 + 6 - 1.5)$$
$$- (2)(153.36)(3 + 6 - 1.5 - 3.2) = 0$$

or

$$D^2 + 15D - 25.43 = 0$$

Hence,

$$D \approx \mathbf{1.6 \text{ m}}$$

Part b
From Eq. (14.82),

$$F = P_1 - \sigma_6 D = 153.36 - (51.86)(1.6) = \mathbf{70.38 \text{ kN/m}} \qquad \blacksquare$$

Example 14.12

Refer to Example 14.11.

    a. Increase the actual depth of penetration as $D_{\text{actual}} = 1.75 D_{\text{theory}}$.
    b. Determine M_{max}.

c. Use Rowe's moment reduction diagram (Figure 14.33) to find an appropriate sheet-pile section. For the sheet pile (Table 14.1), use $E = 207 \times 10^3$ MN/m^2 and $\sigma_{all} = 172{,}500$ kN/m^2.

Solution

Part a

$$D_{actual} = 1.75 D_{theory} = (1.75)(1.6) = \mathbf{2.8\,m}$$

Part b

From Eq. (14.82),

$$F = P_1 - \sigma_6 D_{theory} = 153.36 - (51.86)(1.6) = 70.38\,\text{kN/m}$$

From Eq. (14.69), for zero shear, use

$$\frac{1}{2}\sigma_1' L_1 - F + \sigma_1'(z - L_1) + \frac{1}{2}K_a \gamma'(z - L_1)^2 = 0$$

Let $z - L_1 = x$, So,

$$\frac{1}{2}\sigma_1' L_1 - F + \sigma_1' x + \frac{1}{2}K_a \gamma' x^2 = 0$$

or

$$\left(\frac{1}{2}\right)(13.82)(3) - 70.38 + (13.82)(x) + \left(\frac{1}{2}\right)(0.271)(20 - 9.81)x^2 = 0$$

$$1.38x^2 + 13.82x - 49.65 = 0$$

From the above equation, $x \approx 3$ m. Taking the moment about the zero shear point,

$$M_{max} = -\frac{1}{2}\sigma_1' L_1\left(x + \frac{L_1}{3}\right) + F(x + l_2) - \frac{\sigma_1' x^2}{2} - \frac{1}{2}K_a \gamma' x^2\left(\frac{x}{3}\right)$$

or

$$M_{max} = -\left(\frac{1}{2}\right)(13.82)\left(3 + \frac{3}{3}\right) + 70.38(3 + 1.5) - \frac{(13.82)(3)^2}{2} - \left(\frac{1}{2}\right)(0.271)(3)^2\left(\frac{3}{3}\right)$$

$$= \mathbf{225.66\ kN \cdot m/m}$$

Part c

From Eq. (14.84),

$$s_n = 1.25 \frac{c}{\gamma L_1 + \gamma' L_2} = 1.25\left[\frac{41}{(17 \times 3) + (20 - 9.81)(6)}\right] = 0.457$$

From Eq. (14.85),

$$\alpha = \frac{L_1 + L_2}{L_1 + L_2 + D_{actual}} = \frac{3 + 6}{3 + 6 + 2.8} = 0.763$$

Now, referring to Figure 14.33 for $S_n = 0.457$ and $\alpha = 0.763$, we have

log ρ	M_d/M_{max}
−3.1	≈ 0.9
−2.6	≈ 0.9
−2.0	≈ 0.9

Hence, for all log ρ values, $M_d/M_{max} \approx 0.9$. The following table now can be prepared.

Section	I (m⁴/m)	H' (m)	$\rho = (10.91 \times 10^{-7}) \times$ (H'^4/EI)	log ρ	S (m³/m)	$M_d = S\sigma_{all}$	M_d/M_{max}
PZC-12	192.06 $\times 10^{-6}$	11.8	5.93×10^{-4}	−3.2	120.42 $\times 10^{-5}$	207.72	0.92

Note: $H' = L_1 + L_2 + D_{actual} = 3 + 6 + 2.8 = 11.8$ m
$M_{max} = 225.66$ kN $\cdot$ m/m

Figure 14.35 shows the plot of M_d/M_{max} versus log ρ. Section PZC-12 falls above the line of $M_d/M_{max} = 0.9$. So,

PZC-12 will be sufficient.

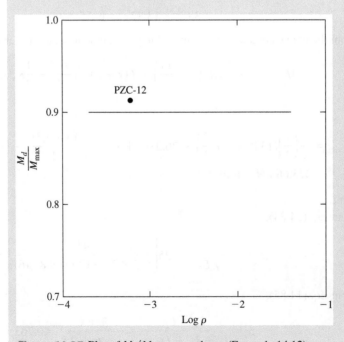

Figure 14.35 Plot of M_d/M_{max} versus log ρ (Example 14.12)

14.15   Anchors

Sections 14.9 through 14.14 gave an analysis of anchored sheet-pile walls and discussed how to obtain the force F per unit length of the sheet-pile wall that has to be sustained by the anchors. The current section covers in more detail the various types of anchor generally used and the procedures for evaluating their ultimate holding capacities.

The general types of anchor used in sheet-pile walls are as follows:

1. Anchor plates and beams (deadman)
2. Tie backs
3. Vertical anchor piles
4. Anchor beams supported by batter (compression and tension) piles

Anchor plates and beams are generally made of cast concrete blocks. (See Figure 14.36a.) The anchors are attached to the sheet pile by *tie rods*. A *wale* is placed at the front or back face of a sheet pile for the purpose of conveniently attaching the tie rod to the wall. To protect the tie rod from corrosion, it is generally coated with paint or asphaltic materials.

In the construction of *tiebacks,* bars or cables are placed in predrilled holes (see Figure 14.36b) with concrete grout (cables are commonly high-strength, prestressed steel tendons). Figures 14.36c and 14.36d show a vertical anchor pile and an anchor beam with batter piles.

Placement of Anchors

The resistance offered by anchor plates and beams is derived primarily from the passive force of the soil located in front of them. Figure 14.36a, in which *AB* is the sheet-pile wall, shows the best location for maximum efficiency of an anchor plate. If the anchor is placed inside wedge *ABC,* which is the Rankine active zone, it would not provide any resistance to failure. Alternatively, the anchor could be placed in zone *CFEH.* Note that line *DFG* is the slip line for the Rankine passive pressure. If part of the passive wedge is located inside the active wedge *ABC,* full passive resistance of the anchor cannot be realized upon failure of the sheet-pile wall. However, if the anchor is placed in zone *ICH,* the Rankine passive zone in front of the anchor slab or plate is located completely outside the Rankine active zone *ABC.* In this case, full passive resistance from the anchor can be realized.

Figures 14.36b, 14.36c, and 14.36d also show the proper locations for the placement of tiebacks, vertical anchor piles, and anchor beams supported by batter piles.

14.16   Holding Capacity of Anchor Plates in Sand

Semi-Empirical Method

Ovesen and Stromann (1972) proposed a semi-empirical method for determining the ultimate resistance of anchors in sand. Their calculations, made in three steps, are carried out as follows:

> *Step 1.* **Basic Case.** Determine the depth of embedment, *H.* Assume that the anchor slab has height *H* and is continuous (i.e., *B* = length of anchor

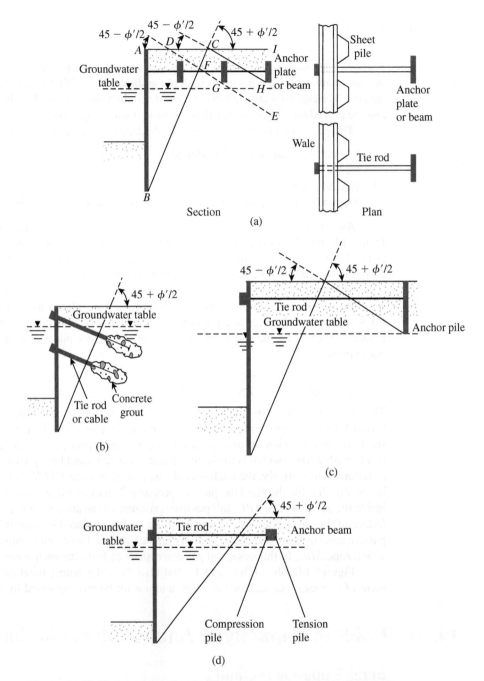

Figure 14.36 Various types of anchoring for sheet-pile walls: (a) anchor plate or beam; (b) tieback; (c) vertical anchor pile; (d) anchor beam with batter piles

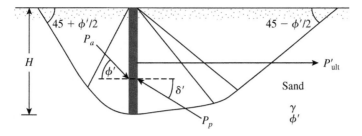

Figure 14.37 Basic case: continuous vertical anchor in granular soil

slab perpendicular to the cross section $= \infty$), as shown in Figure 14.37, in which the following notation is used:

P_p = passive force per unit length of anchor
P_a = active force per unit length of anchor
ϕ' = effective soil friction angle
δ' = friction angle between anchor slab and soil
P'_{ult} = ultimate resistance per unit length of anchor
W = effective weight per unit length of anchor slab

Also,

$$P'_{ult} = \tfrac{1}{2}\gamma H^2 K_p \cos\delta' - P_a \cos\phi' = \tfrac{1}{2}\gamma H^2 K_p \cos\delta' - \tfrac{1}{2}\gamma H^2 K_a \cos\phi'$$
$$= \tfrac{1}{2}\gamma H^2 (K_p \cos\delta' - K_a \cos\phi') \tag{14.86}$$

where

K_a = active pressure coefficient with $\delta' = \phi'$
    (see Figure 14.38a)
K_p = passive pressure coefficient

To obtain $K_p \cos\delta'$, first calculate

$$K_p \sin\delta' = \frac{W + P_a \sin\phi'}{\tfrac{1}{2}\gamma H^2} = \frac{W + \tfrac{1}{2}\gamma H^2 K_a \sin\phi'}{\tfrac{1}{2}\gamma H^2} \tag{14.87}$$

Then use the magnitude of $K_p \sin\delta'$ obtained from Eq. (14.87) to estimate the magnitude of $K_p \cos\delta'$ from the plots given in Figure 14.38b.

Step 2. **Strip Case.** Determine the actual height h of the anchor to be constructed. If a continuous anchor (i.e., an anchor for which $B = \infty$) of height h

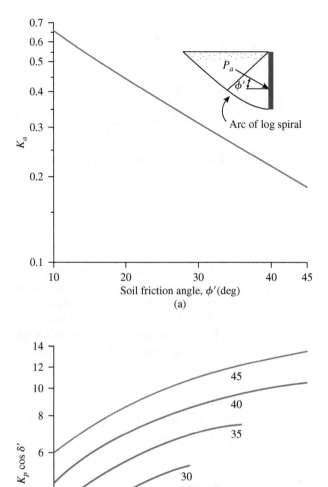

Figure 14.38 (a) Variation of K_a for $\delta' = \phi'$, (b) variation of $K_p \cos \delta'$ with $K_p \sin \delta'$ (Based on Ovesen and Stromann, 1972)

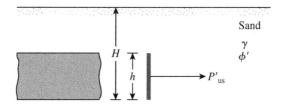

Figure 14.39 Strip case: vertical anchor

is placed in the soil so that its depth of embedment is *H,* as shown in Figure 14.39, the ultimate resistance per unit length is

$$P'_{us} = \left[\frac{C_{ov} + 1}{C_{ov} + \left(\dfrac{H}{h} \right)} \right] \underset{\substack{\uparrow \\ \text{Eq. (14.86)}}}{P'_{ult}} \tag{14.88}$$

where

P'_{us} = ultimate resistance for the *strip case*
C_{ov} = 19 for dense sand and 14 for loose sand

Step 3. **Actual Case.** In practice, the anchor plates are placed in a row with center-to-center spacing *S′*, as shown in Figure 14.40a. The ultimate resistance of each anchor is

$$P_{ult} = P'_{us} B_e \tag{14.89}$$

where B_e = equivalent length.

The equivalent length is a function of *S′*, *B*, *H*, and *h*. Figure 14.40b shows a plot of $(B_e - B)/(H + h)$ against $(S' - B)/(H + h)$ for the cases of loose and dense sand. With known values of *S′*, *B*, *H*, and *h*, the value of B_e can be calculated and used in Eq. (14.89) to obtain P_{ult}.

Stress Characteristic Solution

Neely, Stuart, and Graham (1973) proposed a stress characteristic solution for anchor pull-out resistance using the *equivalent free surface* concept. Figure 14.41 shows the assumed failure surface for a strip anchor. In this figure, *OX* is the equivalent free surface. The shear stress (s_o) mobilized along *OX* can be given as

$$m = \frac{s_o}{\sigma'_o \tan \phi'} \tag{14.90}$$

where

m = shear stress mobilization factor
σ'_o = effective normal stress along *OX*

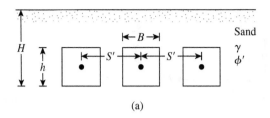

(a)

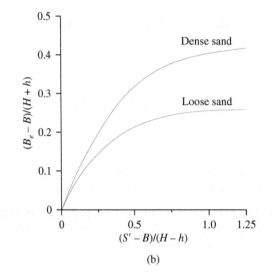

(b)

Figure 14.40 (a) Actual case for row of anchors; (b) variation of $(B_e - B)/(H + h)$ with $(S' - B)/(H + h)$ (Based on Ovesen and Stromann, 1972)

Using this analysis, the ultimate resistance (P_{ult}) of an anchor (length $= B$ and height $= h$) can be given as

$$P_{ult} = M_{\gamma q} (\gamma h^2) B F_s \tag{14.91}$$

where

$M_{\gamma q}$ = force coefficient
F_s = shape factor
γ = effective unit weight of soil

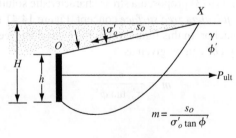

Figure 14.41 Assumed failure surface in soil for stress characteristic solution

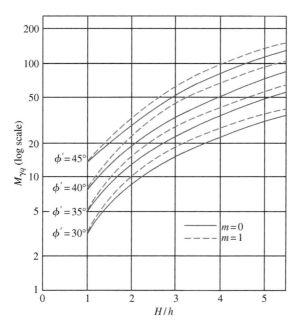

Figure 14.42 Variation of $M_{\gamma q}$ with H/h and ϕ' (Based on Neeley et al., 1973.)

The variations of $M_{\gamma q}$ for $m = 0$ and 1 are shown in Figure 14.42. For conservative design, $M_{\gamma q}$ with $m = 0$ may be used. The shape factor (F_s) determined experimentally is shown in Figure 14.43 as a function of B/h and H/h.

Empirical Correlation Based on Model Tests

Ghaly (1997) used the results of 104 laboratory tests, 15 centrifugal model tests, and 9 field tests to propose an empirical correlation for the ultimate resistance of single anchors. The correlation can be written as

$$P_{ult} = \frac{5.4}{\tan \phi'}\left(\frac{H^2}{A}\right)^{0.28} \gamma AH \qquad (14.92)$$

where A = area of the anchor = Bh.

Ghaly also used the model test results of Das and Seeley (1975) to develop a load–displacement relationship for single anchors. The relationship can be given as

$$\frac{P}{P_{ult}} = 2.2\left(\frac{u}{H}\right)^{0.3} \qquad (14.93)$$

where u = horizontal displacement of the anchor at a load level P.

Equations (14.92) and (14.93) apply to single anchors (i.e., anchors for which $S'/B = \infty$). For all practical purposes, when $S'/B \approx 2$ the anchors behave as single anchors.

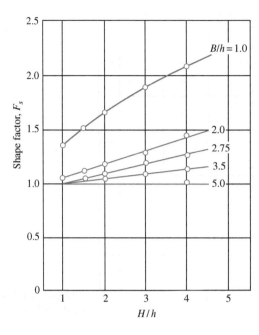

Figure 14.43 Variation of shape factor with H/h and B/h (Based on Neeley et al., 1973.)

Factor of Safety for Anchor Plates

The allowable resistance per anchor plate may be given as

$$P_{all} = \frac{P_{ult}}{FS}$$

where FS = factor of safety.

Generally, a factor of safety of 2 is suggested when the method of Ovesen and Stromann is used. A factor of safety of 3 is suggested for P_{ult} calculated by Eq. (14.92).

Spacing of Anchor Plates

The center-to-center spacing of anchors, S', may be obtained from

$$S' = \frac{P_{all}}{F}$$

where F = force per unit length of the sheet pile.

Example 14.13

Refer to Figure 14.40a. Given: $B = h = 0.4$ m, $S' = 1.2$ m, $H = 1$ m, $\gamma = 16.51$ kN/m³, and $\phi' = 35°$. Determine the ultimate resistance for each anchor plate. The anchor plates are made of concrete and have thicknesses of 0.15 m.

Solution

From Figure 14.38a for $\phi' = 35°$, the magnitude of K_a is about 0.26.

$$W = Ht\gamma_{concrete} = (1 \text{ m})(0.15 \text{ m})(23.5 \text{ kN/m}^3)$$

$$= 3.525 \text{ kN/m}$$

From Eq. (14.87),

$$K_p \sin \delta' = \frac{W + \frac{1}{2}\gamma H^2 K_a \sin \phi'}{\frac{1}{2}\gamma H^2}$$

$$= \frac{3.525 + (0.5)(16.51)(1)^2(0.26)(\sin 35)}{(0.5)(16.51)(1)^2} = 0.576$$

From Figure 14.38b with $\phi' = 35°$ and $K_p \sin \delta' = 0.576$, the value of $K_p \cos \delta'$ is about 4.5. Now, using Eq. (14.86),

$$P'_{ult} = \frac{1}{2}\gamma H^2 (K_p \cos \delta' - K_a \cos \phi')$$

$$= \left(\frac{1}{2}\right)(16.51)(1)^2[4.5 - (0.26)(\cos 35)] = 35.39 \text{ kN/m}$$

In order to calculate P'_{us}, let us assume the sand to be loose. So, C_{ov} in Eq. (14.88) is equal to 14. Hence,

$$P'_{us} = \left[\frac{C_{ov} + 1}{C_{ov} + \left(\dfrac{H}{h}\right)}\right] P'_{ult} = \left[\frac{14 + 1}{14 + \left(\dfrac{1}{0.4}\right)}\right](35.39) = 32.17 \text{ kN/m}$$

$$\frac{S' - B}{H + h} = \frac{1.2 - 0.4}{1 + 0.4} = \frac{0.8}{1.4} = 0.571$$

For $(S' - B)/(H + h) = 0.571$ and loose sand, Figure 14.40b yields

$$\frac{B_e - B}{H - h} = 0.229$$

So

$$B_e = (0.229)(H + h) + B = (0.229)(1 + 0.4) + 0.4$$

$$= 0.72$$

Hence, from Eq. (14.89)

$$P_{ult} = P'_{us} B_e = (32.17)(0.72) = \mathbf{23.16 \text{ kN}}$$

Example 14.14

Refer to a *single anchor* given in Example 14.13 using the stress characteristic solution. Estimate the ultimate anchor resistance. Use $m = 0$ in Figure 14.42.

Solution

Given: $B = h = 0.4$ m and $H = 1$ m.

Thus,

$$\frac{H}{h} = \frac{1 \text{ m}}{0.4 \text{ m}} = 2.5$$

$$\frac{B}{h} = \frac{0.4 \text{ m}}{0.4 \text{ m}} = 1$$

From Eq. (14.91),

$$P_{\text{ult}} = M_{\gamma q} \gamma h^2 \, BF_s$$

From Figure 14.42, with $\phi' = 35°$ and $H/h = 2.5$, $M_{\gamma q} \approx 18.2$. Also, from Figure 14.43, with $H/h = 2.5$ and $B/h = 1$, $F_s \approx 1.8$. Hence,

$$P_{\text{ult}} = (18.2)(16.51)(0.4)^2(0.4)(1.8) \approx \mathbf{34.62 \text{ kN}} \qquad \blacksquare$$

Example 14.15

Solve Example Problem 14.14 using Eq. (14.92).

Solution

From Eq. (14.92),

$$P_{\text{ult}} = \frac{5.4}{\tan \phi'} \left(\frac{H^2}{A} \right)^{0.28} \gamma A H$$

$$H = 1 \text{ m}$$

$$A = Bh = (0.4 \times 0.4) = 0.16 \text{ m}^2$$

$$P_{\text{ult}} = \frac{5.4}{\tan 35} \left[\frac{(1)^2}{0.16} \right]^{0.28} (16.51)(0.16)(1) \approx \mathbf{34.03 \text{ kN}} \qquad \blacksquare$$

14.17 Holding Capacity of Anchor Plates in Clay ($\phi = 0$ Condition)

Relatively few studies have been conducted on the ultimate resistance of anchor plates in clayey soils ($\phi = 0$). Mackenzie (1955) and Tschebotarioff (1973) identified the nature of variation of the ultimate resistance of strip anchors and beams as a function of H, h, and c (undrained cohesion based on $\phi = 0$) in a nondimensional form based on laboratory model test results. This is shown in the form of a nondimensional plot in Figure 14.44 (P_{ult}/hBc versus H/h) and can be used to estimate the ultimate resistance of anchor plates in saturated clay ($\phi = 0$).

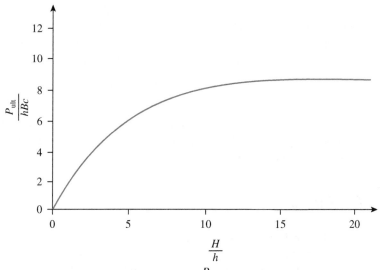

Figure 14.44 Experimental variation of $\dfrac{P_{\text{ult}}}{hBc}$ with H/h for plate anchors in clay (Based on Mackenzie (1955) and Tschebotarioff (1973))

14.18 Ultimate Resistance of Tiebacks

According to Figure 14.45, the ultimate resistance offered by a tieback in sand is

$$P_{\text{ult}} = \pi d l \overline{\sigma}_o' K \tan \phi' \tag{14.94}$$

where

ϕ' = effective angle of friction of soil
$\overline{\sigma}_o'$ = average effective vertical stress ($= \gamma z$ in dry sand)
K = earth pressure coefficient

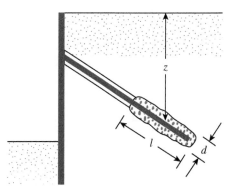

Figure 14.45 Parameters for defining the ultimate resistance of tiebacks

The magnitude of K can be taken to be equal to the earth pressure coefficient at rest (K_o) if the concrete grout is placed under pressure (Littlejohn, 1970). The lower limit of K can be taken to be equal to the Rankine active earth pressure coefficient.

In clays, the ultimate resistance of tiebacks may be approximated as

$$P_{ult} = \pi dl c_a \tag{14.95}$$

where c_a = adhesion.

The value of c_a may be approximated as $\frac{2}{3}c_u$ (where c_u = undrained cohesion). A factor of safety of 1.5 to 2 may be used over the ultimate resistance to obtain the allowable resistance offered by each tieback.

Problems

14.1 Figure P14.1 shows a cantilever sheet-pile wall penetrating a granular soil. Here, $L_1 = 4$ m, $L_2 = 8$ m, $\gamma = 16.1$ kN/m³, $\gamma_{sat} = 18.2$ kN/m³, and $\phi' = 32°$.
   a. What is the theoretical depth of embedment, D?
   b. For a 30% increase in D, what should be the total length of the sheet piles?
   c. Determine the theoretical maximum moment of the sheet pile.

14.2 Redo Problem 14.1 with the following: $L_1 = 3$ m, $L_2 = 6$ m, $\gamma = 17.3$ kN/m³, $\gamma_{sat} = 19.4$ kN/m³, and $\phi' = 30°$.

14.3 Refer to Figure 14.11. Given: $L = 3$ m, $\gamma = 16.7$ kN/m³, and $\phi' = 30°$. Calculate the theoretical depth of penetration, D, and the maximum moment.

14.4 Refer to Figure P14.4, for which $L_1 = 2.4$ m, $L_2 = 4.6$ m, $\gamma = 15.7$ kN/m³, $\gamma_{sat} = 17.3$ kN/m³, and $\phi' = 30°$, and $c = 29$ kN/m².
   a. What is the theoretical depth of embedment, D?
   b. Increase D by 40%. What length of sheet piles is needed?
   c. Determine the theoretical maximum moment in the sheet pile.

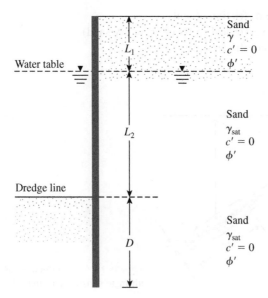

Figure P14.1

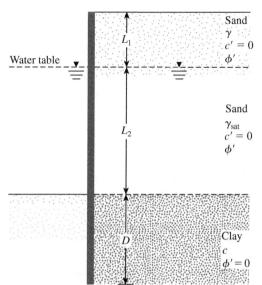

Figure P14.4

14.5 Refer to Figure 14.15. Given: $L = 4$ m; for sand, $\gamma = 16$ kN/m³; $\phi' = 35°$; and, for clay, $\gamma_{sat} = 19.2$ kN/m³ and $c = 45$ kN/m². Determine the theoretical value of D and the maximum moment.

14.6 An anchored sheet-pile bulkhead is shown in Figure P14.6. Let $L_1 = 4$ m, $L_2 = 9$ m, $l_1 = 2$ m, $\gamma = 17$ kN/m³, $\gamma_{sat} = 19$ kN/m³, and $\phi' = 34°$.
   a. Calculate the theoretical value of the depth of embedment, D.
   b. Draw the pressure distribution diagram.
   c. Determine the anchor force per unit length of the wall.
   Use the free earth-support method.

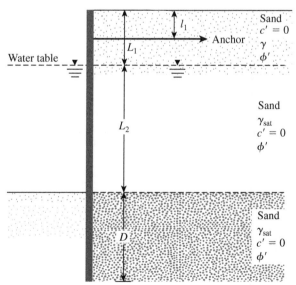

Figure P14.6

14.7 In Problem 14.6, assume that $D_{actual} = 1.3 D_{theory}$.
   a. Determine the theoretical maximum moment.
   b. Using Rowe's moment reduction technique, choose a sheet-pile section. Take $E = 210 \times 10^3$ MN/m^2 and $\sigma_{all} = 210,000$ kN/m^2.

14.8 Refer to Figure P14.6. Given: $L_1 = 4$ m, $L_2 = 8$ m, $l_1 = l_2 = 2$ m, $\gamma = 16$ kN/m^3, $\gamma_{sat} = 18.5$ kN/m^3, and $\phi' = 35°$. Use the charts presented in Section 14.10 and determine:
   a. Theoretical depth of penetration
   b. Anchor force per unit length
   c. Maximum moment in the sheet pile

14.9 Refer to Figure P14.6, for which $L_1 = 4$ m, $L_2 = 7$ m, $l_1 = 1.5$ m, $\gamma = 17.5$ kN/m^3, $\gamma_{sat} = 19.5$ kN/m^3, and $\phi' = 30°$. Use the computational diagram method (Section 14.12) to determine D, F, and M_{max}. Assume that $C = 0.68$ and $R = 0.6$.

14.10 An anchored sheet-pile bulkhead is shown in Figure P14.10. Let $L_1 = 2$ m, $L_2 = 6$ m, $l_1 = 1$ m, $\gamma = 16$ kN/m^3, $\gamma_{sat} = 18.86$ kN/m^3, $\phi' = 32°$, and $c = 27$ kN/m^2.
   a. Determine the theoretical depth of embedment, D.
   b. Calculate the anchor force per unit length of the sheet-pile wall.
      Use the free earth support method.

14.11 In Figure 14.40a, for the anchor slab in sand, $H = 1.52$ m, $h = 0.91$ m, $B = 1.22$ m, $S' = 2.13$ m, $\phi' = 30°$, and $\gamma = 17.3$ kN/m^3. The anchor plates are made of concrete and have a thickness of 76 mm. Using Ovesen and Stromann's method, calculate the ultimate holding capacity of each anchor. Take $\gamma_{concrete} = 23.58$ kN/m^3.

14.12 A single anchor slab is shown in Figure P14.12. Here, $H = 0.9$ m, $h = 0.3$ m, $\gamma = 17$ kN/m^3, and $\phi' = 32°$. Calculate the ultimate holding capacity of the anchor slab if the width B is (a) 0.3 m, (b) 0.6 m, and (c) = 0.9 m.
   (*Note:* center-to-center spacing, $S' = \infty$.) Use the empirical correlation given in Section 14.16 [Eq. (14.92)].

14.13 Repeat Problem 14.12 using Eq. (14.91). Use $m = 0$ in Figure 14.42.

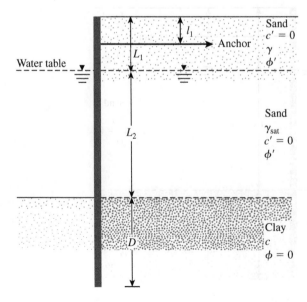

Figure P14.10

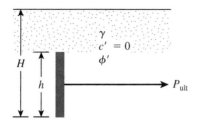

Figure P14.12

References

CASAGRANDE, L. (1973). "Comments on Conventional Design of Retaining Structures," *Journal of the Soil Mechanics and Foundations Division,* ASCE, Vol. 99, No. SM2, pp. 181–198.

DAS, B. M. and SEELEY, G. R. (1975). "Load–Displacement Relationships for Vertical Anchor Plates," *Journal of the Geotechnical Engineering Division,* American Society of Civil Engineers, Vol. 101, No, GT7, pp. 711–715.

GHALY, A. M. (1997). "Load–Displacement Prediction for Horizontally Loaded Vertical Plates." *Journal of Geotechnical and Geoenvironmental Engineering,* ASCE, Vol. 123, No. 1, pp. 74–76.

HAGERTY, D. J. and NOFAL, M. M. (1992). "Design Aids: Anchored Bulkheads in Sand," *Canadian Geotechnical Journal,* Vol. 29, No. 5, pp. 789–795.

LITTLEJOHN, G. S. (1970). "Soil Anchors," *Proceedings, Conference on Ground Engineering,* Institute of Civil Engineers, London, pp. 33–44.

MACKENZIE, T. R. (1955). *Strength of Deadman Anchors in Clay,* M.S. Thesis, Princeton University, Princeton, N. J.

NATARAJ, M. S. and HOADLEY, P. G. (1984). "Design of Anchored Bulkheads in Sand," *Journal of Geotechnical Engineering,* American Society of Civil Engineers, Vol. 110, No. GT4, pp. 505–515.

NEELEY, W. J., STUART, J. G., and GRAHAM, J. (1973). "Failure Loads of Vertical Anchor Plates in Sand," *Journal of the Soil Mechanics and Foundations Division,* American Society of Civil Engineers, Vol. 99, No. SM9, pp. 669–685.

OVESEN, N. K. and STROMANN, H. (1972). "Design Methods for Vertical Anchor Slabs in Sand," *Proceedings, Specialty Conference on Performance of Earth and Earth-Supported Structures.* American Society of Civil Engineers, Vol. 2.1, pp. 1481–1500.

ROWE, P. W. (1952). "Anchored Sheet-Pile Walls," *Proceedings, Institute of Civil Engineers,* Vol. 1, Part 1, pp. 27–70.

ROWE, P. W. (1957). "Sheet-Pile Walls in Clay," *Proceedings, Institute of Civil Engineers,* Vol. 7, pp. 654–692.

TSCHEBOTARIOFF, G. P. (1973). *Foundations, Retaining and Earth Structures,* 2nd ed., McGraw-Hill, New York.

TSINKER, G. P. (1983). "Anchored Street Pile Bulkheads: Design Practice," *Journal of Geotechnical Engineering,* American Society of Civil Engineers, Vol. 109, No. GT8, pp. 1021–1038.

7 Braced Cuts

15.1 Introduction

Sometimes construction work requires ground excavations with vertical or near-vertical faces—for example, basements of buildings in developed areas or underground transportation facilities at shallow depths below the ground surface (a cut-and-cover type of construction). The vertical faces of the cuts need to be protected by temporary bracing systems to avoid failure that may be accompanied by considerable settlement or by bearing capacity failure of nearby foundations.

Figure 15.1 shows two types of braced cut commonly used in construction work. One type uses the *soldier beam* (Figure 15.1a), which is driven into the ground before excavation and is a vertical steel or timber beam. *Laggings,* which are horizontal timber planks, are placed between soldier beams as the excavation proceeds. When the excavation reaches the desired depth, *wales* and *struts* (horizontal steel beams) are installed. The struts are compression members. Figure 15.1b shows another type of braced excavation. In this case, interlocking *sheet piles* are driven into the soil before excavation. Wales and struts are inserted immediately after excavation reaches the appropriate depth.

Figure 15.2 shows the braced-cut construction used for the Chicago subway in 1940. Timber lagging, timber struts, and steel wales were used. Figure 15.3 shows a braced cut made during the construction of the Washington, DC, metro in 1974. In this cut, timber lagging, steel H-soldier piles, steel wales, and pipe struts were used.

To design braced excavations (i.e., to select wales, struts, sheet piles, and soldier beams), an engineer must estimate the lateral earth pressure to which the braced cuts will be subjected. The theoretical aspects of the lateral earth pressure on a braced cut is discussed in Section 15.2. The total active force per unit length of the wall (P_a) can be calculated by using the general wedge theory. However, that analysis will not provide the relationships required for estimating the variation of lateral pressure with depth, which is a function of several factors, such as the type of soil, the experience of the construction crew, the type of construction equipment used, and so forth. For that reason, empirical pressure envelopes developed from field observations are used for the design of braced cuts. This procedure is discussed in the following sections.

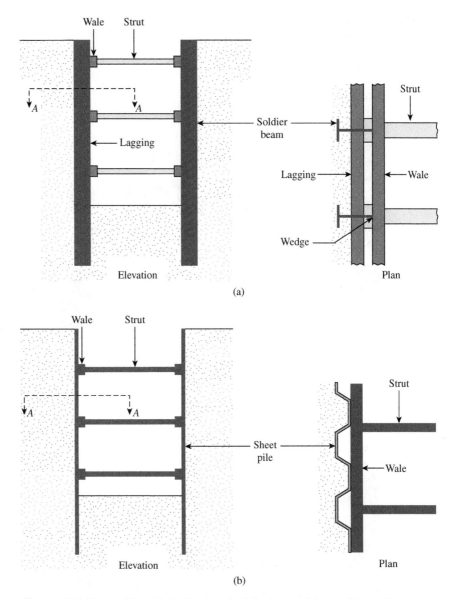

Figure 15.1 Types of braced cut: (a) use of soldier beams; (b) use of sheet piles

15.2 Braced Cut Analysis Based on General Wedge Theory

Figure 15.4 shows a braced cut of height H. Let us assume that AB is a *frictionless* wall retaining a granular soil. During the excavation process followed by the placement of struts, the upper portion of the soil mass next to the cut does not undergo sufficient

Figure 15.2 Braced cut in Chicago Subway construction, January 1940 (*Courtesy of Ralph B. Peck*)

Figure 15.3 Braced cut in the construction of Washington, D.C. Metro, May 1974 (*Courtesy of Ralph B. Peck*)

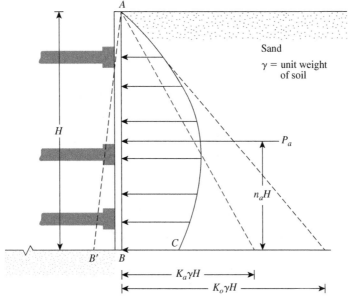

Note: K_o = earth pressure coefficient at-rest
$\quad\ K_a$ = Rankine active earth pressure coefficient

Figure 15.4 Earth pressure behind a frictionless wall retaining sand—wall rotation about the top

lateral deformation. However, as the depth of excavation increases, the time lag between the excavation and placement of struts increases, also resulting in a gradual increase in the lateral deformation of wall AB. Ideally, at the end of excavation, wall AB will be deformed to the shape AB'. The lateral earth-pressure distribution along wall AB will be of the nature shown in Figure 15.4. It is important to note the following:

- The wall AB rotates about A (i.e., rotation about the top).
- At A, the lateral earth pressure will be close to the at-rest earth pressure (practically no lateral deformation of the wall).
- At B, the lateral earth pressure may be less than the Rankine active earth pressure. (The deformation of the wall is large, and the soil may be in a state well past the plastic equilibrium.)
- Hence, the lateral earth-pressure diagram will approximate to the form ACB, as shown in Figure 15.4.

With this type of pressure distribution, the point of application of the resultant active thrust, P_a, will be at a height $n_a H$ measured from the bottom of the wall. The magnitude of n_a will be greater than 1/3.

The magnitude of the active thrust P_a can be determined by considering several trial failure surfaces in soil based on the general wedge theory (Terzaghi, 1943). Figure 15.5 shows a braced cut AB in $c'-\phi'$ soil. Bb_1 is assumed to be a failure surface in the soil

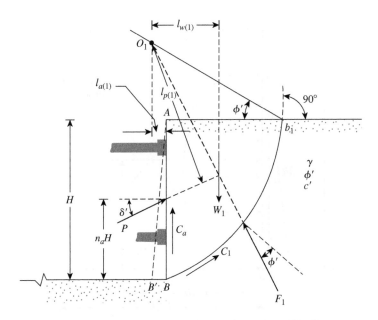

Figure 15.5 Determination of active force on bracing system of open cut in a $c' - \phi'$ soil—the general wedge theory

behind the wall that is an arc of a logarithmic spiral with its center located at O_1. The equation to the logarithmic spiral can be given as

$$r = r_o e^{\theta \tan \phi'} \tag{15.1}$$

where r = radius of the spiral
   r_o = starting radius at $\theta = 0$
   ϕ' = angle of friction of soil
   θ = angle between r and r_o

An important property of the logarithmic spiral defined by Eq. (15.1) is that any radial line makes an angle ϕ' with the normal to the curve drawn at the point where the radial line and the spiral intersect.

The forces *per unit length of the cut* acting on the soil wedge are

- Weight of the wedge = W_1
- Resultant of the normal and shear forces along $Bb_1 = F_1$
- Cohesive force along $Bb = C_1$
- Adhesive force along $AB = C_a = c_a'H$ (where c_a' = unit adhesion)
- Active force P inclined at an angle δ' to the normal drawn to the wall AB

Now, taking the moment of all the forces about O_1,

$$W_1 l_{w(1)} + F_1(0) - C_a l_{a(1)} - M_{C(1)} - P l_{p(1)} = 0 \tag{15.2}$$

where

$M_{C(1)}$ = moment of the cohesive force

$$C_1 = \frac{c'}{2 \tan \phi'}(r^2 - r_o^2) \tag{15.3}$$

$r_o = \overline{O_1 b_1}$
$r = \overline{O_1 B}$
c' = unit cohesion

Thus, from Eq. (15.2),

$$P = \frac{W_1 l_{w(1)} - C_a l_{a(1)} - M_{C(1)}}{l_{p(1)}} \tag{15.4}$$

If this procedure for determining P is repeated for several trial wedges, the maximum value of $P = P_a$ (active thrust) can be obtained.

Table 15.1 provides the variation of $P_a/(0.5\gamma H^2)$ with ϕ', δ'/ϕ', and n_a for granular soil ($c' = 0$) determined by using the general wedge theory described previously. The values of $P_a/(0.5\gamma H^2)$ may be, in general, 15 to 20% more than those obtained from Coulomb's theory for similar wall and soil parameters. However, the analysis does not provide the distribution of the lateral earth pressure with depth. Also, the magnitude of n_a will depend on several factors that may be soil and site specific.

Table 15.1 Active Pressure for Wall Rotation—General Wedge Theory (Granular Soil Backfill)

Soil friction angle, ϕ' (deg)	δ'/ϕ'	$P_a/0.5\ \gamma H^2$			
		$n_a = 0.3$	$n_a = 0.4$	$n_a = 0.5$	$n_a = 0.6$
25	0	0.371	0.405	0.447	0.499
	1/2	0.345	0.376	0.413	0.460
	2/3	0.342	0.373	0.410	0.457
	1	0.344	0.375	0.413	0.461
30	0	0.304	0.330	0.361	0.400
	1/2	0.282	0.306	0.334	0.386
	2/3	0.281	0.305	0.332	0.367
	1	0.289	0.313	0.341	0.377
35	0	0.247	0.267	0.290	0.318
	1/2	0.231	0.249	0.269	0.295
	2/3	0.232	0.249	0.270	0.296
	1	0.243	0.262	0.289	0.312
40	0	0.198	0.213	0.230	0.252
	1/2	0.187	0.200	0.216	0.235
	2/3	0.190	0.204	0.220	0.239
	1	0.197	0.211	0.228	0.248
45	0	0.205	0.220	0.237	0.259
	1/2	0.149	0.159	0.171	0.185
	2/3	0.153	0.164	0.176	0.196
	1	0.173	0.184	0.198	0.215

15.3 Pressure Envelope for Braced-Cut Design

As mentioned in Section 15.1, the lateral earth pressure in a braced cut is dependent on the type of soil, construction method, and type of equipment used. The lateral earth pressure changes from place to place. Each strut should also be designed for the maximum load to which it may be subjected. Therefore, the braced cuts should be designed using apparent-pressure diagrams that are envelopes of all the pressure diagrams determined from measured strut loads in the field. Figure 15.6 shows the method for obtaining the apparent-pressure diagram at a section from strut loads. In this figure, let $P_1, P_2, P_3, P_4, \ldots$ be the measured strut loads. The apparent horizontal pressure can then be calculated as

$$\sigma_1 = \frac{P_1}{(s)\left(d_1 + \dfrac{d_2}{2}\right)}$$

$$\sigma_2 = \frac{P_2}{(s)\left(\dfrac{d_2}{2} + \dfrac{d_3}{2}\right)}$$

$$\sigma_3 = \frac{P_3}{(s)\left(\dfrac{d_3}{2} + \dfrac{d_4}{2}\right)}$$

$$\sigma_4 = \frac{P_4}{(s)\left(\dfrac{d_4}{2} + \dfrac{d_5}{2}\right)}$$

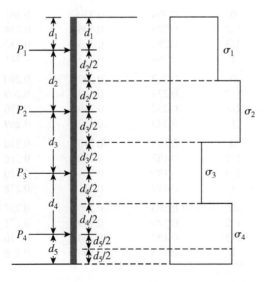

Figure 15.6 Procedure for calculating apparent-pressure diagram from measured strut loads

where

$\sigma_1, \sigma_2, \sigma_3, \sigma_4$ = apparent pressures
s = center-to-center spacing of the struts

Using the procedure just described for strut loads observed from the Berlin subway cut, Munich subway cut, and New York subway cut, Peck (1969) provided the envelope of apparent-lateral-pressure diagrams for design of cuts in *sand*. This envelope is illustrated in Figure 15.7, in which

$$\sigma_a = 0.65\gamma H K_a \qquad (15.5)$$

where

γ = unit weight
H = height of the cut
K_a = Rankine active pressure coefficient = $\tan^2(45 - \phi'/2)$
ϕ' = effective friction angle of sand

Cuts in Clay

In a similar manner, Peck (1969) also provided the envelopes of apparent-lateral-pressure diagrams for cuts in *soft to medium clay* and in *stiff clay*. The pressure envelope for soft to medium clay is shown in Figure 15.8 and is applicable to the condition

$$\frac{\gamma H}{c} > 4$$

where c = undrained cohesion ($\phi = 0$).

The pressure, σ_a, is the larger of

$$\sigma_a = \gamma H\left[1 - \left(\frac{4c}{\gamma H}\right)\right]$$

and $\qquad (15.6)$

$$\sigma_a = 0.3\gamma H$$

where γ = unit weight of clay.

The pressure envelope for cuts in stiff clay is shown in Figure 15.9, in which

$$\sigma_a = 0.2\gamma H \text{ to } 0.4\gamma H \qquad \text{(with an average of } 0.3\gamma H) \qquad (15.7)$$

is applicable to the condition $\gamma H/c \leq 4$.

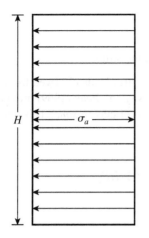

Figure 15.7 Peck's (1969) apparent-pressure envelope for cuts in sand

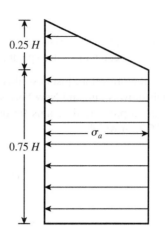

Figure 15.8 Peck's (1969) apparent-pressure envelope for cuts in soft to medium clay

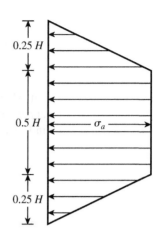

Figure 15.9 Peck's (1969) apparent-pressure envelope for cuts in stiff clay

When using the pressure envelopes just described, keep the following points in mind:

1. They apply to excavations having depths greater than about 6 m ($\approx$20 ft).
2. They are based on the assumption that the water table is below the bottom of the cut.
3. Sand is assumed to be drained with zero pore water pressure.
4. Clay is assumed to be undrained and pore water pressure is not considered.

15.4 Pressure Envelope for Cuts in Layered Soil

Sometimes, layers of both sand and clay are encountered when a braced cut is being constructed. In this case, Peck (1943) proposed that an equivalent value of cohesion ($\phi = 0$) should be determined according to the formula (see Figure 15.10a).

$$c_{av} = \frac{1}{2H} [\gamma_s K_s H_s^2 \tan \phi_s' + (H - H_s)n'q_u] \tag{15.8}$$

where

H = total height of the cut
γ_s = unit weight of sand
H_s = height of the sand layer
K_s = a lateral earth pressure coefficient for the sand layer ($\approx$1)
ϕ_s' = effective angle of friction of sand
q_u = unconfined compression strength of clay
n' = a coefficient of progressive failure (ranging from 0.5 to 1.0; average value 0.75)

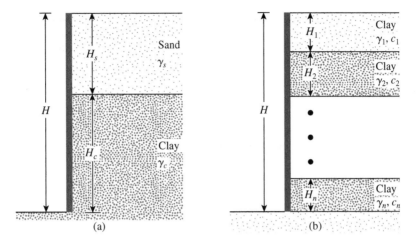

Figure 15.10 Layered soils in braced cuts

The average unit weight of the layers may be expressed as

$$\gamma_a = \frac{1}{H} \left[\gamma_s H_s + (H - H_s)\gamma_c \right] \tag{15.9}$$

where γ_c = saturated unit weight of clay layer.

Once the average values of cohesion and unit weight are determined, the pressure envelopes in clay can be used to design the cuts.

Similarly, when several clay layers are encountered in the cut (Figure 15.10b), the average undrained cohesion becomes

$$c_{\text{av}} = \frac{1}{H}(c_1 H_1 + c_2 H_2 + \cdots + c_n H_n) \tag{15.10}$$

where

$c_1, c_2, \ldots, c_n$ = undrained cohesion in layers $1, 2, \ldots, n$
$H_1, H_2, \ldots, H_n$ = thickness of layers $1, 2, \ldots, n$

The average unit weight is now

$$\gamma_a = \frac{1}{H}(\gamma_1 H_1 + \gamma_2 H_2 + \gamma_3 H_3 + \cdots + \gamma_n H_n) \tag{15.11}$$

15.5 Design of Various Components of a Braced Cut

Struts

In construction work, struts should have a minimum vertical spacing of about 2.75 m (9 ft) or more. Struts are horizontal columns subject to bending. The load-carrying capacity of columns depends on their *slenderness ratio*, which can be reduced by providing vertical and horizontal supports at intermediate points. For wide cuts, splicing the struts may be necessary. For braced cuts in clayey soils, the depth of the first strut below the ground surface should be less than the depth of tensile crack, z_c. From Eq. (12.8),

$$\sigma'_a = \gamma z K_a - 2c'\sqrt{K_a}$$

where K_a = coefficient of Rankine active pressure.

For determining the depth of tensile crack,

$$\sigma'_a = 0 = \gamma z_c K_a - 2c'\sqrt{K_a}$$

or

$$z_c = \frac{2c'}{\sqrt{K_a}\gamma}$$

With $\phi = 0$, $K_a = \tan^2(45 - \phi/2) = 1$, so

$$z_c = \frac{2c}{\gamma}$$

A simplified conservative procedure may be used to determine the strut loads. Although this procedure will vary, depending on the engineers involved in the project, the following is a step-by-step outline of the general methodology (see Figure 15.11):

Step 1. Draw the pressure envelope for the braced cut. (See Figures 15.7, 15.8, and 15.9.) Also, show the proposed strut levels. Figure 15.11a shows a pressure envelope for a sandy soil; however, it could also be for a clay. The strut levels are marked *A, B, C,* and *D*. The sheet piles (or soldier beams) are assumed to be hinged at the strut levels, except for the top and bottom ones. In Figure 15.11a, the hinges are at the level of struts *B* and *C*. (Many designers also assume the sheet piles or soldier beams to be hinged at all strut levels except for the top.)

Step 2. Determine the reactions for the two simple cantilever beams (top and bottom) and all the simple beams between. In Figure 15.11b, these reactions are *A, B_1, B_2, C_1, C_2,* and *D*.

Step 3. The strut loads in the figure may be calculated via the formulas

$$P_A = (A)(s)$$
$$P_B = (B_1 + B_2)(s) \tag{15.12}$$
$$P_C = (C_1 + C_2)(s)$$

and

$$P_D = (D)(s)$$

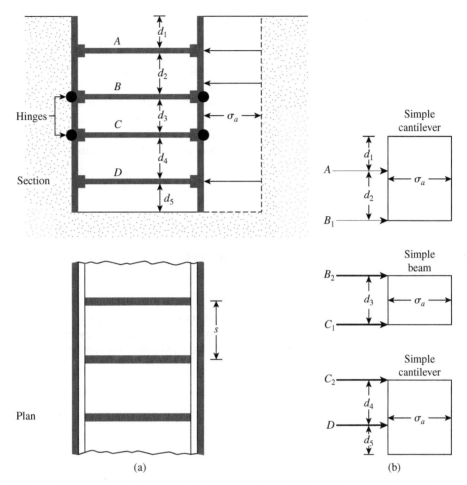

Figure 15.11 Determination of strut loads: (a) section and plan of the cut; (b) method for determining strut loads

where

$$P_A, P_B, P_C, P_D = \text{loads to be taken by the individual struts at levels } A,$$
$$B, C, \text{ and } D, \text{ respectively}$$

$$A, B_1, B_2, C_1, C_2, D = \text{reactions calculated in Step 2 (note the unit:}$$
$$\text{force/unit length of the braced cut)}$$

$$s = \text{horizontal spacing of the struts (see plan in}$$
$$\text{Figure 15.11a)}$$

Step 4. Knowing the strut loads at each level and the intermediate bracing conditions allows selection of the proper sections from the steel construction manual.

Sheet Piles

The following steps are involved in designing the sheet piles:

Step 1. For each of the sections shown in Figure 15.11b, determine the maximum bending moment.

Step 2. Determine the maximum value of the maximum bending moments (M_{max}) obtained in Step 1. Note that the unit of this moment will be, for example, kN-m/m (lb-ft/ft) length of the wall.

Step 3. Obtain the required section modulus of the sheet piles, namely,

$$S = \frac{M_{max}}{\sigma_{all}} \tag{15.13}$$

where σ_{all} = allowable flexural stress of the sheet-pile material.

Step 4. Choose a sheet pile having a section modulus greater than or equal to the required section modulus from a table such as Table 14.1.

Wales

Wales may be treated as continuous horizontal members if they are spliced properly. Conservatively, they may also be treated as though they are pinned at the struts. For the section shown in Figure 15.11a, the maximum moments for the wales (assuming that they are pinned at the struts) are,

$$\text{At level } A, \quad M_{max} = \frac{(A)(s^2)}{8}$$

$$\text{At level } B, \quad M_{max} = \frac{(B_1 + B_2)s^2}{8}$$

$$\text{At level } C, \quad M_{max} = \frac{(C_1 + C_2)s^2}{8}$$

and

$$\text{At level } D, \quad M_{max} = \frac{(D)(s^2)}{8}$$

where A, B_1, B_2, C_1, C_2, and D are the reactions under the struts per unit length of the wall (see Step 2 of strut design).

Now determine the section modulus of the wales:

$$S = \frac{M_{max}}{\sigma_{all}}$$

The wales are sometimes fastened to the sheet piles at points that satisfy the lateral support requirements.

Example 15.1

The cross section of a long braced cut is shown in Figure 15.12a.

    a. Draw the earth-pressure envelope.
    b. Determine the strut loads at levels *A, B,* and *C.*
    c. Determine the section modulus of the sheet-pile section required.
    d. Determine a design section modulus for the wales at level *B.*

(*Note:* The struts are placed at 3 m, center to center, in the plan.) Use

$$\sigma_{all} = 170 \times 10^3 \text{ kN/m}^2$$

Solution

Part a

We are given that $\gamma = 18$ kN/m^2, $c = 35$ kN/m^2, and $H = 7$ m. So,

$$\frac{\gamma H}{c} = \frac{(18)(7)}{35} = 3.6 < 4$$

Thus, the pressure envelope will be like the one in Figure 15.9. The envelope is plotted in Figure 15.12a with maximum pressure intensity, σ_a, equal to $0.3\gamma H = 0.3(18)(7) = \textbf{37.8 kN/m}^2$.

Part b

To calculate the strut loads, examine Figure 15.12b. Taking the moment about B_1, we have $\Sigma M_{B_1} = 0$, and

$$A(2.5) - \left(\frac{1}{2}\right)(37.8)(1.75)\left(1.75 + \frac{1.75}{3}\right) - (1.75)(37.8)\left(\frac{1.75}{2}\right) = 0$$

or

$$A = 54.02 \text{ kN/m}$$

Also, Σ vertical forces = 0. Thus,

$$\tfrac{1}{2}(1.75)(37.8) + (37.8)(1.75) = A + B_1$$

or

$$33.08 + 66.15 - A = B_1$$

So,

$$B_1 = 45.2 \text{ kN/m}$$

Due to symmetry,

$$B_2 = 45.2 \text{ kN/m}$$

and

$$C = 54.02 \text{ kN/m}$$

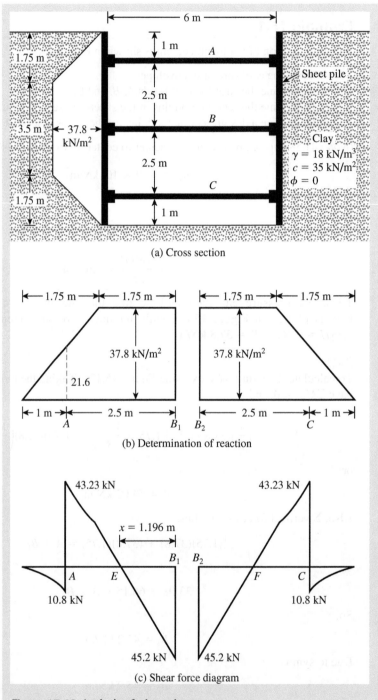

(a) Cross section

(b) Determination of reaction

(c) Shear force diagram

Figure 15.12 Analysis of a braced cut

Hence, the strut loads at the levels indicated by the subscripts are

$$P_A = 54.02 \times \text{horizontal spacing}, s = 54.02 \times 3 = \textbf{162.06 kN}$$

$$P_B = (B_1 + B_2)3 = (45.2 + 45.2)3 = \textbf{271.2 kN}$$

and

$$P_C = 54.02 \times 3 = \textbf{162.06 kN}$$

Part c

At the left side of Figure 15.12b, for the maximum moment, the shear force should be zero. The nature of the variation of the shear force is shown in Figure 15.12c. The location of point E can be given as

$$x = \frac{\text{reaction at } B_1}{37.8} = \frac{45.2}{37.8} = 1.196 \text{ m}$$

Also,

$$\text{Magnitude of moment at } A = \frac{1}{2}(1)\left(\frac{37.8}{1.75} \times 1\right)\left(\frac{1}{3}\right)$$

$$= 3.6 \text{ kN-m/meter of wall}$$

and

$$\text{Magnitude of moment at } E = (45.2 \times 1.196) - (37.8 \times 1.196)\left(\frac{1.196}{2}\right)$$

$$= 54.06 - 27.03 = 27.03 \text{ kN-m/meter of wall}$$

Because the loading on the left and right sections of Figure 15.12b are the same, the magnitudes of the moments at F and C (see Figure 15.12c) will be the same as those at E and A, respectively. Hence, the maximum moment is 27.03 kN-m/meter of wall.

The section modulus of the sheet piles is thus

$$S = \frac{M_{max}}{\sigma_{all}} = \frac{27.03 \text{ kN-m}}{170 \times 10^3 \text{ kN/m}^2} = \textbf{15.9} \times \textbf{10}^{-5} \textbf{ m}^3\textbf{/m of the wall}$$

Part d

The reaction at level B has been calculated in part b. Hence,

$$M_{max} = \frac{(B_1 + B_2)s^2}{8} = \frac{(45.2 + 45.2)3^2}{8} = 101.7 \text{ kN-m}$$

and

$$\text{Section modulus, } S = \frac{101.7}{\sigma_{all}} = \frac{101.7}{(170 \times 1000)}$$

$$= \textbf{0.598} \times \textbf{10}^{-3} \textbf{ m}^3 \qquad \blacksquare$$

Example 15.2

Refer to the braced cut shown in Figure 15.13, for which $\gamma = 112$ lb/ft^3, $\phi' = 32°$, and $c' = 0$. The struts are located 12 ft on center in the plan. Draw the earth-pressure envelope and determine the strut loads at levels A, B, and C.

Solution

For this case, the earth-pressure envelope shown in Figure 15.7 is applicable. Hence,

$$K_a = \tan^2\left(45 - \frac{\phi'}{2}\right) = \tan^2\left(45 - \frac{32}{2}\right) = 0.307$$

From Equation (15.5)

$$\sigma_a = 0.65\,\gamma H K_a = (0.65)(112)(27)(0.307) = 603.44 \text{ lb/ft}^2$$

Figure 15.14a shows the pressure envelope. Refer to Figure 15.14b and calculate B_1:

$$\sum M_{B_1} = 0$$

$$A = \frac{(603.44)(15)\left(\dfrac{15}{2}\right)}{9} = 7543 \text{ lb/ft}$$

$$B_1 = (603.44)(15) - 7543 = 1508.6 \text{ lb/ft} \approx 1509 \text{ lb/ft}$$

Now, refer to Figure 15.14c and calculate B_2:

$$\sum M_{B_2} = 0$$

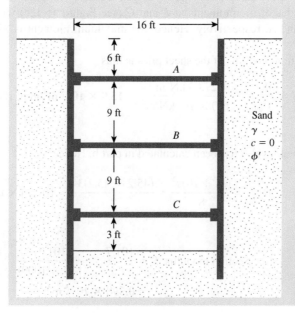

Figure 15.13

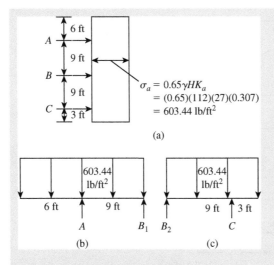

(a)

(b)   (c)   *Figure 15.14* Load diagrams

$$C = \frac{(603.44)(12)\left(\dfrac{12}{2}\right)}{9} = 4827.5 \text{ lb/ft} \approx 4827 \text{ lb/ft}$$

$$B_2 = (603.44)(12) - 4827.5 = 2413.7 \text{ lb/ft} = 2414 \text{ lb/ft}$$

The strut loads are

At A, $(7.543)(\text{spacing}) = (7.543)(12) = $ **90.52 kip**

At B, $(B_1 + B_2)(\text{spacing}) = (1.509 + 2.414)(12) = $ **47.07 kip**

At C, $(4.827)(\text{spacing}) = (4.827)(12) = $ **57.93 kip**   ■

Example 15.3

For the braced cut described in Example 15.2, determine:

  a. The sheet-pile section modulus.
  b. The required section modulus of the wales at level A; assume that $\sigma_{all} = 24$ kip/in^2

Solution
Part a
Refer to the load diagrams shown in Figure 15.14b and 15.14c, Figure 15.15 shows the shear force diagrams based on the load diagrams. First, determine x_1 and x_2:

$$x_1 = \frac{3.923}{0.603} = 6.5 \text{ ft}$$

$$x_2 = \frac{3.017}{0.603} = 5 \text{ ft}$$

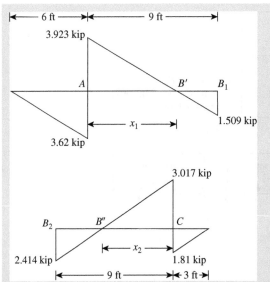

Figure 15.15 Shear force diagrams

Then the moments are

$$\text{At } A, \quad \frac{1}{2}(3.62)(6) = 10.86 \text{ kip-ft}$$

$$\text{At } C, \quad \frac{1}{2}(1.81)(3) = 2.715 \text{ kip-ft}$$

$$\text{At } B', \quad \frac{1}{2}(1.509)(2.5) = 1.89 \text{ kip-ft}$$

$$\text{At } B'', \quad \frac{1}{2}(2.414)(4) = 4.828 \text{ kip-ft}$$

M_A is maximum, so

$$S = \frac{M_{max}}{\sigma_{all}} = \frac{(10.86 \text{ kip-ft})(12)}{24 \text{ kip/in}^2} = \textbf{5.43 in}^3\textbf{/ft}$$

Part b
For the wale at level A,

$$M_{max} = \frac{A(s^2)}{8}$$

$A = 7543$ lb/ft (from Example 15.2). So,

$$M_{max} = \frac{(7.543)(12^2)}{8} = 135.77 \text{ kip-ft/ft}$$

$$S = \frac{M_{max}}{\sigma_{all}} = \frac{(135.77)(12)}{24 \text{ kip/in}^2} = \textbf{67.9 in}^3\textbf{/ft of wall}$$ ∎

15.6 Case Studies of Braced Cuts

The procedure for determining strut loads and the design of sheet piles and wales presented in the preceding sections appears to be fairly straightforward. It is, however, only possible if a proper pressure envelope is chosen for the design, which is difficult. This section describes some case studies of braced cuts and highlights the difficulties and degree of judgment needed for successful completion of various projects.

Subway Extension of the Massachusetts Bay Transportation Authority (MBTA)

Lambe (1970) provided data on the performance of three excavations for the subway extension of the MBTA in Boston (test sections A, B, and D), all of which were well instrumented. Figure 15.16 gives the details of test section B, where the cut was 58 ft, including subsoil conditions. The subsoil consisted of gravel, sand, silt, and clay (fill) to a depth of about 26 ft, followed by a light gray, slightly organic silt to a depth of 46 ft. A layer of coarse sand and gravel with some clay was present from 46 ft to 54 ft below the ground surface. Rock was encountered below 54 ft. The horizontal spacing of the struts was 12 ft center-to-center.

Because the apparent pressure envelopes available (Section 15.3) are for *sand* and *clay* only, questions may arise about how to treat the fill, silt, and till. Figure 15.17 shows the apparent pressure envelopes proposed by Peck (1969), considering the soil as *sand* and also as *clay*, to overcome that problem. For the average soil parameters of the profile, the following values of σ_a were used to develop the pressure envelopes shown in Figure 15.17.

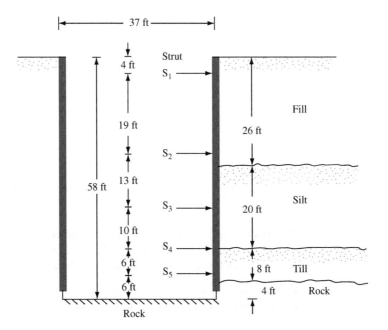

Figure 15.16 Schematic diagram of test section B for subway extension, MTBA

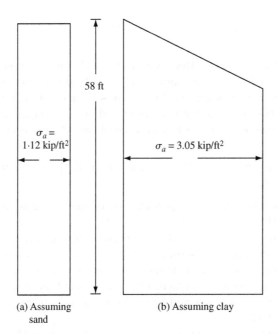

58 ft

$\sigma_a = 1.12$ kip/ft^2

$\sigma_a = 3.05$ kip/ft^2

(a) Assuming sand

(b) Assuming clay

Figure 15.17 Pressure envelopes: (a) assuming sand; (b) assuming clay

Sand

$$\sigma_a = 0.65\gamma H K_a \qquad (15.14)$$

For $\gamma = 114$ lb/ft^3, $H = 58$ ft, and $K_a = 0.26$,

$$\sigma_a = (0.65)(114)(58)(0.26) = 1117 \text{ lb/ft}^2 \approx 1.12 \text{ kip/ft}^2$$

Clay

$$\sigma_a = \gamma H \left[1 - \left(\frac{4c}{\gamma H} \right) \right] \qquad (15.15)$$

For $c = 890$ lb/ft^2,

$$\sigma_a = (114)(58) \left[1 - \frac{(4)(890)}{(114)(58)} \right] = 3052 \text{ lb/ft}^2 \approx 3.05 \text{ kip/ft}^2$$

Table 15.2 shows the variations of the strut load, based on the assumed pressure envelopes shown in Figure 15.17. Also shown in Table 15.2 are the measured strut loads in the field and the design strut loads. This comparison indicates that

1. In most cases the measured strut loads differed widely from those predicted. This result is due primarily to the uncertainties involved in the assumption of the soil parameters.
2. The actual design strut loads were substantially higher than those measured.

Table 15.2 Computed and Measured Strut Loads at Test Section B

Strut number	Computed load (kip)		Measured strut load (kip)
	Envelope based on sand	Envelope based on clay	
S-1	182	230	70.4
S-2	215	580	215
S-3	154	420	304
S-4	108	292	230
S-5	75	219	274

B. Construction of National Plaza (South Half) in Chicago

The construction of the south half of the National Plaza in Chicago required a braced cut 70 ft deep. Swatek et al. (1972) reported the case history for this construction. Figure 15.18 shows a schematic diagram for the braced cut and the subsoil profile. There were six levels of struts. Table 15.3 gives the actual maximum wale and strut loads.

Figure 15.19 presents a lateral earth-pressure envelope based on the maximum wale loads measured. To compare the theoretical prediction to the actual observation

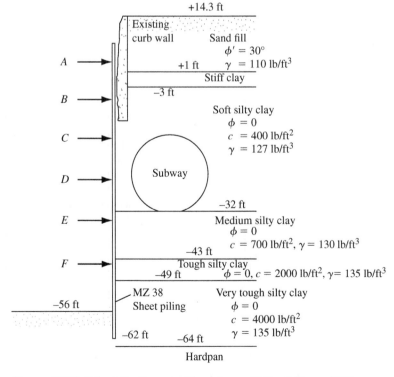

Figure 15.18 Schematic diagram of braced cut—National Plaza of Chicago

Table 15.3 National Plaza Wale and Strut Loads

Strut level	Elevation (ft)	Load measured (kip/ft)
A	+3	16.0
B	−6	26.5
C	−15	29.0
D	−24.5	29.0
E	−34	29.0
F	−44.5	30.7
		Σ160.2

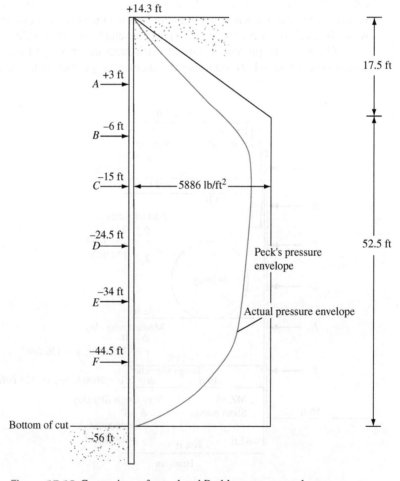

Figure 15.19 Comparison of actual and Peck's pressure envelopes

Table 15.4 Conversion of Soil Layers using Eq. (15.10)

Elevation (ft)	Thickness, H (ft)	c (lb/ft²)	Equivalent c (lb/ft²)
+1 to −32 ft	33	400	$c_{av} = \dfrac{1}{57}[(33)(400) + (11)(700) + (6)(2000)$
			$\qquad + (7)(4000)]$
−32 ft to −43 ft	11	700	$= 1068 \text{ lb/ft}^2$
−43 ft to −49 ft	6	2000	
−49 ft to −56 ft	7	4000	
	Σ57		

requires making an approximate calculation. To do so, we convert the clayey soil layers from Elevation +1 ft to −56 ft to a single equivalent layer in Table 15.4 by using Eq. (15.10).

Now, using Eq. (15.8), we can convert the sand layer located between elevations +14 ft and +1 ft and the equivalent clay layer of 57 ft to one equivalent clay layer with a thickness of 70 ft:

$$c_{av} = \frac{1}{2H}[\gamma_s K_s H_s^2 \tan \phi_s' + (H - H_s)n'q_u]$$

$$= \left[\frac{1}{(2)(70)}\right][(110)(1)(13)^2 \tan 30 + (57)(0.75)(2 \times 1068)] \approx 730 \text{ lb/ft}^2$$

Equation (15.11) gives

$$\gamma_{av} = \frac{1}{H}[\gamma_1 H_1 + \gamma_2 H_2 + \cdots + \gamma_n H_n)$$

$$= \frac{1}{70}[(110)(13) + (127)(33) + (130)(11) + (135)(6) + (135)(7)]$$

$$= 125.8 \text{ lb/ft}^3$$

For the equivalent clay layer of 70 ft,

$$\frac{\gamma_{av} H}{c_{av}} = \frac{(125.8)(70)}{730} = 12.06 > 4$$

Hence the apparent pressure envelope will be of the type shown in Figure 15.8. From Eq. (15.6)

$$\sigma_a = \gamma H\left[1 - \left(\frac{4c_{av}}{\gamma_{av}H}\right)\right] = (125.8)(70)\left[1 - \frac{(4)(730)}{(125.8)(70)}\right] = 5886 \text{ lb/ft}^2$$

The pressure envelope is shown in Figure 15.19. The area of this pressure diagram is 201 kip/ft. Thus Peck's pressure envelope gives a lateral earth pressure of about 1.8 times that actually observed. This result is not surprising because the pressure envelope provided by Figure 15.8 is an envelope developed considering several cuts made at different locations. Under actual field conditions, past experience with the behavior of similar soils can help reduce overdesigning substantially.

15.7 Bottom Heave of a Cut in Clay

Braced cuts in clay may become unstable as a result of heaving of the bottom of the excavation. Terzaghi (1943) analyzed the factor of safety of long braced excavations against bottom heave. The failure surface for such a case in a homogeneous soil is shown in Figure 15.20. In the figure, the following notations are used: B = width of the cut, H = depth of the cut, T = thickness of the clay below the base of excavation, and q = uniform surcharge adjacent to the excavation.

The ultimate bearing capacity at the base of a soil column with a width of B' can be given as

$$q_{ult} = cN_c \tag{15.16}$$

where $N_c = 5.7$ (for a perfectly rough foundation).

The vertical load per unit area along fi is

$$q = \gamma H + q - \frac{cH}{B'} \tag{15.17}$$

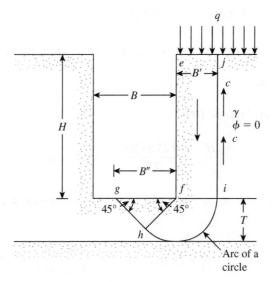

Figure 15.20 Heaving in braced cuts in clay

Hence, the factor of safety against bottom heave is

$$FS = \frac{q_{ult}}{q} = \frac{cN_c}{\gamma H + q - \dfrac{cH}{B'}} = \frac{cN_c}{\left(\gamma + \dfrac{q}{H} - \dfrac{c}{B'}\right)H} \tag{15.18}$$

For excavations of limited length L, the factor of safety can be modified to

$$FS = \frac{cN_c\left(1 + 0.2\,\dfrac{B'}{L}\right)}{\left(\gamma + \dfrac{q}{H} - \dfrac{c}{B'}\right)H} \tag{15.19}$$

where $B' = T$ or $B/\sqrt{2}$ (whichever is smaller).

In 2000, Chang suggested a revision of Eq. (15.19) with the following changes:

1. The shearing resistance along ij may be considered as an increase in resistance rather than a reduction in loading.
2. In Figure 15.20, fg with a width of B'' at the base of the excavation may be treated as a negatively loaded footing.
3. The value of the bearing capacity factor N_c should be 5.14 (not 5.7) for a perfectly smooth footing, because of the restraint-free surface at the base of the excavation.

With the foregoing modifications, Eq. (15.19) takes the form

$$FS = \frac{5.14c\left(1 + \dfrac{0.2B''}{L}\right) + \dfrac{cH}{B'}}{\gamma H + q} \tag{15.20}$$

where

$B' = T$ if $T \leqslant B/\sqrt{2}$
$B' = B/\sqrt{2}$ if $T > B/\sqrt{2}$
$B'' = \sqrt{2}B'$

Bjerrum and Eide (1956) compiled a number of case records for the bottom heave of cuts in clay. Chang (2000) used those records to calculate FS by means of Eq. (15.20); his findings are summarized in Table 15.5. It can be seen from this table that the actual field observations agree well with the calculated factors of safety.

Equation (15.20) is recommended for use in this test. In most cases, a factor of safety of about 1.5 is recommended.

In homogeneous clay, if FS becomes less than 1.5, the sheet pile is driven deeper. (See Figure 15.21.) Usually, the depth d is kept less than or equal to $B/2$, in which case

Table 15.5 Calculated Factors of Safety for Selected Case Records Compiled by Bjerrum and Eide (1956) and Calculated by Chang (2000)

Site	B (m)	B/L	H (m)	H/B	γ (kN/m³)	c (kN/m²)	q (kN/m²)	FS [Eq. (15.20)]	Type of failure
Pumping station, Fornebu, Oslo	5.0	1.0	3.0	0.6	17.5	7.5	0	1.05	Total failure
Storehouse, Drammen	4.8	0	2.4	0.5	19.0	12	15	1.05	Total failure
Sewerage tank, Drammen	5.5	0.69	3.5	0.64	18.0	10	10	0.92	Total failure
Excavation, Grey Wedels Plass, Oslo	5.8	0.72	4.5	0.78	18.0	14	10	1.07	Total failure
Pumping station, Jernbanetorget, Oslo	8.5	0.70	6.3	0.74	19.0	22	0	1.26	Partial failure
Storehouse, Freia, Oslo	5.0	0	5.0	1.00	19.0	16	0	1.10	Partial failure
Subway, Chicago	16	0	11.3	0.70	19.0	35	0	1.00	Near failure

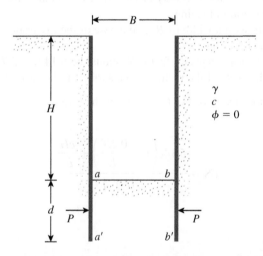

Figure 15.21 Force on the buried length of sheet pile

the force P per unit length of the buried sheet pile (aa' and bb') may be expressed as (U.S. Department of the Navy, 1971)

$$P = 0.7(\gamma HB - 1.4cH - \pi cB) \qquad \text{for } d > 0.47B \qquad (15.21)$$

and

$$P = 1.5d\left(\gamma H - \frac{1.4cH}{B} - \pi c\right) \qquad \text{for } d < 0.47B \qquad (15.22)$$

Example 15.4

In Figure 15.22. for a braced cut in clay, $B = 4\,\text{m}$, $L = 15\,\text{m}$, $H = 6\,\text{m}$, $T = 1.5\,\text{m}$, $\gamma = 17\,\text{kN/m}^3$, $c = 40\,\text{kN/m}^2$, and $q = 0$. Calculate the factor of safety against heave. Use Eq. (15.20).

Solution

From Eq. (15.20),

$$FS = \frac{5.14c\left(1 + \dfrac{0.2B''}{L}\right) + \dfrac{cH}{B'}}{\gamma H + q}$$

with $T = 2\,\text{m}$,

$$\frac{B}{\sqrt{2}} = \frac{4}{\sqrt{2}} = 2.83\,\text{m}$$

So

$$T \le \frac{B}{\sqrt{2}}$$

Hence, $B' = T = 2\,\text{m}$, and it follows that

$$B'' = \sqrt{2}B' = (\sqrt{2})(2) = 2.83\,\text{m}$$

and

$$FS = \frac{(5.14)(40)\left[1 + \dfrac{(0.2)(2.83)}{15}\right] + \dfrac{(40)(6)}{2}}{(17)(6)} = \mathbf{2.55}$$

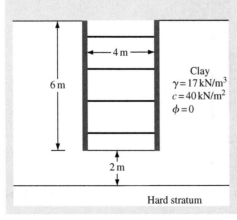

Clay
$\gamma = 17\,\text{kN/m}^3$
$c = 40\,\text{kN/m}^2$
$\phi = 0$

4 m

6 m

2 m

Hard stratum

Figure 15.22 Factor of safety against heaving for a braced cut

15.8 Stability of the Bottom of a Cut in Sand

The bottom of a cut in sand is generally stable. When the water table is encountered, the bottom of the cut is stable as long as the water level inside the excavation is higher than the groundwater level. In case dewatering is needed (see Figure 15.23), the factor of safety against piping should be checked. [*Piping* is another term for failure by heave, as defined in Section 2.12; see Eq. (2.50).] Piping may occur when a high hydraulic gradient is created by water flowing into the excavation. To check the factor of safety, draw flow nets and determine the maximum exit gradient [$i_{max(exit)}$] that will occur at points *A* and *B*. Figure 15.24 shows such a flow net, for which the maximum exit gradient is

$$i_{max(exit)} = \frac{\dfrac{h}{N_d}}{a} = \frac{h}{N_d a} \qquad (15.23)$$

where

a = length of the flow element at *A* (or *B*)
N_d = number of drops (*Note:* in Figure 15.24, $N_d = 8$; see also Section 2.11)

The factor of safety against piping may be expressed as

$$FS = \frac{i_{cr}}{i_{max(exit)}} \qquad (15.24)$$

where i_{cr} = critical hydraulic gradient.

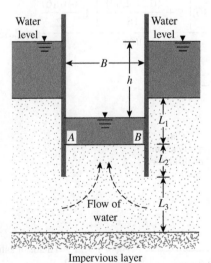

Figure 15.23 Stability of the bottom of a cut in sand

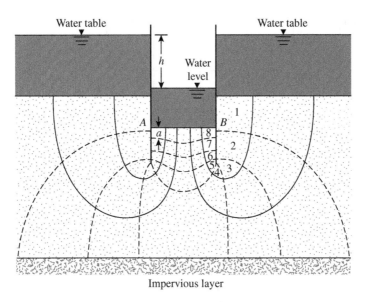

Figure 15.24 Determining the factor of safety against piping by drawing a flow net

The relationship for i_{cr} was given in Chapter 1 as

$$i_{cr} = \frac{G_s - 1}{e + 1}$$

The magnitude of i_{cr} varies between 0.9 and 1.1 in most soils, with an average of about 1. A factor of safety of about 1.5 is desirable.

The maximum exit gradient for sheeted excavations in sands with $L_3 = \infty$ can also be evaluated theoretically (Harr, 1962). (Only the results of these mathematical derivations will be presented here. For further details, see the original work.) To calculate the maximum exit gradient, examine Figures 15.25 and 15.26 and perform the following steps:

1. Determine the modulus, m, from Figure 15.25 by obtaining $2L_2/B$ (or $B/2L_2$) and $2L_1/B$.
2. With the known modulus and $2L_1/B$, examine Figure 15.26 and determine $L_2 i_{exit(max)}/h$. Because L_2 and h will be known, $i_{exit(max)}$ can be calculated.
3. The factor of safety against piping can be evaluated by using Eq. (15.24).

Marsland (1958) presented the results of model tests conducted to study the influence of seepage on the stability of sheeted excavations in sand. The results were summarized by the U.S. Department of the Navy (1971) in NAVFAC DM-7 and are given in Figure 15.27a, b, and c. Note that Figure 15.27b is for the case of determining the sheet pile penetration L_2 needed for the required factor of safety against piping when the sand layer extends to a great depth below the excavation. By contrast, Figure 15.27c represents the case in which an impervious layer lies at a limited depth below the bottom of the excavation.

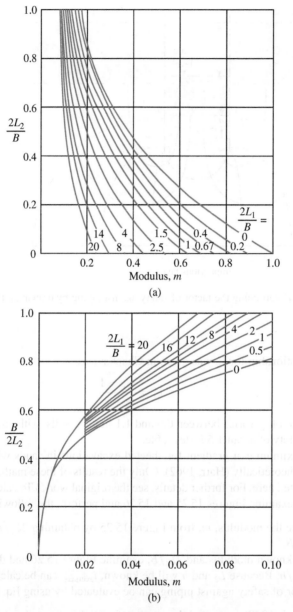

Figure 15.25 Variation of modulus (Based on *Groundwater and Seepage*, by M. E. Harr. McGraw-Hill, 1962.)

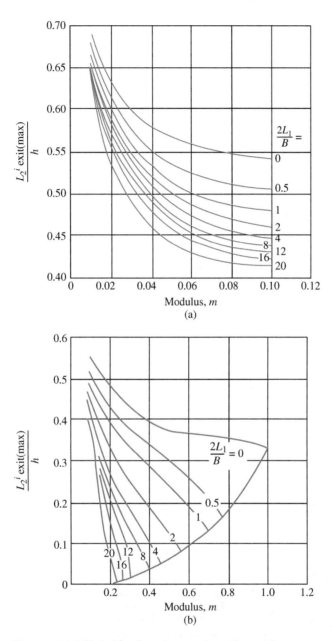

Figure 15.26 Variation of maximum exit gradient with modulus (Based on *Groundwater and Seepage*, by M. E. Harr. McGraw-Hill, 1962.)

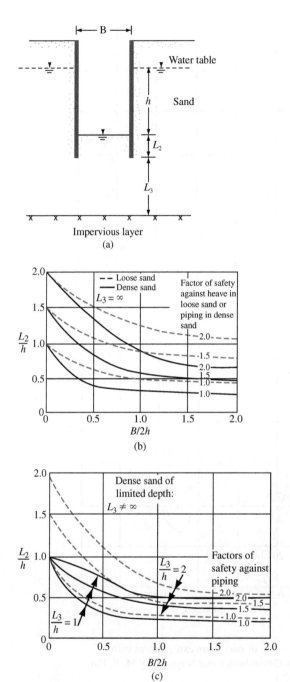

Figure 15.27 Influence of seepage on the stability of sheeted excavation (U.S. Department of the Navy, 1971.)

Example 15.5

In Figure 15.23, let $h = 4.5\,\text{m}$, $L_1 = 5\,\text{m}$, $L_2 = 4\,\text{m}$, $B = 5\,\text{m}$, and $L_3 = \infty$. Determine the factor of safety against piping. Use Figures 15.25 and 15.26.

Solution
We have

$$\frac{2L_1}{B} = \frac{2(5)}{5} = 2$$

and

$$\frac{B}{2L_2} = \frac{5}{2(4)} = 0.625$$

According to Figure 15.25b, for $2L_1/B = 2$ and $B/2L_2 = 0.625$, $m \approx 0.033$. From Figure 15.26a, for $m = 0.033$ and $2L_1/B = 2$, $L_2 i_{\text{exit(max)}}/h = 0.54$. Hence,

$$i_{\text{exit(max)}} = \frac{0.54(h)}{L_2} = 0.54(4.5)/4 = 0.608$$

and

$$\text{FS} = \frac{i_{\text{cr}}}{i_{\text{max(exit)}}} = \frac{1}{0.608} = \mathbf{1.645} \qquad \blacksquare$$

15.9 Lateral Yielding of Sheet Piles and Ground Settlement

In braced cuts, some lateral movement of sheet-pile walls may be expected. (See Figure 15.28.) The amount of lateral yield (δ_H) depends on several factors, the most important of which is the elapsed time between excavation and the placement of wales and struts. As discussed before, in several instances the sheet piles (or the soldier piles, as the case may be) are driven to a certain depth below the bottom of the excavation. The reason is to reduce the lateral yielding of the walls during the last stages of excavation. Lateral yielding of the walls will cause the ground surface surrounding the cut to settle. The degree of lateral yielding, however, depends mostly on the type of soil below the bottom of the cut. If clay below the cut extends to a great depth and $\gamma H/c$ is less than about 6, extension of the sheet piles or soldier piles below the bottom of the cut will help considerably in reducing the lateral yield of the walls.

However, under similar circumstances, if $\gamma H/c$ is about 8, the extension of sheet piles into the clay below the cut does not help greatly. In such circumstances, we may expect a great degree of wall yielding that could result in the total collapse of

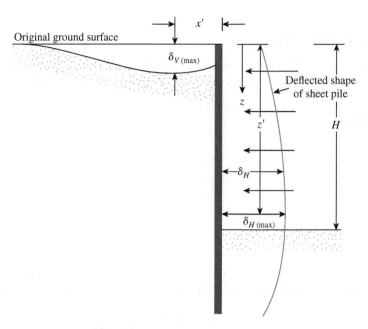

Figure 15.28 Lateral yielding of sheet pile and ground settlement

the bracing systems. If a hard layer of soil lies below a clay layer at the bottom of the cut, the piles should be embedded in the stiffer layer. This action will greatly reduce lateral yield.

The lateral yielding of walls will generally induce ground settlement, δ_V, around a braced cut. Such settlement is generally referred to as *ground loss*. On the basis of several field observations, Peck (1969) provided curves for predicting ground settlement in various types of soil. (See Figure 15.29.) The magnitude of ground loss varies extensively; however, the figure may be used as a general guide.

Moormann (2004) analyzed about 153 case histories dealing mainly with the excavation in soft clay (that is, undrained shear strength, $c \leq 75$ kN/m^2). Following is a summary of his analysis relating to $\delta_{V(max)}$, x', $\delta_{H(max)}$, and z' (see Figure 15.28).

- Maximum Vertical Movement [$\delta_{V(max)}$]

  $\delta_{V(max)}/H \approx 0.1$ to 10.1% with an average of 1.07% (soft clay)
  $\delta_{V(max)}/H \approx 0$ to 0.9% with an average of 0.18% (stiff clay)
  $\delta_{V(max)}/H \approx 0$ to 2.43% with an average of 0.33% (non-cohesive soils)

- Location of $\delta_{V(max)}$, that is x' (Figure 15.28)

  For 70% of all case histories considered, $x' \leq 0.5H$.
  However, in soft clays, x' may be as much as $2H$.

- Maximum Horizontal Deflection of Sheet Piles, $\delta_{H(max)}$

  For 40% of excavation in soft clay, $0.5\% \leq \delta_{H(max)}/H \leq 1\%$.
  The average value of $\delta_{H(max)}/H$ is about 0.87%.

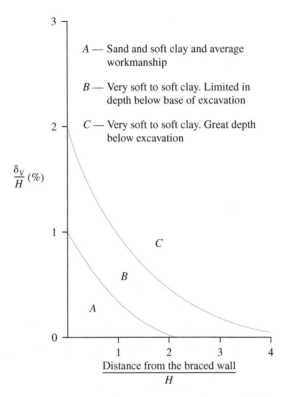

Figure 15.29 Variation of ground settlement with distance (Based on Peck, R. B. (1969). "Deep Excavation and Tunneling in Soft Ground," *Proceedings Seventh International Conference on Soil Mechanics and Foundation Engineering,* Mexico City, State-of-the-Art Volume, pp. 225–290.)

In stiff clays, the average value of $\delta_{H(max)}/H$ is about 0.25%.
In non-cohesive soils, $\delta_{H(max)}/H$ is about 0.27% of the average.

* Location of $\delta_{H(max)}$, that is z' (Figure 15.28)

For deep excavation of soft and stiff cohesive soils, z'/H is about 0.5 to 1.0.

Problems

15.1 Refer to the braced cut shown in Figure P15.1. Given: $\gamma = 17$ kN/m^3, $\phi' = 35°$, and $c' = 0$. The struts are located at 3 m center-to-center in the plan. Draw the earth-pressure envelope and determine the strut loads at levels *A, B,* and *C.*

15.2 For the braced cut described in Problem 15.1, determine the following:
a. The sheet-pile section modulus
b. The section modulus of the wales at level *B*
Assume that $\sigma_{all} = 170$ MN/m^2.

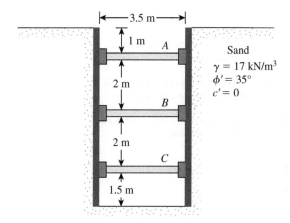

Figure P15.1

15.3 Redo Problem 15.1 with $\gamma = 18$ kN/m³, $\phi' = 40°$, $c' = 0$, and the center-to-center strut spacing in the plan $= 4$ m.

15.4 Determine the sheet-pile section modulus for the braced cut described in Problem 15.3. Given: $\sigma_{all} = 170$ MN/m².

15.5 Refer to Figure 15.10a. For the braced cut, given $H = 8$ m; $H_s = 3$ m; $\gamma_s = 17.5$ kN/m³; angle of friction of sand, $\phi'_s = 34°$; $H_c = 5$ m; $\gamma_c = 18.2$ kN/m³; and unconfined compression strength of clay layer, $q_u = 55$ kN/m².
a. Estimate the average cohesion (c_{av}) and average unit weight (γ_{av}) for the construction of the earth-pressure envelope.
b. Plot the earth-pressure envelope.

15.6 Refer to Figure 15.10b, which shows a braced cut in clay. Given: $H = 25$ ft, $H_1 = 5$ ft, $c_1 = 2125$ lb/ft², $\gamma_1 = 111$ lb/ft³, $H_2 = 10$ ft, $c_2 = 1565$ lb/ft², $\gamma_2 = 107$ lb/ft³, $H_3 = 10$ ft, $c_3 = 1670$ lb/ft², and $\gamma_3 = 109$ lb/ft³.
a. Determine the average cohesion (c_{av}) and average unit weight (γ_{av}) for the construction of the earth-pressure envelope.
b. Plot the earth-pressure envelope.

15.7 Refer to Figure P15.7. Given: $\gamma = 17.5$ kN/m³, $c = 30$ kN/m², and center-to-center spacing of struts in the plan $= 5$ m. Draw the earth-pressure envelope and determine the strut loads at levels A, B, and C.

15.8 Determine the sheet-pile section modulus for the braced cut described in Problem 15.7. Use $\sigma_{all} = 170$ MN/m².

15.9 Redo Problem 15.7 assuming that $c = 60$ kN/m².

15.10 Determine the factor of safety against bottom heave for the braced cut described in Problem 15.7. Use Eq. (15.20) and assume the length of the cut, $L = 18$ m.

15.11 Determine the factor of safety against bottom heave for the braced cut described in Problem 15.9. Use Eq. (15.19). The length of the cut is 12.5.

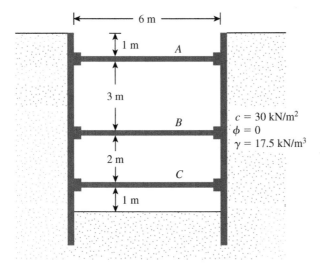

Figure P15.7

References

BJERRUM, L. and EIDE, O. (1956). "Stability of Strutted Excavation in Clay," *Geotechnique,* Vol. 6, No. 1, pp. 32–47.

CHANG, M. F. (2000). "Basal Stability Analysis of Braced Cuts in Clay," *Journal of Geotechnical and Geoenvironmental Engineering,* ASCE, Vol. 126, No. 3, pp. 276–279.

HARR, M. E. (1962). *Groundwater and Seepage,* McGraw-Hill, New York.

LAMBE, T. W. (1970). "Braced Excavations." *Proceedings of the Specialty Conference on Lateral Stresses in the Ground and Design of Earth-Retaining Structures,* American Society of Civil Engineers, pp. 149–218.

MOORMANN, C. (2004). "Analysis of Wall and Ground Movements Due to Deep Excavations in Soft Soil Based on New Worldwide Data Base," *Soils and Foundations,* Vol. 44, No. 1, pp. 87–98.

PECK, R. B. (1943). "Earth Pressure Measurements in Open Cuts, Chicago (ILL.) Subway," *Transactions,* American Society of Civil Engineers, Vol. 108, pp. 1008–1058.

PECK, R. B. (1969). "Deep Excavation and Tunneling in Soft Ground," *Proceedings Seventh International Conference on Soil Mechanics and Foundation Engineering,* Mexico City, State-of-the-Art Volume, pp. 225–290.

SWATEK, E. P., JR., ASROW, S. P., and SEITZ, A. (1972). "Performance of Bracing for Deep Chicago Excavation," *Proceeding of the Specialty Conference on Performance of Earth and Earth Supported Structures,* American Society of Civil Engineers, Vol. 1, Part 2, pp. 1303–1322.

TERZAGHI, K. (1943). *Theoretical Soil Mechanics,* Wiley, New York.

U.S. DEPARTMENT OF THE NAVY (1971). "Design Manual—Soil Mechanics. Foundations, and Earth Structures." NAVFAC DM-7, Washington, D.C.

CHAPTER 8

Seepage

8.1 Introduction

In the preceding chapter, we considered some simple cases for which direct application of Darcy's law was required to calculate the flow of water through soil. In many instances, the flow of water through soil is not in one direction only, nor is it uniform over the entire area perpendicular to the flow. In such cases, the groundwater flow is generally calculated by the use of graphs referred to as *flow nets*. The concept of the flow net is based on *Laplace's equation of continuity*, which governs the steady flow condition for a given point in the soil mass.

In this chapter, we will discuss the following:

- Derivation of Laplace's equation of continuity and some simple applications of the equation
- Procedure to construct flow nets and calculation of seepage in isotropic and anisotropic soils
- Seepage through earth dams

8.2 Laplace's Equation of Continuity

To derive the Laplace differential equation of continuity, let us consider a single row of sheet piles that have been driven into a permeable soil layer, as shown in Figure 8.1a. The row of sheet piles is assumed to be impervious. The steady-state flow of water from the upstream to the downstream side through the permeable layer is a two-dimensional flow. For flow at a point A, we consider an elemental soil block. The block has dimensions dx, dy, and dz (length dy is perpendicular to the plane of the paper); it is shown in an enlarged scale in Figure 8.1b. Let v_x and v_z be the components of the discharge velocity in the horizontal and vertical directions, respectively. The rate of flow of water into the elemental

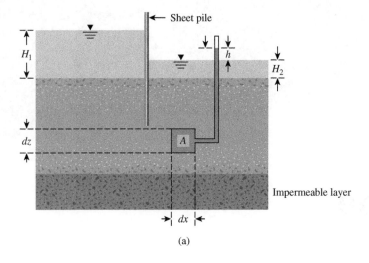

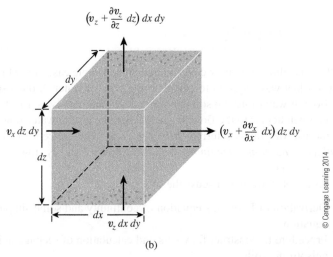

Figure 8.1 (a) Single-row sheet piles driven into permeable layer; (b) flow at A

block in the horizontal direction is equal to $v_x \, dz \, dy$, and in the vertical direction it is v_z $dx \, dy$. The rates of outflow from the block in the horizontal and vertical directions are, respectively,

$$\left(v_x + \frac{\partial v_x}{\partial x} \, dx \right) dz \, dy$$

and

$$\left(v_z + \frac{\partial v_z}{\partial z} \, dz \right) dx \, dy$$

Assuming that water is incompressible and that no volume change in the soil mass occurs, we know that the total rate of inflow should equal the total rate of outflow. Thus,

$$\left[\left(v_x + \frac{\partial v_x}{\partial x}\,dx\right)dz\,dy + \left(v_z + \frac{\partial v_z}{\partial z}\,dz\right)dx\,dy\right] - [v_x\,dz\,dy + v_z\,dx\,dy] = 0$$

or

$$\frac{\partial v_x}{\partial x} + \frac{\partial v_z}{\partial z} = 0 \tag{8.1}$$

With Darcy's law, the discharge velocities can be expressed as

$$v_x = k_x i_x = k_x \frac{\partial h}{\partial x} \tag{8.2}$$

and

$$v_z = k_z i_z = k_z \frac{\partial h}{\partial z} \tag{8.3}$$

where k_x and k_z are the hydraulic conductivities in the horizontal and vertical directions, respectively.

From Eqs. (8.1), (8.2), and (8.3), we can write

$$k_x \frac{\partial^2 h}{\partial x^2} + k_z \frac{\partial^2 h}{\partial z^2} = 0 \tag{8.4}$$

If the soil is isotropic with respect to the hydraulic conductivity—that is, $k_x = k_z$—the preceding continuity equation for two-dimensional flow simplifies to

$$\frac{\partial^2 h}{\partial x^2} + \frac{\partial^2 h}{\partial z^2} = 0 \tag{8.5}$$

8.3 Continuity Equation for Solution of Simple Flow Problems

The continuity equation given in Eq. (8.5) can be used in solving some simple flow problems. To illustrate this, let us consider a one-dimensional flow problem, as shown in Figure 8.2, in which a constant head is maintained across a two-layered soil for the flow of water. The head difference between the top of soil layer no. 1 and the bottom of soil layer no. 2 is h_1. Because the flow is in only the z direction, the continuity equation [Eq. (8.5)] is simplified to the form

$$\frac{\partial^2 h}{\partial z^2} = 0 \tag{8.6}$$

or

$$h = A_1 z + A_2 \tag{8.7}$$

where A_1 and A_2 are constants.

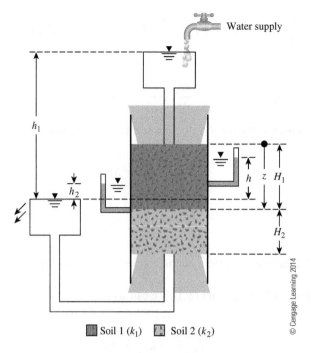

Figure 8.2 Flow through a two-layered soil

To obtain A_1 and A_2 for flow through soil layer no. 1, we must know the boundary conditions, which are as follows:

Condition 1: At $z = 0$, $h = h_1$.
Condition 2: At $z = H_1$, $h = h_2$.

Combining Eq. (8.7) and Condition 1 gives

$$A_2 = h_1 \qquad (8.8)$$

Similarly, combining Eq. (8.7) and Condition 2 with Eq. (8.8) gives

$$h_2 = A_1 H_1 + h_1$$

or

$$A_1 = -\left(\frac{h_1 - h_2}{H_1}\right) \qquad (8.9)$$

Combining Eqs. (8.7), (8.8), and (8.9), we obtain

$$h = -\left(\frac{h_1 - h_2}{H_1}\right)z + h_1 \qquad (\text{for } 0 \leq z \leq H_1) \qquad (8.10)$$

For flow through soil layer no. 2, the boundary conditions are

Condition 1: At $z = H_1$, $h = h_2$.
Condition 2: At $z = H_1 + H_2$, $h = 0$.

From Condition 1 and Eq. (8.7),

$$A_2 = h_2 - A_1 H_1 \qquad (8.11)$$

Also, from Condition 2 and Eqs. (8.7) and (8.11),

$$0 = A_1(H_1 + H_2) + (h_2 - A_1 H_1)$$
$$A_1 H_1 + A_1 H_2 + h_2 - A_1 H_1 = 0$$

or

$$A_1 = -\frac{h_2}{H_2} \qquad (8.12)$$

So, from Eqs. (8.7), (8.11), and (8.12),

$$h = -\left(\frac{h_2}{H_2}\right)z + h_2\left(1 + \frac{H_1}{H_2}\right) \qquad \text{(for } H_1 \leq z \leq H_1 + H_2) \qquad (8.13)$$

At any given time, flow through soil layer no. 1 equals flow through soil layer no. 2, so

$$q = k_1\left(\frac{h_1 - h_2}{H_1}\right)A = k_2\left(\frac{h_2 - 0}{H_2}\right)A$$

where A = area of cross section of the soil
$\quad k_1$ = hydraulic conductivity of soil layer no. 1
$\quad k_2$ = hydraulic conductivity of soil layer no. 2

or

$$h_2 = \frac{h_1 k_1}{H_1\left(\dfrac{k_1}{H_1} + \dfrac{k_2}{H_2}\right)} \qquad (8.14)$$

Substituting Eq. (8.14) into Eq. (8.10), we obtain

$$h = h_1\left(1 - \frac{k_2 z}{k_1 H_2 + k_2 H_1}\right) \qquad \text{(for } 0 \leq z \leq H_1) \qquad (8.15)$$

Similarly, combining Eqs. (8.13) and (8.14) gives

$$h = h_1\left[\left(\frac{k_1}{k_1 H_2 + k_2 H_1}\right)(H_1 + H_2 - z)\right] \qquad \text{(for } H_1 \leq z \leq H_1 + H_2) \qquad (8.16)$$

Example 8.1

Refer to Figure 8.2. Given: $H_1 = 12$ in., $H_2 = 20$ in., $h_1 = 24$ in., $h = 20$ in., $z = 8$ in., $k_1 = 0.026$ in./sec, and diameter of the soil specimen is $D = 3$ in. Determine the rate of flow of water through the two-layered soil (in^3/hr).

Solution

Since $z = 8$ in. is located in soil layer 1, Eq. (8.15) is valid. Thus,

$$h = h_1\left(1 - \frac{k_2 z}{k_1 H_2 + k_2 H_1}\right) = h_1\left[1 - \frac{z}{\left(\dfrac{k_1}{k_2}\right)H_2 + H_1}\right]$$

$$20 = 24\left[1 - \frac{8}{\left(\dfrac{k_1}{k_2}\right)20 + 12}\right]$$

$$\frac{k_1}{k_2} = 1.795 \approx 1.8$$

Given that $k_1 = 0.026$ in./sec,

$$k_2 = \frac{k_1}{1.8} = \frac{0.026}{1.8} = 0.0144 \text{ in./sec}$$

The rate of flow is

$$q = k_{eq}\, iA$$

$$i = \frac{h_1}{H_1 + H_2} = \frac{24}{12 + 20} = 0.75$$

$$A = \frac{\pi}{4}D^2 = \frac{\pi}{4}(3)^2 = 7.069 \text{ in}^2$$

$$k_{eq} = \frac{H_1 + H_2}{\dfrac{H_1}{k_1} + \dfrac{H_2}{k_2}} = \frac{12 + 20}{\dfrac{12}{0.026} + \dfrac{20}{0.0144}} = 0.0173 \text{ in./sec} = 62.28 \text{ in./hr}$$

Thus,

$$q = k_{eq}\, iA = (62.28)(0.75)(7.069) = \mathbf{330.19\ in^3/hr}$$

8.4 Flow Nets

The continuity equation [Eq. (8.5)] in an isotropic medium represents two orthogonal families of curves—that is, the flow lines and the equipotential lines. A *flow line* is a line along which a water particle will travel from upstream to the downstream side in the permeable soil medium. An *equipotential line* is a line along which the potential head at all points is equal. Thus, if piezometers are placed at different points along an equipotential line, the water level will rise to the same elevation in all of them. Figure 8.3a demonstrates

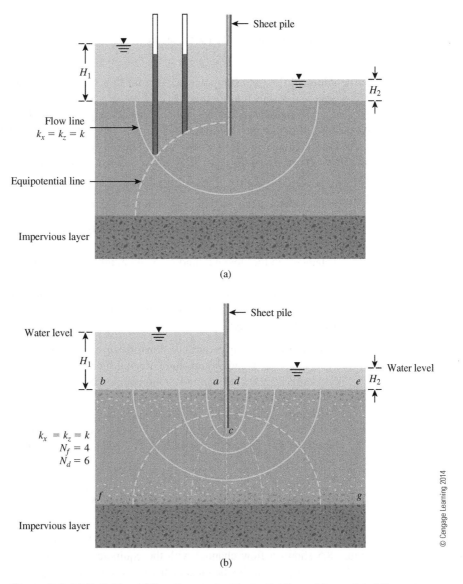

Figure 8.3 (a) Definition of flow lines and equipotential lines; (b) completed flow net

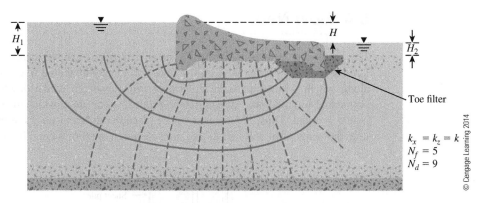

$k_x = k_z = k$
$N_f = 5$
$N_d = 9$

© Cengage Learning 2014

Figure 8.4 Flow net under a dam with toe filter

the definition of flow and equipotential lines for flow in the permeable soil layer around the row of sheet piles shown in Figure 8.1 (for $k_x = k_z = k$).

A combination of a number of flow lines and equipotential lines is called a *flow net.* As mentioned in the introduction, flow nets are constructed for the calculation of ground-water flow and the evaluation of heads in the media. To complete the graphic construction of a flow net, one must draw the flow and equipotential lines in such a way that

1. The equipotential lines intersect the flow lines at right angles.
2. The flow elements formed are approximate squares.

Figure 8.3b shows an example of a completed flow net. One more example of flow net in isotropic permeable layer is given in Figure 8.4. In these figures, N_f is the number of flow channels in the flow net, and N_d is the number of potential drops (defined later in this chapter).

Drawing a flow net takes several trials. While constructing the flow net, keep the boundary conditions in mind. For the flow net shown in Figure 8.3b, the following four boundary conditions apply:

Condition 1: The upstream and downstream surfaces of the permeable layer (lines *ab* and *de*) are equipotential lines.

Condition 2: Because *ab* and *de* are equipotential lines, all the flow lines intersect them at right angles.

Condition 3: The boundary of the impervious layer—that is, line *fg*—is a flow line, and so is the surface of the impervious sheet pile, line *acd*.

Condition 4: The equipotential lines intersect *acd* and *fg* at right angles.

8.5 Seepage Calculation from a Flow Net

In any flow net, the strip between any two adjacent flow lines is called a *flow channel.* Figure 8.5 shows a flow channel with the equipotential lines forming square elements. Let h_1, h_2, h_3, h_4, . . ., h_n be the piezometric levels corresponding to the equipotential lines. The rate of seepage through the flow channel per unit length (perpendicular to the vertical

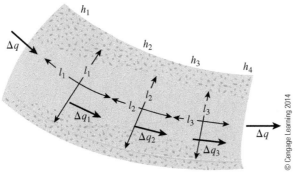

© Cengage Learning 2014

Figure 8.5 Seepage through a flow channel with square elements

section through the permeable layer) can be calculated as follows. Because there is no flow across the flow lines,

$$\Delta q_1 = \Delta q_2 = \Delta q_3 = \cdots = \Delta q \tag{8.17}$$

From Darcy's law, the flow rate is equal to kiA. Thus, Eq. (8.17) can be written as

$$\Delta q = k\left(\frac{h_1 - h_2}{l_1}\right)l_1 = k\left(\frac{h_2 - h_3}{l_2}\right)l_2 = k\left(\frac{h_3 - h_4}{l_3}\right)l_3 = \cdots \tag{8.18}$$

Equation (8.18) shows that if the flow elements are drawn as approximate squares, the drop in the piezometric level between any two adjacent equipotential lines is the same. This is called the *potential drop*. Thus,

$$h_1 - h_2 = h_2 - h_3 = h_3 - h_4 = \cdots = \frac{H}{N_d} \tag{8.19}$$

and

$$\Delta q = k\frac{H}{N_d} \tag{8.20}$$

where H = head difference between the upstream and downstream sides
  N_d = number of potential drops

In Figure 8.3b, for any flow channel, $H = H_1 - H_2$ and $N_d = 6$.

If the number of flow channels in a flow net is equal to N_f, the total rate of flow through all the channels per unit length can be given by

$$q = k\frac{HN_f}{N_d} \tag{8.21}$$

Although drawing square elements for a flow net is convenient, it is not always necessary. Alternatively, one can draw a rectangular mesh for a flow channel, as shown in Figure 8.6, provided that the width-to-length ratios for all the rectangular elements in the flow net are the same. In this case, Eq. (8.18) for rate of flow through the channel can be modified to

$$\Delta q = k\left(\frac{h_1 - h_2}{l_1}\right)b_1 = k\left(\frac{h_2 - h_3}{l_2}\right)b_2 = k\left(\frac{h_3 - h_4}{l_3}\right)b_3 = \cdots \tag{8.22}$$

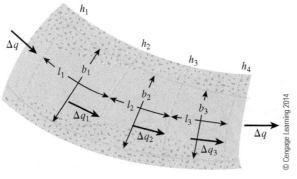

© Cengage Learning 2014

Figure 8.6 Seepage through a flow channel with rectangular elements

If $b_1/l_1 = b_2/l_2 = b_3/l_3 = \cdots = n$ (i.e., the elements are not square), Eqs. (8.20) and (8.21) can be modified to

$$\Delta q = kH\left(\frac{n}{N_d}\right) \tag{8.23}$$

and

$$q = kH\left(\frac{N_f}{N_d}\right)n \tag{8.24}$$

Figure 8.7 shows a flow net for seepage around a single row of sheet piles. Note that flow channels 1 and 2 have square elements. Hence, the rate of flow through these two channels can be obtained from Eq. (8.20):

$$\Delta q_1 + \Delta q_2 = \frac{k}{N_d}H + \frac{k}{N_d}H = \frac{2kH}{N_d}$$

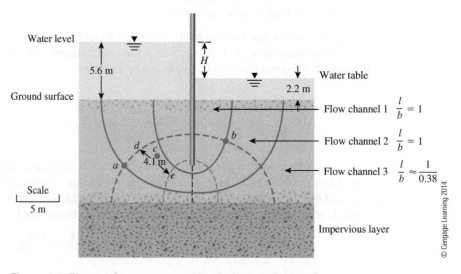

Figure 8.7 Flow net for seepage around a single row of sheet piles

However, flow channel 3 has rectangular elements. These elements have a width-to-length ratio of about 0.38; hence, from Eq. (8.23),

$$\Delta q_3 = \frac{k}{N_d} H(0.38)$$

So, the total rate of seepage can be given as

$$q = \Delta q_1 + \Delta q_2 + \Delta q_3 = 2.38 \frac{kH}{N_d} \tag{8.25}$$

Example 8.2

A flow net for flow around a single row of sheet piles in a permeable soil layer is shown in Figure 8.7. Given that $k_x = k_z = k = 5 \times 10^{-3}$ cm/sec, determine

a. How high (above the ground surface) the water will rise if piezometers are placed at points a and b
b. The total rate of seepage through the permeable layer per unit length
c. The approximate average hydraulic gradient at c

Solution

Part a

From Figure 8.7, we have $N_d = 6$, $H_1 = 5.6$ m, and $H_2 = 2.2$ m. So the head loss of each potential drop is

$$\Delta H = \frac{H_1 - H_2}{N_d} = \frac{5.6 - 2.2}{6} = 0.567 \text{ m}$$

At point a, we have gone through one potential drop. So the water in the piezometer will rise to an elevation of

$$(5.6 - 0.567) = \textbf{5.033 m above the ground surface}$$

At point b, we have five potential drops. So the water in the piezometer will rise to an elevation of

$$[5.6 - (5)(0.567)] = \textbf{2.765 m above the ground surface}$$

Part b

From Eq. (8.25),

$$q = 2.38 \frac{k(H_1 - H_2)}{N_d} = \frac{(2.38)(5 \times 10^{-5} \text{ m/sec})(5.6 - 2.2)}{6}$$

$$= \textbf{6.74} \times \textbf{10}^{-5} \textbf{m}^3\textbf{/sec/m}$$

Part c

The average hydraulic gradient at c can be given as

$$i = \frac{\text{head loss}}{\text{average length of flow between } d \text{ and } e} = \frac{\Delta H}{\Delta L} = \frac{0.567 \text{ m}}{4.1 \text{ m}} = \textbf{0.138}$$

(*Note:* The average length of flow has been scaled.)

8.6 Flow Nets in Anisotropic Soil

The flow-net construction described thus far and the derived Eqs. (8.21) and (8.24) for seepage calculation have been based on the assumption that the soil is isotropic. However, in nature, most soils exhibit some degree of anisotropy. To account for soil anisotropy with respect to hydraulic conductivity, we must modify the flow net construction.

The differential equation of continuity for a two-dimensional flow [Eq. (8.4)] is

$$k_x\frac{\partial^2 h}{\partial x^2} + k_z\frac{\partial^2 h}{\partial z^2} = 0$$

For anisotropic soils, $k_x \neq k_z$. In this case, the equation represents two families of curves that do not meet at $90°$. However, we can rewrite the preceding equation as

$$\frac{\partial^2 h}{(k_z/k_x)\,\partial x^2} + \frac{\partial^2 h}{\partial z^2} = 0 \tag{8.26}$$

Substituting $x' = \sqrt{k_z/k_x}\,x$, we can express Eq. (8.26) as

$$\frac{\partial^2 h}{\partial x'^2} + \frac{\partial^2 h}{\partial z^2} = 0 \tag{8.27}$$

Now Eq. (8.27) is in a form similar to that of Eq. (8.5), with x replaced by x', which is the new transformed coordinate. To construct the flow net, use the following procedure:

Step 1: Adopt a vertical scale (that is, z axis) for drawing the cross section.

Step 2: Adopt a horizontal scale (that is, x axis) such that horizontal scale $= \sqrt{k_z/k_x} \times$ vertical scale.

Step 3: With scales adopted as in steps 1 and 2, plot the vertical section through the permeable layer parallel to the direction of flow.

Step 4: Draw the flow net for the permeable layer on the section obtained from step 3, with flow lines intersecting equipotential lines at right angles and the elements as approximate squares.

The rate of seepage per unit length can be calculated by modifying Eq. (8.21) to

$$q = \sqrt{k_x k_z}\,\frac{HN_f}{N_d} \tag{8.28}$$

where $H = $ total head loss
N_f and $N_d = $ number of flow channels and potential drops, respectively (from flow net drawn in step 4)

Note that when flow nets are drawn in transformed sections (in anisotropic soils), the flow lines and the equipotential lines are orthogonal. However, when they are redrawn in a true section, these lines are not at right angles to each other. This fact is shown in Figure 8.8. In this figure, it is assumed that $k_x = 6k_z$. Figure 8.8a shows a flow element in a transformed section. The flow element has been redrawn in a true section in Figure 8.8b.

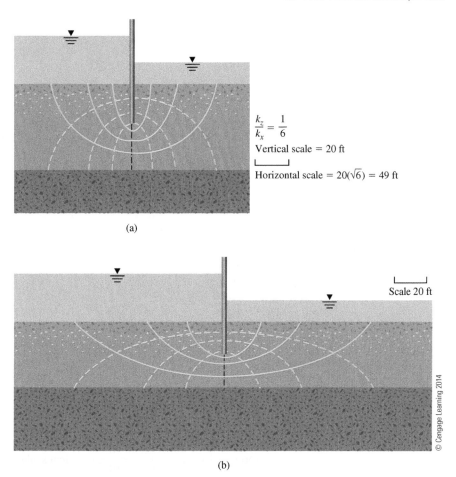

(a)

$$\frac{k_z}{k_x} = \frac{1}{6}$$

Vertical scale = 20 ft

Horizontal scale = $20(\sqrt{6})$ = 49 ft

(b)

Figure 8.8 A flow element in anisotropic soil: (a) in transformed section; (b) in true section

Example 8.3

A dam section is shown in Figure 8.9a. The hydraulic conductivity of the permeable layer in the vertical and horizontal directions are 2×10^{-2} mm/s and 4×10^{-2} mm/s, respectively. Draw a flow net and calculate the seepage loss of the dam in ft^3/day/ft

Solution
From the given data,

$$k_z = 2 \times 10^{-2} \text{ mm/s} = 5.67 \text{ ft/day}$$

$$k_x = 4 \times 10^{-2} \text{ mm/s} = 11.34 \text{ ft/day}$$

and $H = 20$ ft. For drawing the flow net,

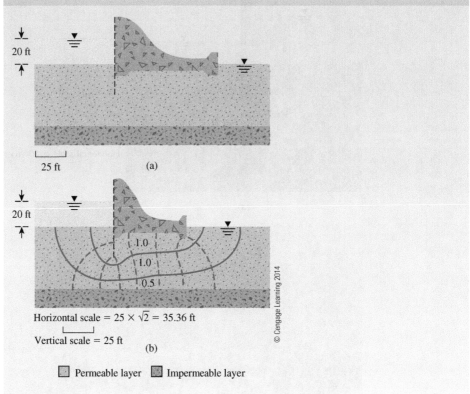

Figure 8.9

$$\text{Horizontal scale} = \sqrt{\frac{2 \times 10^{-2}}{4 \times 10^{-2}}}\,(\text{vertical scale})$$

$$= \frac{1}{\sqrt{2}}\,(\text{vertical scale})$$

On the basis of this, the dam section is replotted, and the flow net drawn as in Figure 8.9b. The rate of seepage is given by $q = \sqrt{k_x k_z}\, H(N_f/N_d)$. From Figure 8.9b, $N_d = 8$ and $N_f = 2.5$ (the lowermost flow channel has a width-to-length ratio of 0.5). So,

$$q = \sqrt{(5.67)(11.34)}\,(20)(2.5/8) = \textbf{50.12 ft}^3\textbf{/day/ft}$$

8.7 Mathematical Solution for Seepage

The seepage under several simple hydraulic structures can be solved mathematically. Harr (1962) has analyzed many such conditions. Figure 8.10 shows a nondimensional plot for the rate of seepage around a single row of sheet piles. In a similar manner, Figure 8.11 is a nondimensional plot for the rate of seepage under a dam. In Figures 8.10 and 8.11, the depth of penetration of the sheet pile is S, and the thickness of the permeable soil layer is T'.

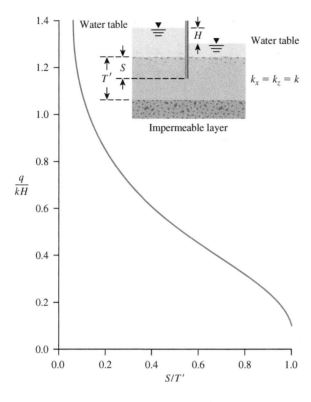

Figure 8.10 Plot of q/kH against S/T' for flow around a single row of sheet piles (*After Harr, 1962. By permission of Dover Publications, Inc.*)

Example 8.4

Refer to Figure 8.11. Given; the width of the dam, $B = 6$ m; length of the dam, $L = 120$ m; $S = 3$ m; $T' = 6$ m; $x = 2.4$ m; and $H_1 - H_2 = 5$ m. If the hydraulic conductivity of the permeable layer is 0.008 cm/sec, estimate the seepage under the dam (Q) in m³/day/m.

Solution

Given that $B = 6$ m, $T' = 6$ m, and $S = 3$ m, so $b = B/2 = 3$ m.

$$\frac{b}{T'} = \frac{3}{6} = 0.5$$

$$\frac{S}{T'} = \frac{3}{6} = 0.5$$

$$\frac{x}{b} = \frac{2.4}{3} = 0.8$$

From Figure 8.11, for $b/T' = 0.5$, $S/T' = 0.5$, and $x/b = 0.8$, the value of $q/kH \approx 0.378$.

Thus,

$$Q = q\,L = 0.378\,k\,HL = (0.378)(0.008 \times 10^{-2} \times 60 \times 60 \times 24 \text{ m/day})(5)(120)$$
$$= \textbf{1567.64 m}^3\textbf{/day}$$

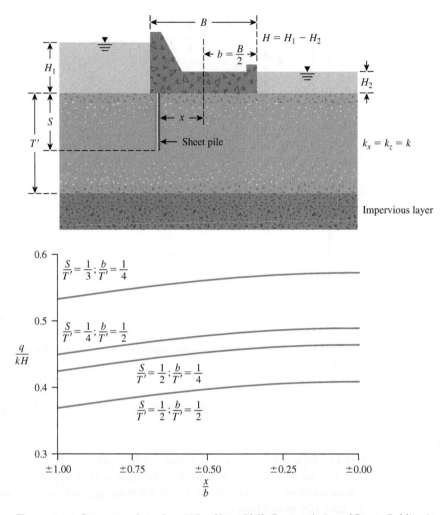

Figure 8.11 Seepage under a dam (*After Harr, 1962. By permission of Dover Publications, Inc.*)

8.8 Uplift Pressure under Hydraulic Structures

Flow nets can be used to determine the uplift pressure at the base of a hydraulic structure. This general concept can be demonstrated by a simple example. Figure 8.12a shows a weir, the base of which is 2 m below the ground surface. The necessary flow net also has been drawn (assuming that $k_x = k_z = k$). The pressure distribution diagram at the base of the weir can be obtained from the equipotential lines as follows.

There are seven equipotential drops (N_d) in the flow net, and the difference in the water levels between the upstream and downstream sides is $H = 7$ m. The head loss for each potential drop is $H/7 = 7/7 = 1$ m. The uplift pressure at

$$a \text{ (left corner of the base)} = \text{(Pressure head at } a\text{)} \times (\gamma_w)$$
$$= [(7 + 2) - 1]\gamma_w = 8\gamma_w$$

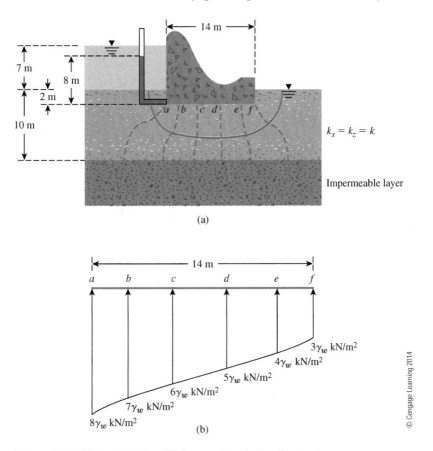

Figure 8.12 (a) A weir; (b) uplift force under a hydraulic structure

Similarly, the uplift pressure at

$$b = [9 - (2)(1)]\gamma_w = 7\gamma_w$$

and at

$$f = [9 - (6)(1)]\gamma_w = 3\gamma_w$$

The uplift pressures have been plotted in Figure 8.12b. The uplift force per unit length measured along the axis of the weir can be calculated by finding the area of the pressure diagram.

8.9 Seepage through an Earth Dam on an Impervious Base

Figure 8.13 shows a homogeneous earth dam resting on an impervious base. Let the hydraulic conductivity of the compacted material of which the earth dam is made be equal to k. The free surface of the water passing through the dam is given by $abcd$. It is assumed that $a'bc$ is parabolic. The slope of the free surface can be assumed to be equal to the

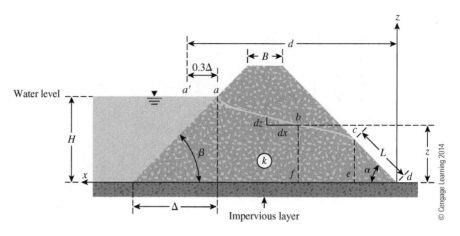

Figure 8.13 Flow through an earth dam constructed over an impervious base

hydraulic gradient. It is also is assumed that, because this hydraulic gradient is constant with depth (Dupuit, 1863),

$$i \approx \frac{dz}{dx} \qquad (8.29)$$

Considering the triangle *cde*, we can give the rate of seepage per unit length of the dam (at right angles to the cross section shown in Figure 8.13) as

$$q = kiA$$

$$i = \frac{dz}{dx} = \tan \alpha$$

$$A = (\overline{ce})(1) = L \sin \alpha$$

So

$$q = k(\tan \alpha)(L \sin \alpha) = kL \tan \alpha \sin \alpha \qquad (8.30)$$

Again, the rate of seepage (per unit length of the dam) through the section *bf* is

$$q = kiA = k\left(\frac{dz}{dx}\right)(z \times 1) = kz\frac{dz}{dx} \qquad (8.31)$$

For continuous flow,

$$q_{\text{Eq. (8.30)}} = q_{\text{Eq. (8.31)}}$$

or

$$kz\frac{dz}{dx} = kL \tan \alpha \sin \alpha$$

or

$$\int_{z=L \sin \alpha}^{z=H} kz \, dz = \int_{x=L \cos \alpha}^{x=d} (kL \tan \alpha \sin \alpha) \, dx$$

$$\tfrac{1}{2}(H^2 - L^2 \sin^2 \alpha) = L \tan \alpha \sin \alpha (d - L \cos \alpha)$$

$$\frac{H^2}{2} - \frac{L^2 \sin^2 \alpha}{2} = Ld\left(\frac{\sin^2 \alpha}{\cos \alpha}\right) - L^2 \sin^2 \alpha$$

$$\frac{H^2 \cos \alpha}{2 \sin^2 \alpha} - \frac{L^2 \cos \alpha}{2} = Ld - L^2 \cos \alpha$$

or

$$L^2 \cos \alpha - 2Ld + \frac{H^2 \cos \alpha}{\sin^2 \alpha} = 0$$

So,

$$L = \frac{d}{\cos \alpha} - \sqrt{\frac{d^2}{\cos^2 \alpha} - \frac{H^2}{\sin^2 \alpha}} \tag{8.32}$$

Following is a step-by-step procedure to obtain the seepage rate q (per unit length of the dam):

Step 1: Obtain α.
Step 2: Calculate Δ (see Figure 8.13) and then 0.3Δ.
Step 3: Calculate d.
Step 4: With known values of α and d, calculate L from Eq. (8.32).
Step 5: With known value of L, calculate q from Eq. (8.30).

The preceding solution generally is referred to as Schaffernak's solution (1917) with Casagrande's (1937) correction, since Casagrande experimentally showed that the parabolic free surface starts from a', not a (Figure 8.13).

Example 8.5

Refer to the earth dam shown in Figure 8.13. Given that $\beta = 45°$, $\alpha = 30°$, $B = 10$ ft, $H = 20$ ft, height of dam $= 25$ ft, and $k = 2 \times 10^{-4}$ ft/min, calculate the seepage rate, q, in ft³/day/ft length.

Solution
We know that $\beta = 45°$ and $\alpha = 30°$. Thus,

$$\Delta = \frac{H}{\tan \beta} = \frac{20}{\tan 45°} = 20 \text{ ft} \quad 0.3\Delta = (0.3)(20) = 6 \text{ ft}$$

$$d = 0.3\Delta + \frac{(25 - 20)}{\tan \beta} + B + \frac{25}{\tan \alpha}$$

$$= 6 + \frac{(25 - 20)}{\tan 45°} + 10 + \frac{25}{\tan 30} = 64.3 \text{ ft}$$

From Eq. (8.32),

$$L = \frac{d}{\cos \alpha} - \sqrt{\frac{d^2}{\cos^2 \alpha} - \frac{H^2}{\sin^2 \alpha}}$$

$$= \frac{64.3}{\cos 30} - \sqrt{\left(\frac{64.3}{\cos 30}\right)^2 - \left(\frac{20}{\sin 30}\right)^2} = 11.7 \text{ ft}$$

From Eq. (8.30)

$$q = kL \tan \alpha \sin \alpha = (2 \times 10^{-4})(11.7)(\tan 30)(\sin 30)$$

$$= 6.754 \times 10^{-4} \text{ ft}^3/\text{min/ft} = \textbf{0.973 ft}^3\textbf{/day/ft}$$

8.10 L. Casagrande's Solution for Seepage through an Earth Dam

Equation (8.32) is derived on the basis of Dupuit's assumption (i.e., $i \approx dz/dx$). It was shown by Casagrande (1932) that, when the downstream slope angle α in Figure 8.13 becomes greater than 30°, deviations from Dupuit's assumption become more noticeable. Thus (see Figure 8.13), L. Casagrande (1932) suggested that

$$i = \frac{dz}{ds} = \sin \alpha \tag{8.33}$$

where $ds = \sqrt{dx^2 + dz^2}$.

So Eq. (8.30) can now be modified as

$$q = kiA = k \sin \alpha(L \sin \alpha) = kL \sin^2 \alpha \tag{8.34}$$

Again,

$$q = kiA = k\left(\frac{dz}{ds}\right)(1 \times z) \tag{8.35}$$

Combining Eqs. (8.34) and (8.35) yields

$$\int_{L \sin \alpha}^{H} z \, dz = \int_{L}^{s} L \sin^2 \alpha \, ds \tag{8.36}$$

where s = length of curve $a'bc$

$$\frac{1}{2}(H^2 - L^2 \sin^2 \alpha) = L \sin^2 \alpha (s - L)$$

or

$$L = s - \sqrt{s^2 - \frac{H^2}{\sin^2 \alpha}} \tag{8.37}$$

With about 4 to 5% error, we can write

$$s = \sqrt{d^2 + H^2} \tag{8.38}$$

Combining Eqs. (8.37) and (8.38) yields

$$L = \sqrt{d^2 + H^2} - \sqrt{d^2 - H^2 \cot^2 \alpha} \tag{8.39}$$

Once the magnitude of L is known, the rate of seepage can be calculated from Eq. (8.34) as

$$q = kL \sin^2 \alpha$$

In order to avoid the approximation introduced in Eqs. (8.38) and (8.39), a solution was provided by Gilboy (1934). This is shown in a graphical form in Figure 8.14. Note, in this graph,

$$m = \frac{L \sin \alpha}{H} \tag{8.40}$$

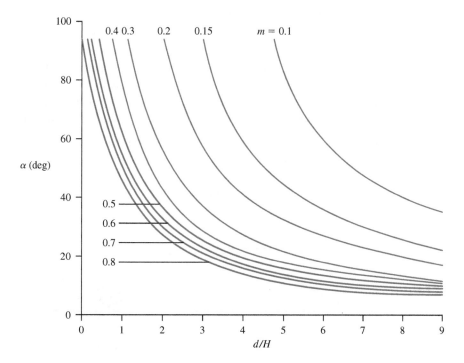

Figure 8.14 Chart for solution by L. Casagrande's method based on Gilboy's solution

In order to use the graph,

> *Step 1:* Determine d/H.
> *Step 2:* For a given d/H and α, determine m.
> *Step 3:* Calculate $L = \dfrac{mH}{\sin \alpha}$.
> *Step 4:* Calculate $kL \sin^2 \alpha$.

8.11 Filter Design

When seepage water flows from a soil with relatively fine grains into a coarser material, there is danger that the fine soil particles may wash away into the coarse material. Over a period of time, this process may clog the void spaces in the coarser material. Hence, the grain-size distribution of the coarse material should be properly manipulated to avoid this situation. A properly designed coarser material is called a *filter.* Figure 8.15 shows the steady-state seepage condition in an earth dam which has a toe filter. For proper selection of the filter material, two conditions should be kept in mind:

> *Condition 1:* The size of the voids in the filter material should be small enough to hold the larger particles of the protected material in place.
> *Condition 2:* The filter material should have a high hydraulic conductivity to prevent buildup of large seepage forces and hydrostatic pressures in the filters.

It can be shows that, if three perfect spheres have diameters greater than 6.5 times the diameter of a smaller sphere, the small sphere can move through the void spaces of the larger ones (Figure 8.16a). Generally speaking, in a given soil, the sizes of the grains vary over a wide range. If the pore spaces in a filter are small enough to hold D_{85} of the soil to be protected, then the finer soil particles also will be protected (Figure 8.16b). This means that the effective diameter of the pore spaces in the filter should be less than D_{85} of the soil to be protected. The effective pore diameter is about $\frac{1}{5} D_{15}$ of the filter. With this in mind and based on the experimental investigation of filters, Terzaghi and Peck (1948) provided the following criteria to satisfy Condition 1:

$$\frac{D_{15(F)}}{D_{85(S)}} \leq 4 \text{ to } 5 \qquad (\text{to satisfy Condition 1}) \tag{8.41}$$

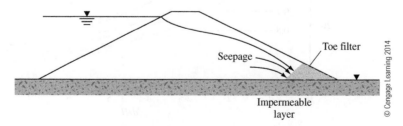

Figure 8.15 Steady-state seepage in an earth dam with a toe filter

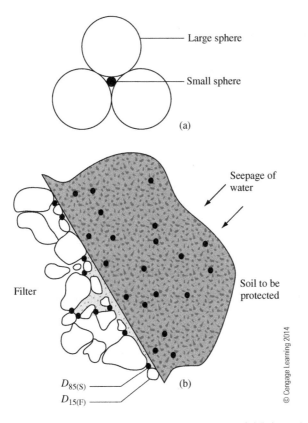

Figure 8.16 (a) Large spheres with diameters of 6.5 times the diameter of the small sphere; (b) boundary between a filter and the soil to be protected

In order to satisfy Condition 2, they suggested that

$$\frac{D_{15(F)}}{D_{15(S)}} \geq 4 \text{ to } 5 \qquad (\text{to satisfy Condition 2}) \qquad (8.42)$$

where $D_{15(F)}$ = diameter through which 15% of filter material will pass
$\qquad D_{15(S)}$ = diameter through which 15% of soil to be protected will pass
$\qquad D_{85(S)}$ = diameter through which 85% of soil to be protected will pass

The proper use of Eqs. (8.41) and (8.42) to determine the grain-size distribution of soils used as filters is shown in Figure 8.17. Consider the soil used for the construction of the earth dam shown in Figure 8.15. Let the grain-size distribution of this soil be given by curve *a* in Figure 8.17. We can now determine $5D_{85(S)}$ and $5D_{15(S)}$ and plot them as shown in Figure 8.17. The acceptable grain-size distribution of the filter material will have to lie in the shaded zone. (*Note:* The shape of curves *b* and *c* are approximately the same as curve *a.*)

The U.S. Navy (1971) requires the following conditions for the design of filters.

Condition 1: For avoiding the movement of the particles of the protected soil:

$$\frac{D_{15(F)}}{D_{85(S)}} < 5$$

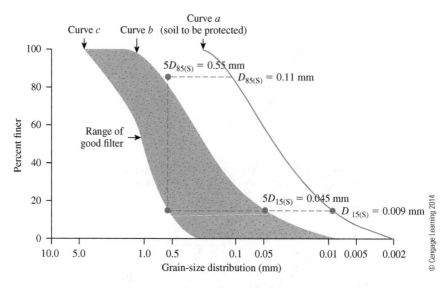

Figure 8.17 Determination of grain-size distribution of filter using Eqs. (8.41) and (8.42)

$$\frac{D_{50\,(F)}}{D_{50\,(S)}} < 25$$

$$\frac{D_{15(F)}}{D_{15(S)}} < 20$$

If the uniformity coefficient C_u of the protected soil is less than 1.5, $D_{15(F)}/D_{85(S)}$ may be increased to 6. Also, if C_u of the protected soil is greater than 4, $D_{15(F)}/D_{15(S)}$ may be increased to 40.

Condition 2:   For avoiding buildup of large seepage force in the filter:

$$\frac{D_{15(F)}}{D_{15(S)}} > 4$$

Condition 3:   The filter material should not have grain sizes greater than 76.2 mm (3 in.). (This is to avoid segregation of particles in the filter.)

Condition 4:   To avoid internal movement of fines in the filter, it should have no more than 5% passing a No. 200 sieve.

Condition 5:   When perforated pipes are used for collecting seepage water, filters also are used around the pipes to protect the fine-grained soil from being washed into the pipes. To avoid the movement of the filter material into the drain-pipe perforations, the following additional conditions should be met:

$$\frac{D_{85(F)}}{\text{slot width}} > 1.2 \text{ to } 1.4$$

$$\frac{D_{85(F)}}{\text{hole diameter}} > 1.0 \text{ to } 1.2$$

8.12 Summary

Following is a summary of the subjects covered in this chapter.

- In an *isotropic soil,* Laplace's equation of continuity for two-dimensional flow is given as [Eq. (8.5)]:

$$\frac{\partial^2 h}{\partial x^2} + \frac{\partial^2 h}{\partial z^2} = 0$$

- A flow net is a combination of flow lines and equipotential lines that are two orthogonal families of lines (Section 8.4).
- In an isotropic soil, seepage (q) for unit length of the structure in unit time can be expressed as [Eq. (8.24)]

$$q = kH\left(\frac{N_f}{N_d}\right)n$$

- The construction of flow nets in *anisotropic soil* was outlined in Section 8.6. For this case, the seepage for unit length of the structure in unit time is [Eq. (8.28)]

$$q = \sqrt{k_x k_z}\,\frac{HN_f}{N_d}$$

- Seepage through an earth dam on an impervious base was discussed in Section 8.9 (Schaffernak's solution with Casagrande's correction) and Section 8.10 (L. Casagrande solution).
- The criteria for filter design are given in Section 8.11 [Eqs. (8.41) and (8.42)], according to which

$$\frac{D_{15(F)}}{D_{85(S)}} \leq 4 \text{ to } 5$$

and

$$\frac{D_{15(F)}}{D_{15(S)}} \geq 4 \text{ to } 5$$

Problems

8.1 Refer to the constant-head permeability test arrangement in a two-layered soil as shown in Figure 8.2. During the test, it was seen that when a constant head of $h_1 = 200$ mm was maintained, the magnitude of h_2 was 80 mm. If k_1 is 0.004 cm/sec, determine the value of k_2 given $H_1 = 100$ mm and $H_2 = 150$ mm.

8.2 Refer to Figure 8.18. Given:

- $H_1 = 6$ m
- $H_2 = 1.5$ m
- $D = 3$ m
- $D_1 = 6$ m

draw a flow net. Calculate the seepage loss per meter length of the sheet pile (at a right angle to the cross section shown).

8.3 Draw a flow net for the single row of sheet piles driven into a permeable layer as shown in Figure 8.18. Given:

- $H_1 = 3$ m
- $H_2 = 0.5$ m
- $D = 1.5$ m
- $D_1 = 3.75$ m

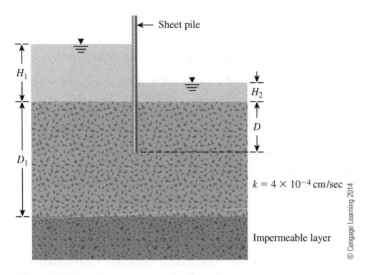

Figure 8.18

calculate the seepage loss per meter length of the sheet pile (at right angles to the cross section shown).

8.4   Refer to Figure 8.18. Given:

- $H_1 = 4$ m
- $D_1 = 6$ m
- $H_2 = 1.5$ m
- $D = 3.6$ m

calculate the seepage loss in m^3/day per meter length of the sheet pile (at right angles to the cross section shown). Use Figure 8.10.

8.5   For the hydraulic structure shown in Figure 8.19, draw a flow net for flow through the permeable layer and calculate the seepage loss in m^3/day/m.

8.6   Refer to Problem 8.5. Using the flow net drawn, calculate the hydraulic uplift force at the base of the hydraulic structure per meter length (measured along the axis of the structure).

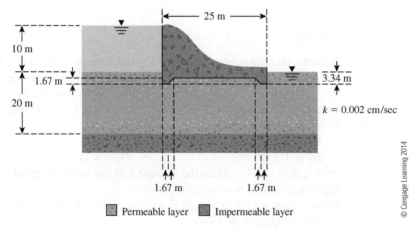

Figure 8.19

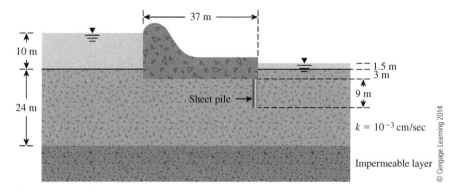

Figure 8.20

8.7 Draw a flow net for the weir shown in Figure 8.20. Calculate the rate of seepage under the weir.

8.8 For the weir shown in Figure 8.21, calculate the seepage in the permeable layer in $m^3/day/m$ for (a) $x' = 1$ m and (b) $x' = 2$ m. Use Figure 8.11.

8.9 An earth dam is shown in Figure 8.22. Determine the seepage rate, q, in $m^3/day/m$ length. Given: $\alpha_1 = 35°$, $\alpha_2 = 40°$, $L_1 = 5$ m, $H = 7$ m, $H_1 = 10$ m, and $k = 3 \times 10^{-4}$ cm/sec. Use Schaffernak's solution.

8.10 Repeat Problem 8.9 using L. Casagrande's method.

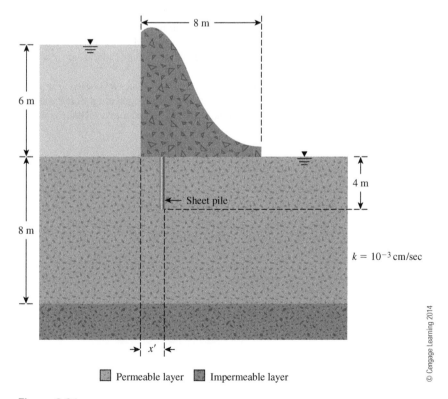

Figure 8.21

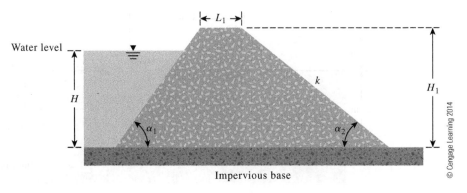

Figure 8.22

References

CASAGRANDE, A. (1937). "Seepage Through Dams," in *Contribution to Soil Mechanics 1925–1940*, Boston Society of Civil Engineers, Boston.

CASAGRANDE, L. (1932). "Naeherungsmethoden zur Bestimmurg von Art und Menge der Sickerung durch geschuettete Daemme," Thesis, Technische Hochschule, Vienna.

DUPUIT, J. (1863). *Etudes Theoriques et Practiques sur le Mouvement des Eaux dans les Canaux Decouverts et a Travers les Terrains Permeables*, Dunod, Paris.

GILBOY, G. (1934). "Mechanics of Hydraulic Fill Dams," in *Contributions to Soil Mechanics 1925–1940*, Boston Society of Civil Engineers, Boston.

HARR, M. E. (1962). *Ground Water and Seepage*, McGraw-Hill, New York.

SCHAFFERNAK, F. (1917). "Über die Standicherheit durchlaessiger geschuetteter Dämme," *Allgem. Bauzeitung.*

TERZAGHI, K., and PECK, R. B. (1948). *Soil Mechanics in Engineering Practice*, Wiley, New York.

U.S. DEPARTMENT OF THE NAVY, Naval Facilities Engineering Command (1971). "Design Manual— Soil Mechanics, Foundations, and Earth Structures," *NAVFAC DM-7*, Washington, D.C.

Appendix - A

A Generalized Case for Rankine Active and Passive Pressure—Granular Backfill

In Sections 13.5, 13.6, and 13.7, we discussed the Rankine active and passive pressure cases for a frictionless wall with a vertical back and a horizontal backfill of granular soil. This can be extended to general cases of frictionless wall with inclined back and inclined backfill (granular soil) as shown in Figure A.1 (Chu, 1991).

Rankine Active Case

For the Rankine active case, the lateral earth pressure (σ_a') at a depth z can be given as

$$\sigma_a' = \frac{\gamma z \cos \alpha \sqrt{1 + \sin^2\phi' - 2\sin\phi' \cos\psi_a}}{\cos \alpha + \sqrt{\sin^2\phi' - \sin^2\alpha}} \tag{A.1}$$

where

$$\psi_a = \sin^{-1}\left(\frac{\sin\alpha}{\sin\phi'}\right) - \alpha + 2\theta \tag{A.2}$$

The pressure σ_a' will be inclined at an angle β_a with the plane drawn at right angle to the back face of the wall, and

$$\beta_a = \tan^{-1}\left(\frac{\sin\phi' \sin\psi_a}{1 - \sin\phi' \cos\psi_a}\right) \tag{A.3}$$

The active force P_a for unit length of the wall can then be calculated as

$$P_a = \frac{1}{2}\gamma H^2 K_{a(R)} \tag{A.4}$$

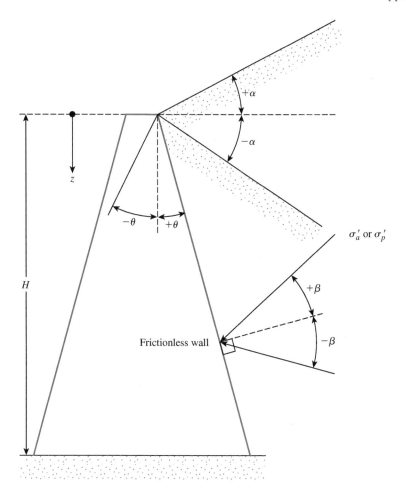

Figure A.1 General case for Rankine active and passive pressures

where

$$K_{a(R)} = \frac{\cos(\alpha - \theta)\sqrt{1 + \sin^2\phi' - 2\sin\phi'\cos\psi_a}}{\cos^2\theta(\cos\alpha + \sqrt{\sin^2\phi' - \sin^2\alpha})}$$

= Rankine active earth-pressure coefficient for generalized case   (A.5)

The location and direction of the resultant force P_a is shown in Figure A.2a. Also shown in this figure is the failure wedge, *ABC*. Note that *BC* will be inclined at an angle η_a. Or

$$\eta_a = \frac{\pi}{4} + \frac{\phi'}{2} + \frac{\alpha}{2} - \frac{1}{2}\sin^{-1}\left(\frac{\sin\alpha}{\sin\phi'}\right)$$ (A.6)

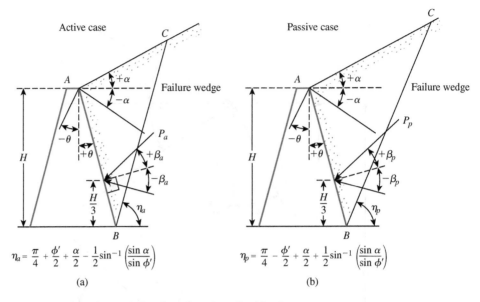

Figure A.2 Location and direction of resultant Rankine force

As a special case, for a vertical back face of the wall (that is, $\theta = 0$) as shown in Figure 13.10, Eqs. (A.4) and (A.5) simplify to the following:

$$P_a = \frac{1}{2} K_{a(R)} \gamma H^2 \tag{13.24}$$

where

$$K_{a(R)} = \cos \alpha \frac{\cos \alpha - \sqrt{\cos^2 \alpha - \cos^2 \phi'}}{\cos \alpha + \sqrt{\cos^2 \alpha - \cos^2 \phi'}} \tag{13.23}$$

Tables A.1 and A.2 give the variations of $K_{a(R)}$ and β_a, respectively, for various values of α, θ, and ϕ'.

Rankine Passive Case

Similar to the active case, for the Rankine passive case, we can obtain the following relationships.

$$\sigma'_p = \frac{\gamma z \cos \alpha \sqrt{1 + \sin^2 \phi' + 2 \sin \phi' \cos \psi_p}}{\cos \alpha - \sqrt{\sin^2 \phi' - \sin^2 \alpha}} \tag{A.7}$$

where

$$\psi_p = \sin^{-1}\left(\frac{\sin \alpha}{\sin \phi'}\right) + \alpha - 2\theta \tag{A.8}$$

Table A.1 Variation of $K_{a(R)}$ [Eq. (A.5)]

		$K_{a(R)}$						
		ϕ' (deg)						
α (deg)	θ (deg)	28	30	32	34	36	38	40
	0	0.361	0.333	0.307	0.283	0.26	0.238	0.217
	2	0.363	0.335	0.309	0.285	0.262	0.240	0.220
	4	0.368	0.341	0.315	0.291	0.269	0.248	0.228
0	6	0.376	0.350	0.325	0.302	0.280	0.260	0.242
	8	0.387	0.362	0.338	0.316	0.295	0.276	0.259
	10	0.402	0.377	0.354	0.333	0.314	0.296	0.280
	15	0.450	0.428	0.408	0.390	0.373	0.358	0.345
	0	0.366	0.337	0.311	0.286	0.262	0.240	0.219
	2	0.373	0.344	0.317	0.292	0.269	0.247	0.226
	4	0.383	0.354	0.328	0.303	0.280	0.259	0.239
5	6	0.396	0.368	0.342	0.318	0.296	0.275	0.255
	8	0.412	0.385	0.360	0.336	0.315	0.295	0.276
	10	0.431	0.405	0.380	0.358	0.337	0.318	0.300
	15	0.490	0.466	0.443	0.423	0.405	0.388	0.373
	0	0.380	0.350	0.321	0.294	0.270	0.246	0.225
	2	0.393	0.362	0.333	0.306	0.281	0.258	0.236
	4	0.408	0.377	0.348	0.322	0.297	0.274	0.252
10	6	0.426	0.395	0.367	0.341	0.316	0.294	0.273
	8	0.447	0.417	0.389	0.363	0.339	0.317	0.297
	10	0.471	0.441	0.414	0.388	0.365	0.344	0.324
	15	0.542	0.513	0.487	0.463	0.442	0.422	0.404
	0	0.409	0.373	0.341	0.311	0.283	0.258	0.235
	2	0.427	0.391	0.358	0.328	0.300	0.274	0.250
	4	0.448	0.411	0.378	0.348	0.320	0.294	0.271
15	6	0.472	0.435	0.402	0.371	0.344	0.318	0.295
	8	0.498	0.461	0.428	0.398	0.371	0.346	0.323
	10	0.527	0.490	0.457	0.428	0.400	0.376	0.353
	15	0.610	0.574	0.542	0.513	0.487	0.463	0.442
	0	0.461	0.414	0.374	0.338	0.306	0.277	0.250
	2	0.486	0.438	0.397	0.360	0.328	0.298	0.271
	4	0.513	0.465	0.423	0.386	0.353	0.323	0.296
20	6	0.543	0.495	0.452	0.415	0.381	0.351	0.324
	8	0.576	0.527	0.484	0.446	0.413	0.383	0.355
	10	0.612	0.562	0.518	0.481	0.447	0.417	0.390
	15	0.711	0.660	0.616	0.578	0.545	0.515	0.488

Table A.2 Variation of β_a [Eq. (A.3)]

| | | β_a | | | | | | |
| | | ϕ' (deg) | | | | | | |
α (deg)	θ (deg)	28	30	32	34	36	38	40
	0	0.000	0.000	0.000	0.000	0.000	0.000	0.000
	2	3.525	3.981	4.484	5.041	5.661	6.351	7.124
	4	6.962	7.848	8.821	9.893	11.075	12.381	13.827
0	6	10.231	11.501	12.884	14.394	16.040	17.837	19.797
	8	13.270	14.861	16.579	18.432	20.428	22.575	24.876
	10	16.031	17.878	19.850	21.951	24.184	26.547	29.039
	15	21.582	23.794	26.091	28.464	30.905	33.402	35.940
	0	5.000	5.000	5.000	5.000	5.000	5.000	5.000
	2	8.375	8.820	9.311	9.854	10.455	11.123	11.870
	4	11.553	12.404	13.336	14.358	15.482	16.719	18.085
5	6	14.478	15.679	16.983	18.401	19.942	21.618	23.441
	8	17.112	18.601	20.203	21.924	23.773	25.755	27.876
	10	19.435	21.150	22.975	24.915	26.971	29.144	31.434
	15	23.881	25.922	28.039	30.227	32.479	34.787	37.140
	0	10.000	10.000	10.000	10.000	10.000	10.000	10.000
	2	13.057	13.491	13.967	14.491	15.070	15.712	16.426
	4	15.839	16.657	17.547	18.519	19.583	20.751	22.034
10	6	18.319	19.460	20.693	22.026	23.469	25.032	26.726
	8	20.483	21.888	23.391	24.999	26.720	28.559	30.522
	10	22.335	23.946	25.653	27.460	29.370	31.385	33.504
	15	25.683	27.603	29.589	31.639	33.747	35.908	38.114
	0	15.000	15.000	15.000	15.000	15.000	15.000	15.000
	2	17.576	18.001	18.463	18.967	19.522	20.134	20.812
	4	19.840	20.631	21.485	22.410	23.417	24.516	25.719
15	6	21.788	22.886	24.060	25.321	26.677	28.139	29.716
	8	23.431	24.778	26.206	27.722	29.335	31.052	32.878
	10	24.783	26.328	27.950	29.654	31.447	33.332	35.310
	15	27.032	28.888	30.793	32.747	34.751	36.802	38.894
	0	20.000	20.000	20.000	20.000	20.000	20.000	20.000
	2	21.925	22.350	22.803	23.291	23.822	24.404	25.045
	4	23.545	24.332	25.164	26.054	27.011	28.048	29.175
20	6	24.876	25.966	27.109	28.317	29.604	30.980	32.455
	8	25.938	27.279	28.669	30.124	31.657	33.276	34.989
	10	26.755	28.297	29.882	31.524	33.235	35.021	36.886
	15	27.866	29.747	31.638	33.552	35.498	37.478	39.491

The inclination β_p of σ'_p, as shown in Figure A.1, is

$$\beta_p = \tan^{-1}\left(\frac{\sin\phi'\sin\psi_p}{1 + \sin\phi'\cos\psi_p}\right) \tag{A.9}$$

The passive force per unit length of the wall is

$$P_p = \frac{1}{2}\gamma H^2 K_{p(R)} \tag{A.10}$$

where

$$K_{p(R)} = \frac{\cos(\alpha - \theta)\sqrt{1 + \sin^2\phi' + 2\sin\phi'\cos\psi_p}}{\cos^2\theta(\cos\alpha - \sqrt{\sin^2\phi' - \sin^2\alpha})} \tag{A.11}$$

The location and direction of P_p along with the failure wedge is shown in Figure A.2b. For walls with vertical backface, $\theta = 0$,

$$P_p = \frac{1}{2}K_{p(R)}\gamma H^2 \tag{13.25}$$

where

$$K_{p(R)} = \cos\alpha\frac{\cos\alpha + \sqrt{\cos^2\alpha - \cos^2\phi'}}{\cos\alpha - \sqrt{\cos^2\alpha - \cos^2\phi'}} \tag{13.26}$$

Tables A.3 and A.4 give the variation of $K_{p(R)}$ and β_p, respectively, for various values of α, θ, and ϕ'.

Table A.3 Variation of $K_{p(R)}$ [Eq. (A.11)]

		$K_{p(R)}$						
				ϕ' (deg)				
α (deg)	θ (deg)	28	30	32	34	36	38	40
	0	2.77	3	3.255	3.537	3.852	4.204	4.599
	2	2.770	3.000	3.255	3.537	3.852	4.204	4.599
	4	2.771	3.001	3.255	3.538	3.852	4.204	4.599
0	6	2.772	3.002	3.256	3.539	3.853	4.205	4.600
	8	2.773	3.003	3.258	3.540	3.854	4.206	4.601
	10	2.775	3.005	3.259	3.542	3.856	4.207	4.602
	15	2.783	3.012	3.266	3.547	3.861	4.212	4.607
	0	2.715	2.943	3.196	3.476	3.788	4.136	4.527
	2	2.734	2.964	3.218	3.500	3.814	4.165	4.558
	4	2.755	2.986	3.242	3.525	3.841	4.194	4.589
5	6	2.776	3.009	3.266	3.551	3.868	4.223	4.621
	8	2.798	3.032	3.290	3.577	3.896	4.253	4.654
	10	2.820	3.055	3.315	3.604	3.925	4.284	4.687
	15	2.880	3.119	3.382	3.675	4.000	4.365	4.774
	0	2.551	2.775	3.022	3.295	3.598	3.937	4.316
	2	2.589	2.815	3.065	3.342	3.649	3.991	4.376
	4	2.627	2.857	3.110	3.389	3.700	4.047	4.436
10	6	2.667	2.899	3.155	3.438	3.753	4.104	4.497
	8	2.707	2.942	3.201	3.488	3.806	4.161	4.560
	10	2.749	2.987	3.249	3.539	3.861	4.220	4.624
	15	2.860	3.104	3.374	3.672	4.004	4.375	4.790
	0	2.284	2.502	2.740	3.003	3.293	3.615	3.977
	2	2.336	2.558	2.801	3.068	3.364	3.693	4.061
	4	2.389	2.616	2.863	3.135	3.436	3.771	4.146
15	6	2.444	2.675	2.926	3.204	3.510	3.851	4.233
	8	2.501	2.735	2.991	3.274	3.586	3.933	4.322
	10	2.559	2.797	3.058	3.345	3.663	4.017	4.413
	15	2.712	2.961	3.234	3.534	3.867	4.237	4.651
	0	1.918	2.132	2.362	2.612	2.886	3.189	3.526
	2	1.979	2.199	2.435	2.691	2.972	3.283	3.629
	4	2.042	2.267	2.509	2.772	3.060	3.378	3.733
20	6	2.107	2.338	2.585	2.854	3.150	3.476	3.840
	8	2.174	2.410	2.664	2.939	3.242	3.577	3.949
	10	2.243	2.485	2.745	3.027	3.337	3.680	4.061
	15	2.426	2.683	2.958	3.258	3.587	3.951	4.357

Table A.4 Variation of β_p [Eq. (A.9)]

		β_p						
		ϕ' (deg)						
α (deg)	θ (deg)	28	30	32	34	36	38	40
	0	0.000	0.000	0.000	0.000	0.000	0.000	0.000
	2	−1.278	−1.333	−1.385	−1.434	−1.481	−1.524	−1.565
	4	−2.554	−2.665	−2.769	−2.867	−2.960	−3.047	−3.129
0	6	−3.827	−3.993	−4.150	−4.298	−4.437	−4.568	−4.691
	8	−5.095	−5.318	−5.527	−5.725	−5.910	−6.085	−6.250
	10	−6.357	−6.636	−6.899	−7.146	−7.379	−7.598	−7.804
	15	−9.474	−9.896	−10.293	−10.668	−11.021	−11.353	−11.666
	0	5.000	5.000	5.000	5.000	5.000	5.000	5.000
	2	3.731	3.674	3.621	3.571	3.524	3.480	3.439
	4	2.458	2.345	2.239	2.139	2.046	1.958	1.875
5	6	1.182	1.013	0.854	0.705	0.565	0.434	0.310
	8	−0.096	−0.320	−0.531	−0.729	−0.915	−1.090	−1.255
	10	−1.374	−1.653	−1.916	−2.163	−2.395	−2.614	−2.819
	15	−4.557	−4.974	−5.367	−5.737	−6.084	−6.411	−6.719
	0	10.000	10.000	10.000	10.000	10.000	10.000	10.000
	2	8.766	8.704	8.647	8.593	8.543	8.497	8.453
	4	7.521	7.398	7.285	7.179	7.080	6.987	6.900
10	6	6.266	6.084	5.915	5.757	5.610	5.472	5.343
	8	5.003	4.763	4.539	4.331	4.136	3.953	3.782
	10	3.734	3.437	3.159	2.900	2.658	2.431	2.219
	15	0.546	0.107	−0.302	−0.684	−1.043	−1.378	−1.693
	0	15.000	15.000	15.000	15.000	15.000	15.000	15.000
	2	13.835	13.763	13.696	13.636	13.580	13.528	13.480
	4	12.650	12.507	12.377	12.258	12.147	12.045	11.951
15	6	11.448	11.237	11.044	10.867	10.704	10.553	10.413
	8	10.229	9.952	9.699	9.466	9.251	9.052	8.868
	10	8.998	8.656	8.344	8.056	7.791	7.545	7.317
	15	5.871	5.375	4.921	4.502	4.114	3.754	3.420
	0	20.000	20.000	20.000	20.000	20.000	20.000	20.000
	2	18.962	18.866	18.782	18.708	18.641	18.580	18.525
	4	17.891	17.703	17.539	17.393	17.262	17.143	17.034
20	6	16.790	16.514	16.273	16.059	15.866	15.690	15.530
	8	15.662	15.302	14.987	14.707	14.454	14.224	14.014
	10	14.509	14.069	13.684	13.340	13.029	12.747	12.488
	15	11.540	10.912	10.360	9.865	9.418	9.010	8.637

References

CHU, S. C. (1991). "Rankine Analysis of Active and Passive Pressures on Dry Sand," *Soils and Foundations*, Vol. 31, No. 4, 115–120.

9 Soil Improvement and Ground Modification

16.1 Introduction

The soil at a construction site may not always be totally suitable for supporting structures such as buildings, bridges, highways, and dams. For example, in granular soil deposits, the *in situ* soil may be very loose and indicate a large elastic settlement. In such a case, the soil needs to be densified to increase its unit weight and thus its shear strength.

Sometimes the top layers of soil are undesirable and must be removed and replaced with better soil on which the structural foundation can be built. The soil used as fill should be well compacted to sustain the desired structural load. Compacted fills may also be required in low-lying areas to raise the ground elevation for construction of the foundation.

Soft saturated clay layers are often encountered at shallow depths below foundations. Depending on the structural load and the depth of the layers, unusually large consolidation settlement may occur. Special soil-improvement techniques are required to minimize settlement.

In Chapter 11, we mentioned that the properties of expansive soils could be altered substantially by adding stabilizing agents such as lime. Improving *in situ* soils by using additives is usually referred to as *stabilization.*

Various techniques are used to

1. Reduce the settlement of structures
2. Improve the shear strength of soil and thus increase the bearing capacity of shallow foundations
3. Increase the factor of safety against possible slope failure of embankments and earth dams
4. Reduce the shrinkage and swelling of soils

This chapter discusses some of the general principles of soil improvement, such as compaction, vibroflotation, precompression, sand drains, wick drains, stabilization by admixtures, jet grouting, and deep mixing, as well as the use of stone columns and sand compaction piles in weak clay to construct foundations.

16.2 General Principles of Compaction

If a small amount of water is added to a soil that is then compacted, the soil will have a certain unit weight. If the moisture content of the same soil is gradually increased and the energy of compaction is the same, the dry unit weight of the soil will gradually increase. The reason is that water acts as a lubricant between the soil particles, and under compaction it helps rearrange the solid particles into a denser state. The increase in dry unit weight with increase of moisture content for a soil will reach a limiting value beyond which the further addition of water to the soil will result in a *reduction* in dry unit weight. The moisture content at which the *maximum dry unit weight* is obtained is referred to as the *optimum moisture content*.

The standard laboratory tests used to evaluate maximum dry unit weights and optimum moisture contents for various soils are

- The Standard Proctor test (ASTM designation D-698)
- The Modified Proctor test (ASTM designation D-1557)

The soil is compacted in a mold in several layers by a hammer. The moisture content of the soil, w, is changed, and the dry unit weight, γ_d, of compaction for each test is determined. The maximum dry unit weight of compaction and the corresponding optimum moisture content are determined by plotting a graph of γ_d against w (%). The standard specifications for the two types of Proctor test are given in Tables 16.1 and 16.2.

Table 16.1 Specifications for Standard Proctor Test (Based on ASTM Designation 698)

Item	Method A	Method B	Method C
Diameter of mold	101.6 mm (4 in.)	101.6 mm (4 in.)	152.4 mm (6 in.)
Volume of mold	944 cm^3 $\left(\frac{1}{30} \text{ ft}^3\right)$	944 cm^3 $\left(\frac{1}{30} \text{ ft}^3\right)$	2124 cm^3(0.075 ft^3)
Mass (weight) of hammer	2.5 kg (5.5 lb)	2.5 kg (5.5 lb)	2.5 kg (5.5 lb)
Height of hammer drop	304.8 mm (12 in.)	304.8 mm (12 in.)	304.8 mm (12 in.)
Number of hammer blows per layer of soil	25	25	56
Number of layers of compaction	3	3	3
Energy of compaction	600 kN · m/m^3 (12,400 ft · lb/ft^3)	600 kN · m/m^3 (12,400 ft · lb/ft^3)	600 kN · m/m^3 (12,400 ft · lb/ft^3)
Soil to be used	Portion passing No. 4 (4.57-mm) sieve. May be used if 20% *or less* by weight of material is retained on No. 4 sieve.	Portion passing $\frac{3}{8}$-in. (9.5-mm) sieve. May be used if soil retained on No. 4 sieve *is more* than 20% and 20% *or less* by weight is retained on 9.5-mm $\left(\frac{3}{8}\text{-in.}\right)$ sieve.	Portion passing $\frac{3}{4}$-in. (19.0-mm) sieve. May be used if *more than* 20% by weight of material is retained on 9.5 mm $\left(\frac{3}{8}\text{-in.}\right)$ sieve and *less than* 30% by weight is retained on 19.00-mm $\left(\frac{3}{4}\text{-in.}\right)$ sieve.

Table 16.2 Specifications for Modified Proctor Test (Based on ASTM Designation 1557)

Item	Method A	Method B	Method C
Diameter of mold	101.6 mm (4 in.)	101.6 mm (4 in.)	152.4 mm (6 in.)
Volume of mold	944 cm^3 $\left(\frac{1}{30}\text{ ft}^3\right)$	944 cm^3 $\left(\frac{1}{30}\text{ ft}^3\right)$	2124 cm^3 (0.075 ft^3)
Mass (weight) of hammer	4.54 kg (10 lb)	4.54 kg (10 lb)	4.54 kg (10 lb)
Height of hammer drop	457.2 mm (18 in.)	457.2 mm (18 in.)	457.2 mm (18 in.)
Number of hammer blows per layer of soil	25	25	56
Number of layers of compaction	5	5	5
Energy of compaction	2700 kN · m/m^3 (56,000 ft · lb/ft^3)	2700 kN · m/m^3 (56,000 ft · lb/ft^3)	2700 kN · m/m^3 (56,000 ft · lb/ft^3)
Soil to be used	Portion passing No. 4 (4.57-mm) sieve. May be used if 20% *or less* by weight of material is retained on No. 4 sieve.	Portion passing 9.5-mm $\left(\frac{3}{8}\text{-in.}\right)$ sieve. May be used if soil retained on No. 4 sieve *is more* than 20% and 20% *or less* by weight is retained on 9.5-mm $\left(\frac{3}{8}\text{-in.}\right)$ sieve.	Portion passing 19.0-mm $\left(\frac{3}{4}\text{-in.}\right)$ sieve. May be used if *more than* 20% by weight of material is retained on 9.5-mm $\left(\frac{3}{8}\text{-in.}\right)$ sieve and *less than* 30% by weight is retained on 19-mm $\left(\frac{3}{4}\text{-in.}\right)$ sieve.

Figure 16.1 shows a plot of γ_d against w (%) for a clayey silt obtained from standard and modified Proctor tests (method A). The following conclusions may be drawn:

1. The maximum dry unit weight and the optimum moisture content depend on the degree of compaction.
2. The higher the energy of compaction, the higher is the maximum dry unit weight.
3. The higher the energy of compaction, the lower is the optimum moisture content.
4. No portion of the compaction curve can lie to the right of the zero-air-void line. The zero-air-void dry unit weight, γ_{zav}, at a given moisture content is the theoretical maximum value of γ_d, which means that all the void spaces of the compacted soil are filled with water, or

$$\gamma_{zav} = \frac{\gamma_w}{\dfrac{1}{G_s} + w} \tag{16.1}$$

where

γ_w = unit weight of water
G_s = specific gravity of the soil solids
w = moisture content of the soil

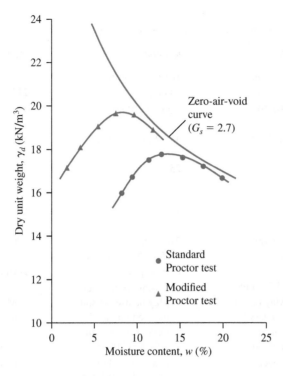

Figure 16.1 Standard and modified Proctor compaction curves for a clayey silt (method A)

5. The maximum dry unit weight of compaction and the corresponding optimum moisture content will vary from soil to soil.

Using the results of laboratory compaction (γ_d versus w), specifications may be written for the compaction of a given soil in the field. In most cases, the contractor is required to achieve a relative compaction of 90% or more on the basis of a specific laboratory test (either the standard or the modified Proctor compaction test). The relative compaction is defined as

$$RC = \frac{\gamma_{d(\text{field})}}{\gamma_{d(\text{max})}} \tag{16.2}$$

Chapter 2 introduced the concept of relative density (for the compaction of granular soils), defined as

$$D_r = \left[\frac{\gamma_d - \gamma_{d(\text{min})}}{\gamma_{d(\text{max})} - \gamma_{d(\text{min})}} \right] \frac{\gamma_{d(\text{max})}}{\gamma_d}$$

where

γ_d = dry unit weight of compaction in the field
$\gamma_{d(max)}$ = maximum dry unit weight of compaction as determined in the laboratory
$\gamma_{d(min)}$ = minimum dry unit weight of compaction as determined in the laboratory

For granular soils in the field, the degree of compaction obtained is often measured in terms of relative density. Comparing the expressions for relative density and relative compaction reveals that

$$RC = \frac{A}{1 - D_r(1 - A)} \tag{16.3}$$

where $A = \dfrac{\gamma_{d(min)}}{\gamma_{d(max)}}$.

16.3 Empirical Relationships for Compaction

Omar et al. (2003) recently presented the results of modified Proctor compaction tests on 311 soil samples. Of these samples, 45 were gravelly soil (GP, GP-GM, GW, GW-GM, and GM), 264 were sandy soil (SP, SP-SM, SW-SM, SW, SC-SM, SC, and SM), and two were clay with low plasticity (CL). All compaction tests were conducted using ASTM 1557 method C to avoid over-size correction. Based on the tests, the following correlations were developed.

$$\rho_{d(max)}(\text{kg/m}^3) = [4{,}804{,}574G_S - 195.55(\text{LL})^2 + 156.971(\text{R\# 4})^{0.5}$$
$$- 9{,}527{,}830]^{0.5} \tag{16.4}$$

$$\ln(w_{opt}) = 1.195 \times 10^{-4}(\text{LL})^2 - 1.964G_s - 6.617 \times 10^{-5}(\text{R\# 4})$$
$$+ 7.651 \tag{16.5}$$

where

$\rho_{d(max)}$ = maximum dry density
w_{opt} = optimum moisture content
G_s = specific gravity of soil solids
LL = liquid limit, in percent
R # 4 = percent retained on No. 4 sieve

It needs to be pointed out that Eqs. (16.4 and 16.5) contain the term for liquid limit. This is because the soils that were considered included silty and clayey sands.

For granular soils with less than 12% fines (i.e., finer than No. 200 sieve), relative density may be a better indicator for end product compaction specification in the field. Based on laboratory compaction tests on 55 clean sands (less than 5% finer than No. 200 sieve), Patra et al. (2010) provided the following relationships

$$D_r = AD_{50}^{-B} \tag{16.6}$$
$$A = 0.216 \ln E - 0.850 \tag{16.7}$$
$$B = -0.03 \ln E + 0.306 \tag{16.8}$$

where D_r = maximum relative density of compaction achieved with compaction energy E (kN-m/m^3)

D_{50} = median grain size (mm)

Gurtug and Sridharan (2004) proposed correlations for optimum moisture content and maximum dry unit weight with the plastic limit (PL) of cohesive soils. These correlations can be expressed as

$$w_{opt}(\%) = [1.95 - 0.38(\log E)](PL) \tag{16.9}$$

$$\gamma_{d(max)}(kN/m^3) = 22.68e^{-0.0183w_{opt}(\%)} \tag{16.10}$$

where PL = plastic limit (%)

E = compaction energy (kN-m/m^3)

For modified Proctor test, $E = 2700$ kN/m^3. Hence,

$$w_{opt}(\%) \approx 0.65(PL)$$

and

$$\gamma_{d(max)}(kN/m^3) = 22.68e^{-0.012(PL)}$$

Osman et al. (2008) analyzed a number of laboratory compaction-test results on fine-grained (cohesive) soil. Based on this study, the following correlations were developed:

$$w_{opt} = (1.99 - 0.165 \ln E)(PI) \tag{16.11}$$

and

$$\gamma_{d(max)} = L - Mw_{opt} \tag{16.12}$$

where

$$L = 14.34 + 1.195 \ln E \tag{16.13}$$
$$M = -0.19 + 0.073 \ln E \tag{16.14}$$

w_{opt} = optimum moisture content (%)

PI = plasticity index (%)

$\gamma_{d(max)}$ = maximum dry unit weight (kN/m^3)

E = compaction energy (kN-m/m^3)

Matteo et al. (2009) analyzed the results of 71 fine-grained soils and provided the following correlations for optimum moisture content (w_{opt}) and maximum dry unit weight [$\gamma_{d(max)}$] for modified Proctor test ($E = 2700$ kN-m/m^3):

$$w_{opt}(\%) = -0.86(LL) + 3.04\left(\frac{LL}{G_s}\right) + 2.2 \tag{16.15}$$

and

$$\gamma_{d(max)}(kN/m^3) = 40.316(w_{opt}^{-0.295})(PI^{0.032}) - 2.4 \tag{16.16}$$

where LL = liquid limit (%)

PI = plasticity index (%)

G_s = specific gravity of soil solids

Example 16.1

For a granular soil, the following are given:

- $G_s = 2.6$
- Liquid limit on the fraction passing No. 40 sieve = 20
- Percent retained on No. 4 sieve = 20

Using Eqs. (16.4) and (16.5), estimate the maximum dry density of compaction and the optimum moisture content based on the modified Proctor test.

Solution
From Eq. (16.4),

$$
\begin{aligned}
\rho_{d(\max)} (\text{kg/m}^3) &= [4{,}804{,}574 G_s - 195.55(\text{LL})^2 + 156{,}971(R\#4)^{0.5} - 9{,}527{,}830]^{0.5} \\
&= [4{,}804{,}574(2.6) - 195.55(20)^2 + 156{,}971(20)^{0.5} - 9{,}527{,}830]^{0.5} \\
&= \textbf{1894 kg/m}^3
\end{aligned}
$$

From Eq. (16.5),

$$
\begin{aligned}
\ln(w_{\text{opt}}) &= 1.195 \times 10^{-4}(\text{LL})^2 - 1.964 G_s - 6.617 \times 10^{-5}(R\#4) + 7{,}651 \\
&= 1.195 \times 10^{-4}(20)^2 - 1.964(2.6) - 6.617 \times 10^{-5}(20) + 7{,}651 \\
&= 2.591 \\
w_{\text{opt}} &= \textbf{13.35\%}
\end{aligned}
$$

∎

Example 16.2

For a sand with 4% finer than No. 200 sieve, estimate the maximum relative density of compaction that may be obtained from a modified Proctor test. Given $D_{50} = 1.4$ mm.

Solution
For modified Proctor test, $E = 2696$ kN-m/m^3

From Eq. (16.7)

$$
A = 0.216 \ln E - 0.850 = (0.216)(\ln 2696) - 0.850 = 0.856
$$

From Eq. (16.8)

$$
B = -0.03 \ln E + 0.306 = -(0.03)(\ln 2696) + 0.306 = 0.069
$$

From Eq. (16.6)

$$
D_r = A D_{50}^{-B} = (0.856)(1.4)^{-0.069} = 0.836 = \textbf{83.6\%}
$$

∎

Example 16.3

For a silty clay soil given LL = 43 and PL = 18. Estimate the maximum dry unit weight of compaction that can be achieved by conducting a modified Proctor test. Use Eq. (16.12).

Solution

For modified Proctor test, $E = 2696$ kN-m/m^3

From Eqs. (16.13) and (16.14),

$$L = 14.34 + 1.195 \ln E = 14.34 + 1.195 \ln (2696) = 23.78$$
$$M = -0.19 + 0.073 \ln E = -0.19 + 0.073 \ln (2696) = 0.387$$

From Eq. (16.11),

$$w_{opt}(\%) = (1.99 - 0.165 \ln E)(PI)$$
$$= [1.99 - 0.165 \ln (2696)](43 - 18)$$
$$= 17.16\%$$

From Eq. (16.12),

$$\gamma_{d(max)} = L - Mw_{opt} = 23.78 - (0.387)(17.16) = \mathbf{17.14 \ kN/m^3} \qquad \blacksquare$$

16.4   Field Compaction

Ordinary compaction in the field is done by rollers. Of the several types of roller used, the most common are

1. Smooth-wheel rollers (or smooth drum rollers)
2. Pneumatic rubber-tired rollers
3. Sheepsfoot rollers
4. Vibratory rollers

Figure 16.2 shows a *smooth-wheel roller* that can also create vertical vibration during compaction. Smooth-wheel rollers are suitable for proof-rolling subgrades and for finishing the construction of fills with sandy or clayey soils. They provide 100% coverage under the wheels, and the contact pressure can be as high as 300 to 400 kN/m^2 ($\approx$45 to 60 lb/in^2). However, they do not produce a uniform unit weight of compaction when used on thick layers.

Pneumatic rubber-tired rollers (Figure 16.3) are better in many respects than smooth-wheel rollers. Pneumatic rollers, which may weigh as much as 2000 kN (450 kip), consist of a heavily loaded wagon with several rows of tires. The tires are closely spaced—four to six in a row. The contact pressure under the tires may range up to 600 to 700 kN/m^2 ($\approx$85 to 100 lb/in^2), and they give about 70 to 80% coverage. Pneumatic rollers, which can be used for sandy and clayey soil compaction, produce a combination of pressure and kneading action.

Figure 16.2 Vibratory smooth-wheel rollers (*Dmitry Kalinovsky/Shutterstock.com*)

Figure 16.3 Pneumatic rubber-tired roller (*Vadim Ratnikov/Shutterstock.com*)

Figure 16.4 Vibratory sheepsfoot roller (*Artit Thongchuea/Shutterstock.com*)

Sheepsfoot rollers (Figure 16.4) consist basically of drums with large numbers of projections. The area of each of the projections may be 25 to 90 cm^2 (4 to 14 in^2). These rollers are *most effective in compacting cohesive soils.* The contact pressure under the projections may range from 1500 to 7500 kN/m^2($\approx$215 to 1100 lb/in^2). During compaction in the field, the initial passes compact the lower portion of a lift. Later, the middle and top of the lift are compacted.

Vibratory rollers are efficient in compacting granular soils. Vibrators can be attached to smooth-wheel, pneumatic rubber-tired or sheepsfoot rollers to send vibrations into the soil being compacted. Figures 16.2 and 16.4 show vibratory smooth-wheel rollers and a vibratory sheepsfoot roller, respectively.

In general, compaction in the field depends on several factors, such as the type of compactor, type of soil, moisture content, lift thickness, towing speed of the compactor, and number of passes the roller makes.

Figure 16.5 shows the variation of the unit weight of compaction with depth for a poorly graded dune sand compacted by a vibratory drum roller. Vibration was produced by mounting an eccentric weight on a single rotating shaft within the drum cylinder. The weight of the roller used for this compaction was 55.7 kN (12.5 kip), and the drum diameter was 1.19 m (3.9 ft). The lifts were kept at 2.44 m (8 ft). Note that, at any depth, the dry unit weight of compaction increases with the number of passes the roller makes. However, the rate of increase in unit weight gradually decreases after about 15 passes. Note also the variation of dry unit weight with depth by the number of roller passes. The dry unit weight and hence the relative density, D_r, reach maximum values at a depth of about 0.5 m ($\approx$1.6 ft) and then gradually decrease as the depth increases. The reason is the lack of confining pressure toward the surface. Once the relation between depth and

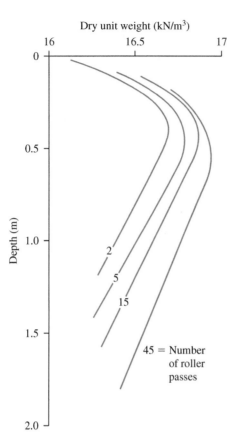

Figure 16.5 Vibratory compaction of a sand: Variation of dry unit weight with depth and number of roller passes; lift thickness = 2.44 m (Based on D'Appolonia, D. J., Whitman, R. V. and D'Appolonia, E. (1969). "Sand Compaction with Vibratory Rollers," *Journal of the Soil Mechanics and Foundations Division,* American Society of Civil Engineers, Vol. 95, N. SM1, pp. 263–284.)

relative density (or dry unit weight) for a soil for a given number of passes is determined, for satisfactory compaction based on a given specification, the approximate thickness of each lift can be easily estimated.

Hand-held vibrating plates can be used for effective compaction of granular soils over a limited area. Vibrating plates are also gang-mounted on machines. These can be used in less restricted areas.

16.5 Compaction Control for Clay Hydraulic Barriers

Compacted clays are commonly used as hydraulic barriers in cores of earth dams, liners and covers of landfills, and liners of surface impoundments. Since the primary purpose of a barrier is to minimize flow, the hydraulic conductivity, *k,* is the controlling factor. In many cases, it is desired that the hydraulic conductivity be less than 10^{-7}cm/s. This can be achieved by controlling the minimum degree of saturation during compaction, a relation that can be explained by referring to the compaction characteristics of three soils described in Table 16.3 (Othman and Luettich, 1994).

Figures 16.6, 16.7, and 16.8 show the standard and modified Proctor test results and the hydraulic conductivities of compacted specimens. Note that the solid symbols

Table 16.3 Characteristics of Soils Reported in Figures 16.6, 16.7, and 16.8

Soil	Classification	Liquid limit	Plasticity index	Percent finer than No. 200 sieve (0.075 mm)
Wisconsin A	CL	34	16	85
Wisconsin B	CL	42	19	99
Wisconsin C	CH	84	60	71

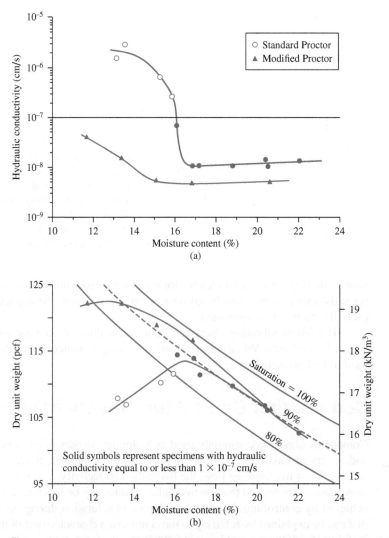

Figure 16.6 Standard and Modified Proctor test results and hydraulic conductivity of Wisconsin A soil (Based on Othman, M. A., and S. M. Luettich, "Compaction Control Criteria for Clay Hydraulic Barriers," *Transportation Research Record 1462*, Transportation Research Board, National Research Council, Washington, D.C., 1994.)

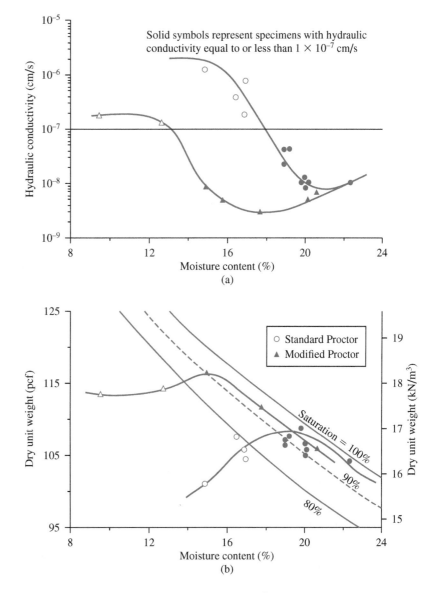

Figure 16.7 Standard and Modified Proctor test results and hydraulic conductivity of Wisconsin B soil (Based on Othman, M. A., and S. M. Luettich, "Compaction Control Criteria for Clay Hydraulic Barriers," *Transportation Research Record 1462*, Transportation Research Board, National Research Council, Washington, D.C., 1994.)

represent specimens with hydraulic conductivities of 10^{-7} cm/s or less. As can be seen from these figures, the data points plot generally parallel to the line of full saturation. Figure 16.9 shows the effect of the degree of saturation during compaction on the hydraulic conductivity of the three soils. It is evident from the figure that, if it is desired that the maximum hydraulic conductivity be 10^{-7} cm/s, then all soils should be compacted at a minimum degree of saturation of 88%.

In field compaction at a given site, soils of various composition may be encountered. Small changes in the content of fines will change the magnitude of hydraulic conductivity. Hence, considering the various soils likely to be encountered at a given site the procedure just described aids in developing a minimum-degree-of-saturation criterion for compaction to construct hydraulic barriers.

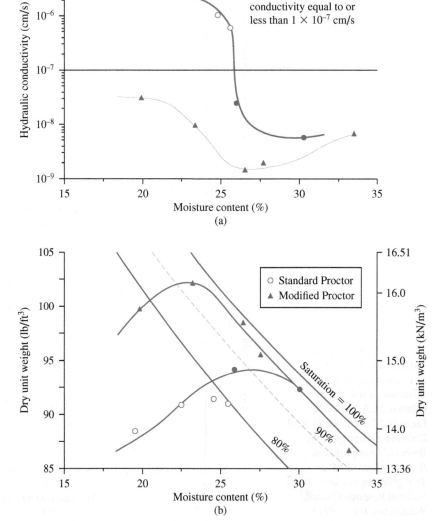

Figure 16.8 Standard and Modified Proctor test results and hydraulic conductivity of Wisconsin C soil (Based on Othman, M. A., and S. M. Luettich, "Compaction Control Criteria for Clay Hydraulic Barriers," *Transportation Research Record 1462*, Transportation Research Board, National Research Council, Washington, D.C., 1994.)

16.6 Vibroflotation

Vibroflotation is a technique developed in Germany in the 1930s for *in situ* densification of thick layers of loose granular soil deposits. Vibroflotation was first used in the United States about 10 years later. The process involves the use of a *vibroflot* (called the *vibrating unit*), as shown in Figure 16.10. The device is about 2 m (6 ft) in length. This vibrating unit has an eccentric weight inside it and can develop a centrifugal force. The weight enables the unit to vibrate horizontally. Openings at the bottom and top of the unit are for water jets. The

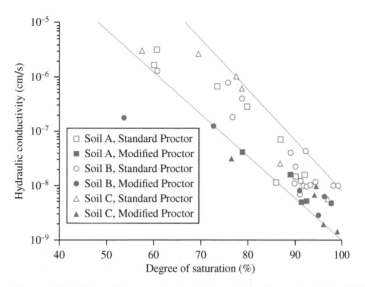

Figure 16.9 Effect of degree of saturation on hydraulic conductivity of Wisconsin A, B, and C soils (After Othman and Luettich, 1994) (Based on Othman, M. A., and S. M. Luettich, "Compaction Control Criteria for Clay Hydraulic Barriers," *Transportation Research Record 1462*, Transportation Research Board, National Research Council, Washington, D.C., 1994.)

vibrating unit is attached to a follow-up pipe. The figure shows the vibroflotation equipment necessary for compaction in the field.

The entire compaction process can be divided into four steps (see Figure 16.11):

Step 1. The jet at the bottom of the vibroflot is turned on, and the vibroflot is lowered into the ground.

Step 2. The water jet creates a quick condition in the soil, which allows the vibrating unit to sink.

Step 3. Granular material is poured into the top of the hole. The water from the lower jet is transferred to the jet at the top of the vibrating unit. This water carries the granular material down the hole.

Step 4. The vibrating unit is gradually raised in about 0.3-m (1-ft) lifts and is held vibrating for about 30 seconds at a time. This process compacts the soil to the desired unit weight.

Table 16.4 gives the details of various types of vibroflot unit used in the United States. The 30-HP electric units have been used since the latter part of the 1940s. The 100-HP units were introduced in the early 1970s. The zone of compaction around a single probe will vary according to the type of vibroflot used. The cylindrical zone of compaction will have a radius of about 2 m (6 ft) for a 30-HP unit and about 3 m (10 ft) for a 100-HP unit. Compaction by vibroflotation involves various probe spacings, depending on the zone of compaction. (See Figure 16.12.) Mitchell (1970) and Brown (1977) reported several successful cases of foundation design that used vibroflotation.

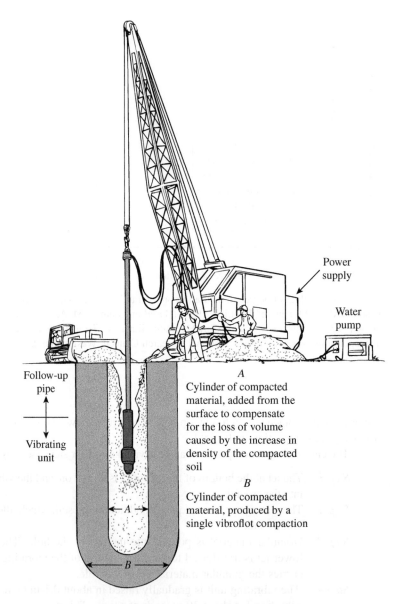

Figure 16.10 Vibroflotation unit (Based on Brown, 1977.)

Follow-up pipe

Vibrating unit

Power supply

Water pump

A
Cylinder of compacted material, added from the surface to compensate for the loss of volume caused by the increase in density of the compacted soil

B
Cylinder of compacted material, produced by a single vibroflot compaction

←*A*→

←—— *B* ——→

The success of densification of *in situ* soil depends on several factors, the most important of which are the grain-size distribution of the soil and the nature of the backfill used to fill the holes during the withdrawal period of the vibroflot. The range of the grain-size distribution of *in situ* soil marked Zone 1 in Figure 16.13 is most suitable for compaction by vibroflotation. Soils that contain excessive amounts of fine sand and silt-size particles are difficult to compact; for such soils, considerable effort is needed to reach the proper relative density of compaction. Zone 2 in Figure 16.13 is the approximate lower limit of grain-size

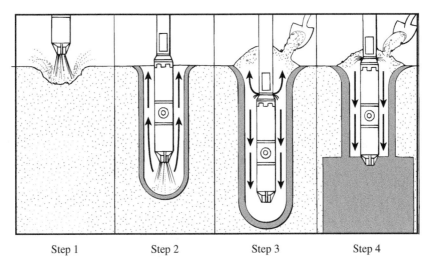

| Step 1 | Step 2 | Step 3 | Step 4 |

Figure 16.11 Compaction by the vibroflotation process (Based on Brown, 1977.)

Table 16.4 Types of Vibrating Units[a]

	100-HP electric and hydraulic motors	30-HP electric motors
	(a) Vibrating tip	
Length	2.1 m	1.86 m
	(7 ft)	(6.11 ft)
Diameter	406 mm	381 mm
	(16 in.)	(15 in.)
Weight	18 kN	18 kN
	(4000 lb)	(4000 lb)
Maximum movement when free	12.5 mm	7.6 mm
	(0.49 in.)	(0.3 in.)
Centrifugal force	160 kN	90 kN
	(18 ton)	(10 ton)
	(b) Eccentric	
Weight	1.16 kN	0.76 kN
	(260 lb)	(170 lb)
Offset	38 mm	32 mm
	(1.5 in.)	(1.25 in.)
Length	610 mm	387 mm
	(24 in.)	(15.3 in.)
Speed	1800 rpm	1800 rpm
	(c) Pump	
Operating flow rate	0–1.6 m^3/min	0–0.6 m^3/min
	(0–400 gal/min)	(0–150 gal/min)
Pressure	690–1035 kN/m^2	690–1035 kN/m^2
	(100–150 lb/in^2)	(100–150 lb/in^2)

(continued)

Table 16.4 (*continued*)

	(d) Lower follow-up pipe and extensions	
Diameter	305 mm	305 mm
	(12 in.)	(12 in.)
Weight	3.65 kN/m	3.65 kN/m
	(250 lb/ft)	(250 lb/ft)

[a]Based on data from Brown, E. E. (1977), "Vibroflotation Compaction of Cohensionless Soils," *Journal of the Geotechnical Engineering Divison,* Vol. 103, No. GT12

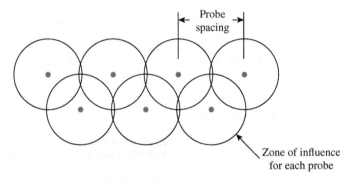

Figure 16.12 Nature of probe spacing for vibroflotation

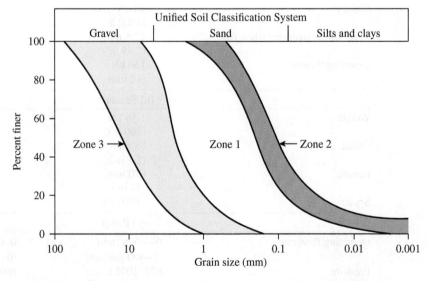

Figure 16.13 Effective range of grain-size distribution of soil for vibroflotation

distribution for compaction by vibroflotation. Soil deposits whose grain-size distribution falls into Zone 3 contain appreciable amounts of gravel. For these soils, the rate of probe penetration may be rather slow, so compaction by vibroflotation might prove to be uneconomical in the long run.

The grain-size distribution of the backfill material is one of the factors that control the rate of densification. Brown (1977) defined a quantity called *suitability number* for rating a backfill material. The suitability number is given by the formula

$$S_N = 1.7 \sqrt{\frac{3}{(D_{50})^2} + \frac{1}{(D_{20})^2} + \frac{1}{(D_{10})^2}}$$

(16.17)

where D_{50}, D_{20}, and D_{10} are the diameters (in mm) through which 50%, 20%, and 10%, respectively, of the material is passing. The smaller the value of S_N, the more desirable is the backfill material. Following is a backfill rating system proposed by Brown (1977):

Range of S_N	Rating as backfill
0–10	Excellent
10–20	Good
20–30	Fair
30–50	Poor
>50	Unsuitable

An excellent case study that evaluated the benefits of vibroflotation was presented by Basore and Boitano (1969). Densification of granular subsoil was necessary for the construction of a three-story office building at the Treasure Island Naval Station in San Francisco, California. The top 9 m ($\approx$30 ft) of soil at the site was loose to medium-dense sand fill that had to be compacted. Figure 16.14a shows the nature of the layout of the vibroflotation points. Sixteen compaction points were arranged in groups of four, with 1.22 m (4 ft), 1.52 m (5 ft), 1.83 m (6 ft), and 2.44 m (8 ft) spacing. Prior to compaction, standard penetration tests were conducted at the centers of groups of three compaction points. After the completion of compaction by vibroflotation, the variation of the standard penetration resistance with depth was determined at the same points.

Figure 16.14b shows the variation of standard penetration resistance, N_{60}, with depth before and after compaction for vibroflotation point spacings $S' = 1.22$ m (4 ft) and 2.44 m (8 ft). From this figure, the following general conclusions can be drawn:

- For any given S', the magnitude of N_{60} after compaction decreases with an increase in depth.
- An increase in N_{60} indicates an increase in the relative density of sand.
- The degree of compaction decreases with the increase in S'. At $S' = 1.22$ m (4 ft), the degree of compaction at any depth is the largest. However, at $S' = 2.44$ m (8 ft), the vibroflotation had practically no effect in compacting soil.

During the past 30 to 35 years, the vibroflotation technique has been used successfully on large projects to compact granular subsoils, thereby controlling structural settlement.

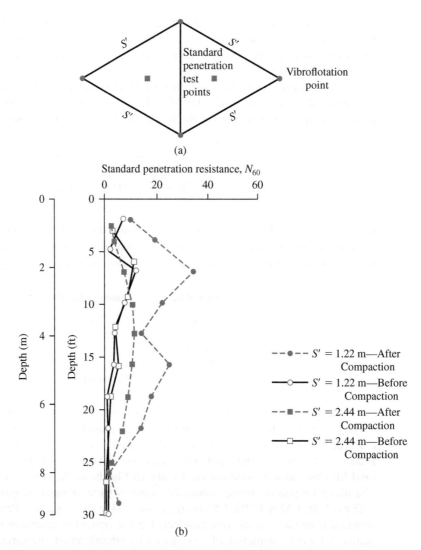

Figure 16.14 (a) Layout of vibroflotation compaction points; (b) variation of standard penetration resistance (N_{60}) before and after compaction (Based on Basore, C. E. and Boitano, J. D. (1969). "Sand Densification by Piles and Vibrofloation," *Journal of Soil Mechanics and Foundation Engineering Division*, American Society of Civil Engineers, Vol. 95, No. 6, pp. 1301–1323, Figure 16.)

16.7 Blasting

Blasting is a technique that has been used successfully in many projects (Mitchell, 1970) for the densification of granular soils. The general soil grain sizes suitable for compaction by blasting are the same as those for compaction by vibroflotation. The process involves the detonation of explosive charges such as 60% dynamite at a certain depth below the ground surface in saturated soil. The lateral spacing of the charges varies from about 3 to 9 m (10 to 30 ft). Three to five successful detonations are usually necessary to achieve the desired compaction. Compaction (up to a relative density of about 80%) up to a depth of about 18 m (60 ft) over a large area can easily be achieved by using this process. Usually, the explosive charges are placed at a depth of about two-thirds of the thickness of the soil layer desired to be compacted. The sphere of influence of compaction by a 60% dynamite charge can be given as follows (Mitchell, 1970):

$$r = \sqrt{\frac{W_{EX}}{C}} \qquad (16.18)$$

where r = sphere of influence
W_{EX} = weight of explosive −60% dynamite
C = 0.0122 when W_{EX} is in kg and r is in m
= 0.0025 when W_{EX} is in lb and r is in ft

Figure 16.15 shows the test results of soil densification by blasting in an area measuring 15 m by 9 m (50 ft by 30 ft)(Mitchell, 1970). For these tests, twenty 2.09-kg (4.6-lb) charges of Gelamite No. 1 (Hercules Powder Company, Wilmington, Delaware) were used.

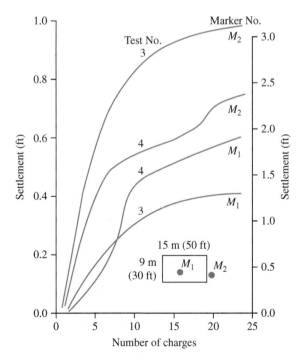

Figure 16.15 Ground settlement as a function of number of explosive charges

16.8 Precompression

When highly compressible, normally consolidated clayey soil layers lie at a limited depth and large consolidation settlements are expected as the result of the construction of large buildings, highway embankments, or earth dams, precompression of soil may be used to minimize postconstruction settlement. The principles of precompression are best explained by reference to Figure 16.16. Here, the proposed structural load per unit area is $\Delta\sigma'_{(p)}$, and the thickness of the clay layer undergoing consolidation is H_c. The maximum primary consolidation settlement caused by the structural load is then

$$S_{c(p)} = \frac{C_c H_c}{1 + e_o} \log \frac{\sigma'_o + \Delta\sigma'_{(p)}}{\sigma'_o} \tag{16.19}$$

The settlement–time relationship under the structural load will be like that shown in Figure 16.16b. However, if a surcharge of $\Delta\sigma'_{(p)} + \Delta\sigma'_{(f)}$ is placed on the ground, the primary consolidation settlement will be

$$S_{c(p+f)} = \frac{C_c H_c}{1 + e_o} \log \frac{\sigma'_o + [\Delta\sigma'_{(p)} + \Delta\sigma'_{(f)}]}{\sigma'_o} \tag{16.20}$$

The settlement–time relationship under a surcharge of $\Delta\sigma'_{(p)} + \Delta\sigma'_{(f)}$ is also shown in Figure 16.16b. Note that a total settlement of $S_{c(p)}$ would occur at time t_2, which is much shorter than t_1. So, if a temporary total surcharge of $\Delta\sigma'_{(p)} + \Delta\sigma'_{(f)}$ is applied on the ground surface for time t_2, the settlement will equal $S_{c(p)}$. At that time, if the surcharge is removed and a structure with a permanent load per unit area of $\Delta\sigma'_{(p)}$ is built, no appreciable settlement will occur. The procedure just described is called *precompression*. The total surcharge $\Delta\sigma'_{(p)} + \Delta\sigma'_{(f)}$ can be applied by means of temporary fills.

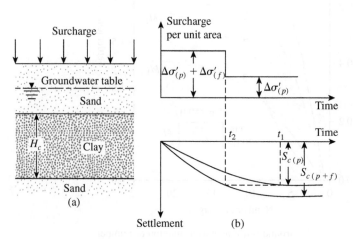

Figure 16.16 Principles of precompression

Derivation of Equations for Obtaining $\Delta\sigma'_{(f)}$ and t_2

Figure 16.16b shows that, under a surcharge of $\Delta\sigma'_{(p)} + \Delta\sigma'_{(f)}$, the degree of consolidation at time t_2 after the application of the load is

$$U = \frac{S_{c(p)}}{S_{c(p+f)}} \qquad (16.21)$$

Substitution of Eqs. (16.19) and (16.20) into Eq. (16.21) yields

$$U = \frac{\log\left[\dfrac{\sigma'_o + \Delta\sigma'_{(p)}}{\sigma'_o}\right]}{\log\left[\dfrac{\sigma'_o + \Delta\sigma'_{(p)} + \Delta\sigma'_{(f)}}{\sigma'_o}\right]} = \frac{\log\left[1 + \dfrac{\Delta\sigma'_{(p)}}{\sigma'_o}\right]}{\log\left\{1 + \dfrac{\Delta\sigma'_{(p)}}{\sigma'_o}\left[1 + \dfrac{\Delta\sigma'_{(f)}}{\Delta\sigma'_{(p)}}\right]\right\}} \qquad (16.22)$$

Figure 16.17 gives magnitudes of U for various combinations of $\Delta\sigma'_{(p)}/\sigma'_o$, and $\Delta\sigma'_{(f)}/\Delta\sigma'_{(p)}$. The degree of consolidation referred to in Eq. (16.22) is actually the average degree of consolidation at time t_2, as shown in Figure 16.17b. However, if the average degree of consolidation is used to determine t_2, some construction problems might occur. The reason is that, after the removal of the surcharge and placement of the structural load, the portion of clay close to the drainage surface will continue to swell, and the soil close to the midplane will continue to settle. (See Figure 16.18.) In some cases, net continuous

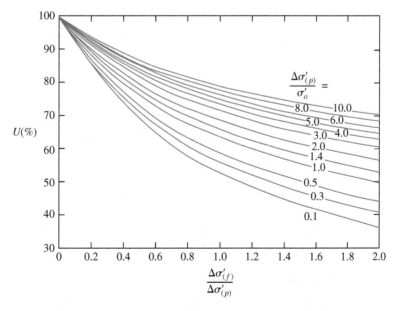

Figure 16.17 Plot of U against $\Delta\sigma'_{(f)}/\Delta\sigma'_{(p)}$ for various values of $\Delta\sigma'_{(p)}/\sigma'_o$—Eq. (16.22)

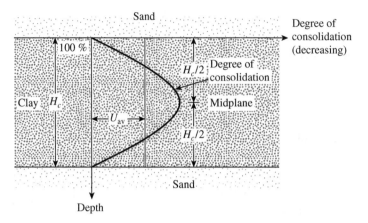

Figure 16.18

settlement might result. A conservative approach may solve the problem; that is, assume that U in Eq. (16.22) is the midplane degree of consolidation (Johnson, 1970a). Now, from Eq. (2.77),

$$U = f(T_v) \qquad (2.77)$$

where

T_v = time factor = $C_v t_2 / H^2$
C_v = coefficient of consolidation
t_2 = time
H = maximum drainage path ($=H_c/2$ for two-way drainage and H_c for one-way drainage)

The variation of U (the midplane degree of consolidation) with T_v is given in Figure 16.19.

Procedure for Obtaining Precompression Parameters

Two problems may be encountered by engineers during precompression work in the field:

1. The value of $\Delta \sigma'_{(f)}$ is known, but t_2 must be obtained. In such a case, obtain σ'_o, $\Delta \sigma_{(p)}$, and solve for U, using Eq. (16.22) or Figure 16.17. For this value of U, obtain T_v from Figure 16.19. Then

$$t_2 = \frac{T_v H^2}{C_v} \qquad (16.23)$$

2. For a specified value of t_2, $\Delta \sigma'_{(f)}$ must be obtained. In such a case, calculate T_v. Then use Figure 16.19 to obtain the midplane degree of consolidation, U. With the estimated value of U, go to Figure 16.17 to get the required value of $\Delta \sigma'_{(f)}/\Delta \sigma'_{(p)}$, and then calculate $\Delta \sigma'_{(f)}$.

   Several case histories on the successful use of precompression techniques for improving foundation soil are available in the literature (for example, Johnson, 1970a).

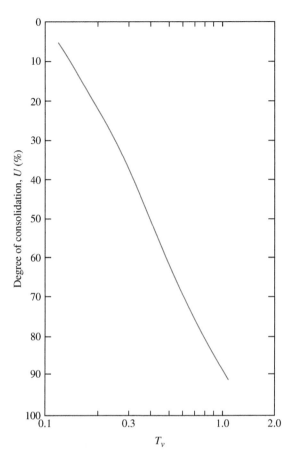

Figure 16.19 Plot of midplane degree of consolidation against T_v

---

Example 16.4

Examine Figure 16.16. During the construction of a highway bridge, the average permanent load on the clay layer is expected to increase by about 115 kN/m². The average effective overburden pressure at the middle of the clay layer is 210 kN/m². Here, $H_c = 6$ m, $C_c = 0.28$, $e_o = 0.9$, and $C_v = 0.36$ m²/mo. The clay is normally consolidated. Determine

    a. The total primary consolidation settlement of the bridge without precompression
    b. The surcharge, $\Delta\sigma'_{(f)}$, needed to eliminate the entire primary consolidation settlement in nine months by precompression.

Solution

Part a

The total primary consolidation settlement may be calculated from Eq. (16.19):

$$S_{c(p)} = \frac{C_c H_c}{1 + e_o} \log\left[\frac{\sigma'_o + \Delta\sigma'_{(p)}}{\sigma'_o}\right] = \frac{(0.28)(6)}{1 + 0.9} \log\left[\frac{210 + 115}{210}\right]$$

$$= 0.1677 \text{ m} = \textbf{167.7 mm}$$

Part b
We have

$$T_v = \frac{C_v t_2}{H^2}$$

$$C_v = 0.36 \text{ m}^2/\text{mo.}$$

$$H = 3 \text{ m (two-way drainage)}$$

$$t_2 = 9 \text{ mo.}$$

Hence,

$$T_v = \frac{(0.36)(9)}{3^2} = 0.36$$

According to Figure 16.19, for $T_v = 0.36$, the value of U is 47%. Now,

$$\Delta\sigma'_{(p)} = 115 \text{ kN/m}^2$$

and

$$\sigma'_o = 210 \text{ kN/m}^2$$

so

$$\frac{\Delta\sigma'_{(p)}}{\sigma'_o} = \frac{115}{210} = 0.548$$

According to Figure 16.17, for $U = 47\%$ and $\Delta\sigma'_{(p)}/\sigma'_o = 0.548$, $\Delta\sigma'_{(f)}/\Delta\sigma'_{(p)} \approx 1.8$; thus,

$$\Delta\sigma'_{(f)} = (1.8)(115) = \mathbf{207 \text{ kN/m}^2} \qquad \blacksquare$$

16.9 Sand Drains

The use of sand drains is another way to accelerate the consolidation settlement of soft, normally consolidated clay layers and achieve precompression before the construction of a desired foundation. Sand drains are constructed by drilling holes through the clay layer(s) in the field at regular intervals. The holes are then backfilled with sand. This can be achieved by several means, such as (a) rotary drilling and then backfilling with sand; (b) drilling by continuous-flight auger with a hollow stem and backfilling with sand (through the hollow steam); and (c) driving hollow steel piles. The soil inside the pile is then jetted out, after which backfilling with sand is done. Figure 16.20 shows a schematic diagram of sand drains. After backfilling the drill holes with sand, a surcharge is applied at the ground surface. The surcharge will increase the pore water pressure in the clay. The excess pore water pressure in the clay will be dissipated by drainage—both vertically and radially to the sand drains—thereby accelerating settlement of the clay layer. In Figure 16.20a, note that the radius of the sand drains is r_w. Figure 16.20b shows the plan of the layout of the sand drains. The effective zone from which the radial drainage will be directed toward a given sand drain is approximately cylindrical, with a diameter of d_e.

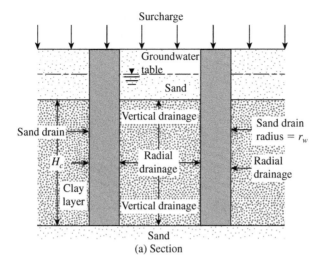

Surcharge

Groundwater table

Sand

Vertical drainage

Sand drain

H_c

Radial drainage

Sand drain radius = r_w

Radial drainage

Clay layer

Vertical drainage

Sand

(a) Section

Sand drain radius = r_w

d_e

(b) Plan

Figure 16.20 Sand drains

To determine the surcharge that needs to be applied at the ground surface and the length of time that it has to be maintained, see Figure 16.16 and use the corresponding equation, Eq. (16.22):

$$U_{v,r} = \frac{\log\left[1 + \dfrac{\Delta\sigma'_{(p)}}{\sigma'_o}\right]}{\log\left\{1 + \dfrac{\Delta\sigma'_{(p)}}{\sigma'_o}\left[1 + \dfrac{\Delta\sigma'_{(f)}}{\Delta\sigma'_{(p)}}\right]\right\}} \qquad (16.24)$$

The notations $\Delta\sigma'_{(p)}$, σ'_o, and $\Delta\sigma'_{(f)}$ are the same as those in Eq. (16.22); however, the left-hand side of Eq. (16.24) is the *average degree* of consolidation instead of the degree of consolidation at midplane. Both *radial* and *vertical* drainage contribute to the average degree of consolidation. If $U_{v,r}$ can be determined for any time t_2 (see Figure 16.16b), the total surcharge $\Delta\sigma'_{(f)} + \Delta\sigma'_{(p)}$ may be obtained easily from Figure 16.17. The procedure for determining the average degree of consolidation $(U_{v,r})$ follows:

For a given surcharge and duration, t_2, the average degree of consolidation due to drainage in the vertical and radial directions is

$$U_{v,r} = 1 - (1 - U_r)(1 - U_v)$$ (16.25)

where

U_r = average degree of consolidation with radial drainage only
U_v = average degree of consolidation with vertical drainage only

The successful use of sand drains has been described in detail by Johnson (1970b). As with precompression, constant field settlement observations may be necessary during the period the surcharge is applied.

Average Degree of Consolidation Due to Radial Drainage Only

Figure 16.21 shows a schematic diagram of a sand drain. In the figure, r_w is the radius of the sand drain and $r_e = d_e/2$ is the radius of the effective zone of drainage. It is also important to realize that, during the installation of sand drains, a certain zone of clay surrounding them is smeared, thereby changing the hydraulic conductivity of the clay. In the figure, r_s is the radial distance from the center of the sand drain to the farthest point of the smeared zone. Now, for the average-degree-of-consolidation relationship, we will use the *theory of equal strain*. Two cases may arise that relate to the nature of the application of surcharge, and they are shown in Figure 16.22. (See the notations shown in Figure 16.16). Either (a) the entire surcharge is applied instantaneously (see Figure 16.22a), or (b) the surcharge is applied in the form of a ramp load (see Figure 16.22b). When the entire surcharge is applied instantaneously (Barron, 1948),

$$U_r = 1 - \exp\left(\frac{-8T_r}{m}\right)$$ (16.26)

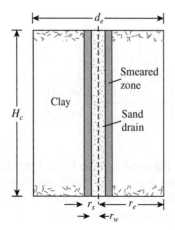

Figure 16.21 Schematic diagram of a sand drain

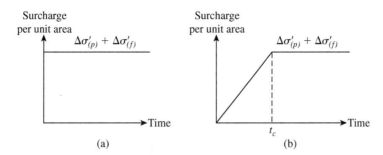

Figure 16.22 Nature of application of surcharge

where

$$m = \frac{n^2}{n^2 - S^2} \ln\left(\frac{n}{S}\right) - \frac{3}{4} + \frac{S^2}{4n^2} + \frac{k_h}{k_s}\left(\frac{n^2 - S^2}{n^2}\right) \ln S \tag{16.27}$$

in which

$$n = \frac{d_e}{2r_w} = \frac{r_e}{r_w} \tag{16.28}$$

$$S = \frac{r_s}{r_w} \tag{16.29}$$

and

k_h = hydraulic conductivity of clay in the horizontal direction in the unsmeared zone

k_s = horizontal hydraulic conductivity in the smeared zone

T_r = nondimensional time factor for radial drainage only = $\dfrac{C_{vr}t_2}{d_e^2}$ \hfill (16.30)

C_{vr} = coefficient of consolidation for radial drainage

$$= \frac{k_h}{\left[\dfrac{\Delta e}{\Delta\sigma'(1 + e_{av})}\right]\gamma_w} \tag{16.31}$$

For a *no-smear case*, $r_s = r_w$ and $k_h = k_s$, so $S = 1$ and Eq. (16.27) becomes

$$m = \left(\frac{n^2}{n^2 - 1}\right) \ln(n) - \frac{3n^2 - 1}{4n^2} \tag{16.32}$$

Table 16.5 gives the values of U_r for various values of T_r and n.

Table 16.5 Variation of U_r for Various Values of T_r and n, No-Smear Case [Eqs. (16.26) and (16.32)]

Degree of consolidation U_r (%)	Time factor T_r for value of n (= r_e/r_w)				
	5	10	15	20	25
0	0	0	0	0	0
1	0.0012	0.0020	0.0025	0.0028	0.0031
2	0.0024	0.0040	0.0050	0.0057	0.0063
3	0.0036	0.0060	0.0075	0.0086	0.0094
4	0.0048	0.0081	0.0101	0.0115	0.0126
5	0.0060	0.0101	0.0126	0.0145	0.0159
6	0.0072	0.0122	0.0153	0.0174	0.0191
7	0.0085	0.0143	0.0179	0.0205	0.0225
8	0.0098	0.0165	0.0206	0.0235	0.0258
9	0.0110	0.0186	0.0232	0.0266	0.0292
10	0.0123	0.0208	0.0260	0.0297	0.0326
11	0.0136	0.0230	0.0287	0.0328	0.0360
12	0.0150	0.0252	0.0315	0.0360	0.0395
13	0.0163	0.0275	0.0343	0.0392	0.0431
14	0.0177	0.0298	0.0372	0.0425	0.0467
15	0.0190	0.0321	0.0401	0.0458	0.0503
16	0.0204	0.0344	0.0430	0.0491	0.0539
17	0.0218	0.0368	0.0459	0.0525	0.0576
18	0.0232	0.0392	0.0489	0.0559	0.0614
19	0.0247	0.0416	0.0519	0.0594	0.0652
20	0.0261	0.0440	0.0550	0.0629	0.0690
21	0.0276	0.0465	0.0581	0.0664	0.0729
22	0.0291	0.0490	0.0612	0.0700	0.0769
23	0.0306	0.0516	0.0644	0.0736	0.0808
24	0.0321	0.0541	0.0676	0.0773	0.0849
25	0.0337	0.0568	0.0709	0.0811	0.0890
26	0.0353	0.0594	0.0742	0.0848	0.0931
27	0.0368	0.0621	0.0776	0.0887	0.0973
28	0.0385	0.0648	0.0810	0.0926	0.1016
29	0.0401	0.0676	0.0844	0.0965	0.1059
30	0.0418	0.0704	0.0879	0.1005	0.1103
31	0.0434	0.0732	0.0914	0.1045	0.1148
32	0.0452	0.0761	0.0950	0.1087	0.1193
33	0.0469	0.0790	0.0987	0.1128	0.1239
34	0.0486	0.0820	0.1024	0.1171	0.1285
35	0.0504	0.0850	0.1062	0.1214	0.1332
36	0.0522	0.0881	0.1100	0.1257	0.1380
37	0.0541	0.0912	0.1139	0.1302	0.1429
38	0.0560	0.0943	0.1178	0.1347	0.1479
39	0.0579	0.0975	0.1218	0.1393	0.1529
40	0.0598	0.1008	0.1259	0.1439	0.1580
41	0.0618	0.1041	0.1300	0.1487	0.1632
42	0.0638	0.1075	0.1342	0.1535	0.1685
43	0.0658	0.1109	0.1385	0.1584	0.1739
44	0.0679	0.1144	0.1429	0.1634	0.1793

Table 16.5 (*continued*)

Degree of consolidation U_r (%)	Time factor T_r for value of n (= r_e/r_w)				
	5	**10**	**15**	**20**	**25**
45	0.0700	0.1180	0.1473	0.1684	0.1849
46	0.0721	0.1216	0.1518	0.1736	0.1906
47	0.0743	0.1253	0.1564	0.1789	0.1964
48	0.0766	0.1290	0.1611	0.1842	0.2023
49	0.0788	0.1329	0.1659	0.1897	0.2083
50	0.0811	0.1368	0.1708	0.1953	0.2144
51	0.0835	0.1407	0.1758	0.2020	0.2206
52	0.0859	0.1448	0.1809	0.2068	0.2270
53	0.0884	0.1490	0.1860	0.2127	0.2335
54	0.0909	0.1532	0.1913	0.2188	0.2402
55	0.0935	0.1575	0.1968	0.2250	0.2470
56	0.0961	0.1620	0.2023	0.2313	0.2539
57	0.0988	0.1665	0.2080	0.2378	0.2610
58	0.1016	0.1712	0.2138	0.2444	0.2683
59	0.1044	0.1759	0.2197	0.2512	0.2758
60	0.1073	0.1808	0.2258	0.2582	0.2834
61	0.1102	0.1858	0.2320	0.2653	0.2912
62	0.1133	0.1909	0.2384	0.2726	0.2993
63	0.1164	0.1962	0.2450	0.2801	0.3075
64	0.1196	0.2016	0.2517	0.2878	0.3160
65	0.1229	0.2071	0.2587	0.2958	0.3247
66	0.1263	0.2128	0.2658	0.3039	0.3337
67	0.1298	0.2187	0.2732	0.3124	0.3429
68	0.1334	0.2248	0.2808	0.3210	0.3524
69	0.1371	0.2311	0.2886	0.3300	0.3623
70	0.1409	0.2375	0.2967	0.3392	0.3724
71	0.1449	0.2442	0.3050	0.3488	0.3829
72	0.1490	0.2512	0.3134	0.3586	0.3937
73	0.1533	0.2583	0.3226	0.3689	0.4050
74	0.1577	0.2658	0.3319	0.3795	0.4167
75	0.1623	0.2735	0.3416	0.3906	0.4288
76	0.1671	0.2816	0.3517	0.4021	0.4414
77	0.1720	0.2900	0.3621	0.4141	0.4546
78	0.1773	0.2988	0.3731	0.4266	0.4683
79	0.1827	0.3079	0.3846	0.4397	0.4827
80	0.1884	0.3175	0.3966	0.4534	0.4978
81	0.1944	0.3277	0.4090	0.4679	0.5137
82	0.2007	0.3383	0.4225	0.4831	0.5304
83	0.2074	0.3496	0.4366	0.4992	0.5481
84	0.2146	0.3616	0.4516	0.5163	0.5668
85	0.2221	0.3743	0.4675	0.5345	0.5868
86	0.2302	0.3879	0.4845	0.5539	0.6081
87	0.2388	0.4025	0.5027	0.5748	0.6311
88	0.2482	0.4183	0.5225	0.5974	0.6558
89	0.2584	0.4355	0.5439	0.6219	0.6827

(*continued*)

Table 16.5 (*continued*)

Degree of consolidation U_r (%)	Time factor T_r for value of n (= r_e/r_w)				
	5	10	15	20	25
90	0.2696	0.4543	0.5674	0.6487	0.7122
91	0.2819	0.4751	0.5933	0.6784	0.7448
92	0.2957	0.4983	0.6224	0.7116	0.7812
93	0.3113	0.5247	0.6553	0.7492	0.8225
94	0.3293	0.5551	0.6932	0.7927	0.8702
95	0.3507	0.5910	0.7382	0.8440	0.9266
96	0.3768	0.6351	0.7932	0.9069	0.9956
97	0.4105	0.6918	0.8640	0.9879	1.0846
98	0.4580	0.7718	0.9640	1.1022	1.2100
99	0.5391	0.9086	1.1347	1.2974	1.4244

If the surcharge is applied in the form of a *ramp* and *there is no smear*, then (Olson, 1977)

$$U_r = \frac{T_r - \frac{1}{A}[1 - \exp(-AT_r)]}{T_{rc}} \quad \text{(for } T_r \leqslant T_{rc}\text{)} \tag{16.33}$$

and

$$U_r = 1 - \frac{1}{AT_{rc}}[\exp(AT_{rc}) - 1]\exp(-AT_{rc}) \quad \text{(for } T_r \geqslant T_{rc}\text{)} \tag{16.34}$$

where

$$T_{rc} = \frac{C_{vr}t_c}{d_e^2} \text{ (see Figure 16.22b for the definition of } t_c\text{)} \tag{16.35}$$

and

$$A = \frac{2}{m} \tag{16.36}$$

Average Degree of Consolidation Due to Vertical Drainage Only

Using Figure 16.22a, for instantaneous application of a surcharge, we may obtain the average degree of consolidation due to vertical drainage only from Eqs. (2.78) and (2.79). We have

$$T_v = \frac{\pi}{4}\left[\frac{U_v(\%)}{100}\right]^2 \quad \text{(for } U_v = 0 \text{ to } 60\%\text{)} \qquad \text{[Eq. (2.78)]}$$

and

$$T_v = 1.781 - 0.933 \log[100 - U_v(\%)] \quad \text{(for } U_v > 60\%\text{)} \qquad \text{[Eq. (2.79)]}$$

where U_v = average degree of consolidation due to vertical drainage only, and

$$T_v = \frac{C_v t_2}{H^2} \qquad \text{[Eq. (2.73)]}$$

where C_v = coefficient of consolidation for vertical drainage.

For the case of ramp loading, as shown in Figure 16.22b, the variation of U_v with T_v can be expressed as (Olson, 1977):

For $T_v \le T_c$:

$$U_v = \frac{T_v}{T_c}\left\{1 - \frac{2}{T_v}\Sigma\frac{1}{M^4}[1 - \exp(-M^2 T_v)]\right\} \qquad (16.37)$$

For $T_v \le T_c$:

$$U_v = 1 - \frac{2}{T_c}\Sigma\frac{1}{M^4}[\exp(-M^2 T_c) - 1]\exp(-M^2 T_v) \qquad (16.38)$$

where

$$M = \frac{\pi}{2}(2m' + 1)$$
$$m' = 0, 1, 2, \ldots$$

$$T_c = \frac{C_v t_c}{H^2} \qquad (16.39)$$

where H = length of maximum vertical drainage path. Figure 16.23 shows the variation of $U_v(\%)$ with T_c and T_v.

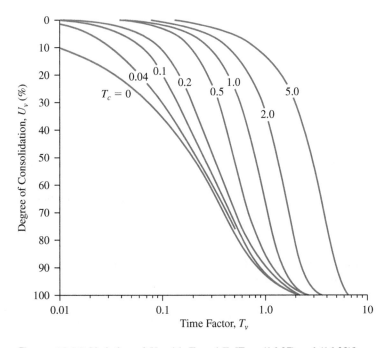

Figure 16.23 Variation of U_v with T_v and T_c [Eqs. (16.37) and (16.38)]

Example 16.5

Redo Example 16.4, with the addition of some sand drains. Assume that $r_w = 0.1$ m, $d_e = 3$ m, $C_v = C_{vr}$, and the surcharge is applied instantaneously. (See Figure 16.22a.) Also assume that this is a no-smear case.

Solution

Part a

The total primary consolidation settlement will be 167.7 mm, as before.

Part b

From Example 16.4, $T_v = 0.36$. Using Eq. (2.78), we obtain

$$T_v = \frac{\pi}{4}\left[\frac{U_v(\%)}{100}\right]^2$$

or

$$U_v = \sqrt{\frac{4T_v}{\pi}} \times 100 = \sqrt{\frac{(4)(0.36)}{\pi}} \times 100 = 67.7\%$$

Also,

$$n = \frac{d_e}{2r_w} = \frac{3}{2 \times 0.1} = 15$$

Again,

$$T_r = \frac{C_{vr}t_2}{d_e^2} = \frac{(0.36)(9)}{(3)^2} = 0.36$$

From Table 16.5 for $n = 15$ and $T_r = 0.36$, the value of U_r is about 77%. Hence,

$$U_{v,r} = 1 - (1 - U_v)(1 - U_r) = 1 - (1 - 0.67)(1 - 0.77)$$

$$= 0.924 = 92.4\%$$

Now, from Figure 16.17, for $\Delta\sigma_p'/\sigma_o' = 0.548$ and $U_{v,r} = 92.4\%$, the value of $\Delta\sigma_f'/\Delta\sigma_p' \approx 0.12$. Hence,

$$\Delta\sigma_{(f)}' = (115)(0.12) = \textbf{13.8 kN/m}^2 \qquad \blacksquare$$

Example 16.6

Suppose that, for the sand drain project of Figure 16.21, the clay is normally consolidated. We are given the following data:

$$
\begin{aligned}
\text{Clay:} \quad H_c &= 15 \text{ ft (two-way drainage)} \\
C_c &= 0.31 \\
e_o &= 1.1
\end{aligned}
$$

Effective overburden pressure at the middle of the clay layer

$$
\begin{aligned}
&= 1000 \text{ lb/ft}^2 \\
C_v &= 0.115 \text{ ft}^2/\text{day}
\end{aligned}
$$

$$
\begin{aligned}
\text{Sand drain:} \quad r_w &= 0.3 \text{ ft} \\
d_e &= 6 \text{ ft} \\
C_v &= C_{\text{vr}}
\end{aligned}
$$

A surcharge is applied as shown in Figure 16.24. Assume this to be a no-smear case. Calculate the degree of consolidation 30 days after the surcharge is first applied. Also, determine the consolidation settlement at that time due to the surcharge.

Solution
From Eq. (16.39),

$$
T_c = \frac{C_v t_c}{H^2} = \frac{(0.115 \text{ ft}^2/\text{day})(60)}{\left(\dfrac{15}{2}\right)^2} = 0.123
$$

and

$$
T_v = \frac{C_v t_2}{H^2} = \frac{(0.115)(30)}{\left(\dfrac{15}{2}\right)^2} = 0.061
$$

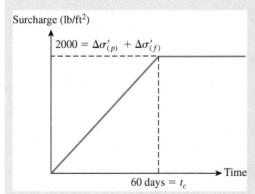

Surcharge (lb/ft²)

$2000 = \Delta\sigma'_{(p)} + \Delta\sigma'_{(f)}$

60 days $= t_c$

Time

Figure 16.24 Ramp load for a sand drain project

Using Figure 16.23 for $T_c = 0.123$ and $T_v = 0.061$, we have $U_v \approx 9\%$. For the sand drain,

$$n = \frac{d_e}{2r_w} = \frac{6}{(2)(0.3)} = 10$$

From Eq. (16.35),

$$T_{rc} = \frac{C_{vr} t_c}{d_e^2} = \frac{(0.115)(60)}{(6)^2} = 0.192$$

and

$$T_r = \frac{C_{vr} t_2}{d_e^2} = \frac{(0.115)(30)}{(6)^2} = 0.096$$

Again, from Eq. (16.33),

$$U_r = \frac{T_r - \dfrac{1}{A}[1 - \exp(-AT_r)]}{T_{rc}}$$

Also, for the no-smear case,

$$m = \frac{n^2}{n^2 - 1} \ln(n) - \frac{3n^2 - 1}{4n^2} = \frac{10^2}{10^2 - 1} \ln(10) - \frac{3(10)^2 - 1}{4(10)^2} = 1.578$$

and

$$A = \frac{2}{m} = \frac{2}{1.578} = 1.267$$

so

$$U_r = \frac{0.096 - \dfrac{1}{1.267}[1 - \exp(-1.267 \times 0.096)]}{0.192} = 0.03 = 3\%$$

From Eq. (16.25),

$$U_{v,r} = 1 - (1 - U_r)(1 - U_v) = 1 - (1 - 0.03)(1 - 0.09) = 0.117 = \textbf{11.7\%}$$

The total primary settlement is thus

$$S_{c(p)} = \frac{C_c H_c}{1 + e_o} \log\left[\frac{\sigma_o' + \Delta\sigma_{(p)}' + \Delta\sigma_f'}{\sigma_o'}\right]$$

$$= \frac{(0.31)(15)}{1 + 1.1} \log\left(\frac{1000 + 2000}{1000}\right) = 1.056 \text{ ft}$$

and the settlement after 30 days is

$$S_{c(p)} U_{v,r} = (1.056)(0.117)(12) = \textbf{1.48 in.} \quad \blacksquare$$

16.10 Prefabricated Vertical Drains

Prefabricated vertical drains (PVDs), also referred to as *wick* or *strip drains,* were originally developed as a substitute for the commonly used sand drain. With the advent of materials science, these drains began to be manufactured from synthetic polymers such as polypropylene and high-density polyethylene. PVDs are normally manufactured with a corrugated or channeled synthetic core enclosed by a geotextile filter, as shown schematically in Figure 16.25. Installation rates reported in the literature are on the order of 0.1 to 0.3 m/s, excluding equipment mobilization and setup time. PVDs have been used extensively in the past for expedient consolidation of low-permeability soils under surface surcharge. The main advantage of PVDs over sand drains is that they do not require drilling; thus, installation is much faster. Figures 16.26a and b are photographs of the installation of PVDs in the field.

Design of PVDs

The relationships for the average degree of consolidation due to radial drainage into sand drains are given in Eqs. (16.26) through (16.31) for equal-strain cases. Yeung (1997) used these relationships to develop design curves for PVDs. The theoretical developments used by Yeung are given next.

Figure 16.27 shows the layout of a square-grid pattern of prefabricated vertical drains. (See also Figure 16.25 for the definition of a and b.) The equivalent diameter of a PVD can be given as

$$d_w = \frac{2(a + b)}{\pi} \tag{16.40}$$

Now, Eq. (16.26) can be rewritten as

$$U_r = 1 - \exp\left(-\frac{8C_{vr}t}{d_w^2} \frac{d_w^2}{d_e^2 m}\right) = 1 - \exp\left(-\frac{8T_r'}{\alpha'}\right) \tag{16.41}$$

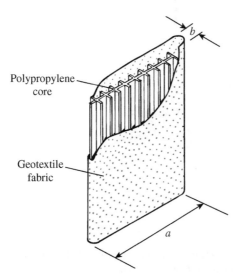

Polypropylene core

Geotextile fabric

Figure 16.25 Prefabricated vertical drain (PVD)

(a)

(b)

Figure 16.26 Installation of PVDs in the field [*Note:* (b) is a closeup view of (a)]
(*Courtesy of E. C. Shin, University of Incheon, Korea*)

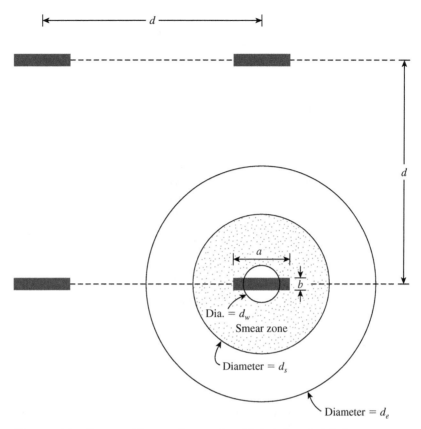

Figure 16.27 Square-grid pattern layout of prefabricated vertical drains

where d_e = diameter of the effective zone of drainage = $2r_e$. Also,

$$T_r' = \frac{C_{vr}t}{d_w^2} \tag{16.42}$$

$$\alpha' = n^2 m = \frac{n^4}{n^2 - S^2} \ln\left(\frac{n}{S}\right) - \left(\frac{3n^2 - S^2}{4}\right) + \frac{k_h}{k_s}(n^2 - S^2) \ln S \tag{16.43}$$

and

$$n = \frac{d_e}{d_w} \tag{16.44}$$

From Eq. (16.41),

$$T_r' = -\frac{\alpha'}{8} \ln(1 - U_r)$$

or

$$(T_r')_1 = \frac{T_r'}{\alpha'} = -\frac{\ln(1 - U_r)}{8} \tag{16.45}$$

Table 16.6 Variation of $(T_r')_1$ with U_r [Eq. (16.45)]

$U_r(\%)$	$(T_r')_1$
0	0
5	0.006
10	0.013
15	0.020
20	0.028
25	0.036
30	0.045
35	0.054
40	0.064
45	0.075
50	0.087
55	0.100
60	0.115
65	0.131
70	0.150
75	0.173
80	0.201
85	0.237
90	0.288
95	0.374

Table 16.6 gives the variation of $(T_r')_1$ with U_r. Also, Figure 16.28 shows plots of α' versus n for $k_h/k_s = 5$ and 10 and $S = 2$ and 3.

Following is a step-by-step procedure for the design of prefabricated vertical drains:

Step 1. Determine time t_2 available for the consolidation process and the $U_{v,r}$ required therefore [Eq. (16.24)]

Step 2. Determine U_r at time t_2 due to vertical drainage. From Eq. (16.25)

$$U_r = 1 - \frac{1 - U_{v,r}}{1 - U_v} \tag{16.46}$$

Step 3. For the PVD that is to be used, calculate d_w from Eq. (16.40).

Step 4. Determine $(T_r')_1$ from Eqs. (16.45) and (16.46).

Step 5. Determine $(T_r')_1$ from Eq. (16.42).

Step 6. Determine

$$\alpha' = \frac{T_r'}{(T_r')_1}.$$

Step 7. Using Figure 16.28 and α' determined from Step 6, determine n.

Step 8. From Eq. (16.44),

$$d_e = n \qquad d_w$$
$$\uparrow \qquad \uparrow$$
$$\text{Step 7} \quad \text{Step 3}$$

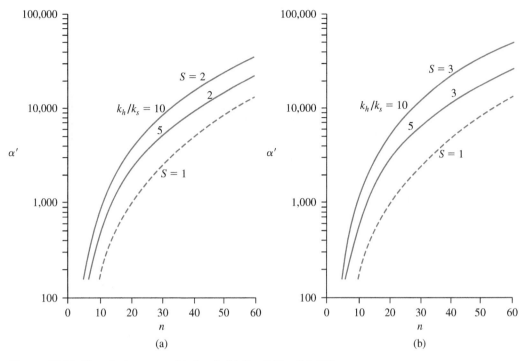

Figure 16.28 Plot of α' versus n: (a) $S = 2$; (b) $S = 3$ [Eq. (16.43)]

Step 9. Choose the drain spacing:

$$d = \frac{d_e}{1.05} \quad \text{(for triangular pattern)}$$

$$d = \frac{d_e}{1.128} \quad \text{(for square pattern)}$$

A Case History

The installation of PVDs combined with preloading is an efficient way to gain strength in soft clays for construction of foundations. An example of a field study can be found in the works of Shibuya and Hanh (2001) which describes a full-scale test embankment 40 m × 40 m in plan constructed over a soft clay layer located at Nong Ngu Hao, Thailand. PVDs were installed in the soft clay layer in a triangular pattern (Figure 16.29a). Figure 16.29b shows the pattern of preloading at the site along with the settlement-time plot at the ground surface below the center of the test embankment. Maximum settlement was reached after about four months. The variation of the undrained shear strength (c_u) with depth in the soft clay layer before and after the soil improvement is shown in Figure 16.29c. The variation of c_u with depth is based on field vane shear tests. The undrained shear strength increases by about 50 to 100% at various depths.

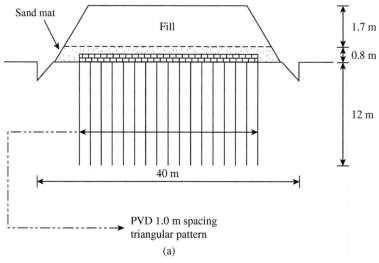

(a)

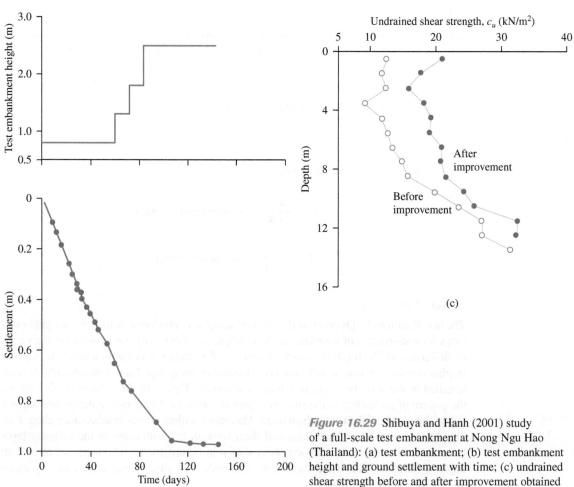

Figure 16.29 Shibuya and Hanh (2001) study of a full-scale test embankment at Nong Ngu Hao (Thailand): (a) test embankment; (b) test embankment height and ground settlement with time; (c) undrained shear strength before and after improvement obtained from vane shear test

16.11 Lime Stabilization

As mentioned in Section 16.1, admixtures are occasionally used to stabilize soils in the field—particularly fine-grained soils. The most common admixtures are lime, cement, and lime–fly ash. The main purposes of stabilizing the soil are to (a) modify the soil, (b) expedite construction, and (c) improve the strength and durability of the soil.

The types of *lime* commonly used to stabilize fine-grained soils are hydrated high-calcium lime $[Ca(OH)_2]$, calcitic quicklime (CaO), monohydrated dolomitic lime $[Ca(OH)_2 \cdot MgO]$, and dolomitic quicklime. The quantity of lime used to stabilize most soils usually is in the range from 5 to 10%. When lime is added to clayey soils, two *pozzolanic* chemical reactions occur: *cation exchange* and *flocculation–agglomeration*. In the cation exchange and flocculation–agglomeration reactions, the *monovalent* cations generally associated with clays are replaced by the *divalent* calcium ions. The cations can be arranged in a series based on their affinity for exchange:

$$Al^{3+} > Ca^{2+} > Mg^{2+} > NH_4^+ > K^+ > Na^+ > Li^+$$

Any cation can replace the ions to its right. For example, calcium ions can replace potassium and sodium ions from a clay. Flocculation–agglomeration produces a change in the texture of clay soils. The clay particles tend to clump together to form larger particles, thereby (a) decreasing the liquid limit, (b) increasing the plastic limit, (c) decreasing the plasticity index, (d) increasing the shrinkage limit, (e) increasing the workability, and (f) improving the strength and deformation properties of a soil. Some examples in which lime influences the plasticity of clayey soils are given in Table 16.7.

Pozzolanic reaction between soil and lime involves a reaction between lime and the silica and alumina of the soil to form cementing material. One such reaction is

$$Ca(OH)_2 + SiO_2 \rightarrow CSH$$
$$\uparrow$$
$$\text{Clay silica}$$

where

$C = CaO$
$S = SiO_2$
$H = H_2O$

The pozzolanic reaction may continue for a long time.

Table 16.7 Influence of Lime on Plasticity of Clay (Based on data from Thompson, 1967)

		0% Lime		5% Lime	
Soil	**AASHTO Classification**	**Liquid limit**	**Plasticity index**	**Liquid limit**	**Plasticity index**
Bryce B	A-7-6(18)	53	29	NP	NP
Cowden B	A-7-6(19)	54	33	NP	NP
Drummer B	A-7-6(19)	54	31	NP	NP
Huey B	A-7-6(17)	46	29	NP	NP

Note: NP—Non-plastic

The first 2 to 3% lime (on the dry-weight basis) substantially influences the workability and the property (such as plasticity) of the soil. The addition of lime to clayey soils also affects their compaction characteristics.

Properties of Cured Lime–Stabilized Soils

The unconfined compression strength (q_u) of fine-grained soils compacted at optimum moisture content may range from 170 kN/m^2 to 2100 kN/m^2 (25 lb/in^2 to 300 lb/in^2), depending upon the nature of the soil. With about 3 to 5% addition of lime and a curing period of 28 days, the unconfined compression strength may increase by 700 kN/m^2 (100 lb/in^2) or more.

The tensile strength (σ_T) of cured fine-grained soils also increases with lime stabilization. Tullock, Hudson, and Kennedy (1970) gave the following relationship between σ_T and q_u:

SI Units

$$\sigma_T \text{ (kN/m}^2) = 47.54 + 50.6 q_u \text{ (MN/m}^2) \tag{16.47a}$$

English Units

$$\sigma_T \text{ (lb/in}^2) = 6.89 + 50.6 q_u \text{ (kip/in}^2) \tag{16.47b}$$

where σ_T is the indirect tensile strength.

Thompson (1966) provided the following relationship to estimate the modulus of elasticity (E_s) of lime-stabilized soils:

SI Units

$$E_s \text{ (MN/m}^2) = 68.86 + 0.124 q_u \text{ (kN/m}^2) \tag{16.48a}$$

English Units

$$E_s \text{ (kip/in}^2) = 9.98 + 0.124 q_u \text{ (lb/in}^2) \tag{16.48b}$$

Poisson's ratio (μ_s) of cured stabilized soils with about 5% lime varies between 0.08 to 0.12 (with an average of 0.11) at a stress level of 25% or less of the ultimate compressive strength. It increases to about 0.27 to 0.37 (with an average of 0.31) at a stress level greater than 50% to 75% of the ultimate compression strength (Transportation Research Board, 1987).

Lime Stabilization in the Field

Lime stabilization in the field can be done in three ways. They are

1. The *in situ* material or the borrowed material can be mixed with the proper amount of lime at the site and then compacted after the addition of moisture.
2. The soil can be mixed with the proper amount of lime and water at a plant and then hauled back to the site for compaction.
3. Lime slurry can be pressure injected into the soil to a depth of 4 to 5 m (12 to 16 ft). Figure 16.30 shows a vehicle used for pressure injection of lime slurry.

The slurry-injection mechanical unit is mounted to the injection vehicle. A common injection unit is a hydraulic-lift mast with crossbeams that contain the injection rods. The rods are pushed into the ground by the action of the lift mast beams. The slurry is generally mixed in a batching tank about 3 m (10 ft) in diameter and 12 m (36 ft) long and is

Figure 16.30 Equipment for pressure injection of lime slurry (*Courtesy of Hayward Baker Inc., Odenton, Maryland*)

pumped at high pressure to the injection rods. Figure 16.31 is a photograph of the lime slurry pressure-injection process. The ratio typically specified for the preparation of lime slurry is 1.13 kg (2.5 lb) of dry lime to a gallon of water.

Because the addition of hydrated lime to soft clayey soils immediately increases the plastic limit, thus changing the soil from plastic to solid and making it appear to "dry up," limited amounts of the lime can be thrown on muddy and troublesome construction sites. This action improves trafficability and may save money and time. Quicklimes have also been successfully used in drill holes having diameters of 100 to 150 mm (4 to 6 in.) for stabilization of subgrades and slopes. For this type of work, holes are drilled in a grid pattern and then filled with quicklime.

16.12 Cement Stabilization

Cement is being increasingly used as a stabilizing material for soil, particularly in the construction of highways and earth dams. The first controlled soil–cement construction in the United States was carried out near Johnsonville, South Carolina, in 1935. Cement can be used to stabilize sandy and clayey soils. As in the case of lime, cement helps decrease the liquid limit and increase the plasticity index and workability of clayey soils. Cement stabilization is effective for clayey soils when the liquid limit is less than 45 to 50 and the plasticity index is less than about 25. The optimum requirements of cement by volume for effective stabilization of various types of soil are given in Table 16.8.

Like lime, cement helps increase the strength of soils, and strength increases with curing time. Table 16.9 presents some typical values of the unconfined compressive

Figure 16.31 Pressure injection of lime slurry (*Courtesy of Hayward Baker Inc., Odenton, Maryland*)

Table 16.8 Cement Requirement by Volume for Effective Stabilization of Various Soils[a]

Soil type		Percent cement by volume
AASHTO classification	**Unified classification**	
A-2 and A-3	GP, SP, and SW	6–10
A-4 and A-5	CL, ML, and MH	8–12
A-6 and A-7	CL, CH	10–14

[a]Based on data from Mitchell, J. K. and Freitag, D. R. (1959). "A Review and Evaluation of Soil-Cement Pavements," *Journal of the Soil Mechanics and Foundations Division,* American Society of Civil Engineers, Vol. 85, No. SM6, pp. 49–73.

Table 16.9 Typical Compressive Strengths of Soils and Soil–Cement Mixtures[a]

Material	Unconfined compressive strength range	
	kN/m²	lb/in²
Untreated soil:		
Clay, peat	Less than 350	Less than 50
Well-compacted sandy clay	70–280	10–40
Well-compacted gravel, sand, and clay mixtures	280–700	40–100
Soil–cement (10% cement by weight):		
Clay, organic soils	Less than 350	Less than 50
Silts, silty clays, very poorly graded sands, slightly organic soils	350–1050	50–150
Silty clays, sandy clays, very poorly graded sands, and gravels	700–1730	100–250
Silty sands, sandy clays, sands, and gravels	1730–3460	250–500
Well-graded sand–clay or gravel–sand–clay mixtures and sands and gravels	3460–10,350	500–1500

[a]Based on data from Mitchell, J. K. and Freitag, D. R. (1959). "A Review and Evaluation of Soil-Cement Pavements," *Journal of the Soil Mechanics and Foundations Division*, American Society of Civil Engineers, Vol. 85, No. SM6, pp. 49–73.

strength of various types of untreated soil and of soil–cement mixtures made with approximately 10% cement by weight.

Granular soils and clayey soils with low plasticity obviously are most suitable for cement stabilization. Calcium clays are more easily stabilized by the addition of cement, whereas sodium and hydrogen clays, which are expansive in nature, respond better to lime stabilization. For these reasons, proper care should be given in the selection of the stabilizing material.

For field compaction, the proper amount of cement can be mixed with soil either at the site or at a mixing plant. If the latter approach is adopted, the mixture can then be carried to the site. The soil is compacted to the required unit weight with a predetermined amount of water.

Similar to lime injection, cement slurry made of portland cement and water (in a water–cement ratio of 0.5:5) can be used for pressure grouting of poor soils under foundations of buildings and other structures. Grouting decreases the hydraulic conductivity of soils and increases their strength and load-bearing capacity. For the design of low-frequency machine foundations subjected to vibrating forces, stiffening the foundation soil by grouting and thereby increasing the resonant frequency is sometimes necessary.

16.13 Fly-Ash Stabilization

Fly ash is a by-product of the pulverized coal combustion process usually associated with electric power-generating plants. It is a fine-grained dust and is composed primarily of silica, alumina, and various oxides and alkalies. Fly ash is pozzolanic in nature and can react with hydrated lime to produce cementitious products. For that reason, lime–fly-ash mixtures can be used to stabilize highway bases and subbases. Effective mixes can be prepared with 10 to 35% fly ash and 2 to 10% lime. Soil–lime–fly-ash mixes are compacted under controlled conditions, with proper amounts of moisture to obtain stabilized soil layers.

A certain type of fly ash, referred to as "Type C" fly ash, is obtained from the burning of coal primarily from the western United States. This type of fly ash contains a fairly large proportion (up to about 25%) of free lime that, with the addition of water, will react with other fly-ash compounds to form cementitious products. Its use may eliminate the need to add manufactured lime.

16.14 Stone Columns

A method now being used to increase the load-bearing capacity of shallow foundations on soft clay layers is the construction of stone columns. This generally consists of water-jetting a vibroflot (see Section 16.6) into the soft clay layer to make a circular hole that extends through the clay to firmer soil. The hole is then filled with an imported gravel. The gravel in the hole is gradually compacted as the vibrator is withdrawn. The gravel used for the stone column has a size range of 6 to 40 mm (0.25 to 1.6 in.). Stone columns usually have diameters of 0.5 to 0.75 m (1.6 to 2.5 ft) and are spaced at about 1.5 to 3 m (5 to 10 ft) center to center. Figure 16.32 shows the construction of a stone column.

After stone columns are constructed, a fill material should always be placed over the ground surface and compacted before the foundation is constructed. The stone columns

Figure 16.32 Construction of a stone column [*DGI-Menard (USA)*.]

tend to reduce the settlement of foundations at allowable loads. Several case histories of construction projects using stone columns are presented in Hughes and Withers (1974), Hughes et al. (1975), Mitchell and Huber (1985), and other works.

Stone columns work more effectively when they are used to stabilize a large area where the undrained shear strength of the subsoil is in the range of 10 to 50 kN/m² (200 to 1000 lb/ft²) than to improve the bearing capacity of structural foundations (Bachus and Barksdale, 1989). Subsoils weaker than that may not provide sufficient lateral support for the columns. For large-site improvement, stone columns are most effective to a depth of 6 to 10 m (20 to 30 ft). However, they have been constructed to a depth of 31 m (100 ft). Bachus and Barksdale provided the following general guidelines for the design of stone columns to stabilize large areas.

Figure 16.33a shows the plan view of several stone columns. The area replacement ratio for the stone columns may be expressed as

$$a_s = \frac{A_s}{A}$$

(16.49)

where

A_s = area of the stone column
A = total area within the unit cell

For an *equilateral triangular pattern* of stone columns,

$$a_s = 0.907\left(\frac{D}{s}\right)^2$$

(16.50)

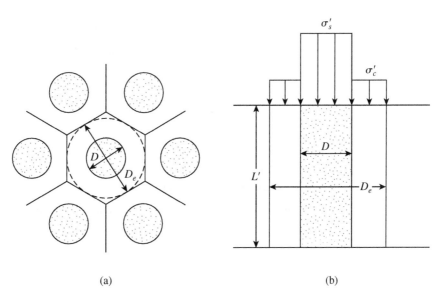

(a)   (b)

Figure 16.33 (a) Stone columns in a triangular pattern; (b) stress concentration due to change in stiffness

where

D = diameter of the stone column
s = spacing between the columns

Combining Eqs. (16.49) and (16.50),

$$\frac{A_s}{A} = \frac{\frac{\pi}{4}D^2}{\frac{\pi}{4}D_e^2} = a_s = 0.907\left(\frac{D}{s}\right)^2$$

or

$$D_e = 1.05s \tag{16.51}$$

Similarly, it can be shown that, for square pattern of stone columns,

$$D_e = 1.13s \tag{16.52}$$

When a uniform stress by means of a fill operation is applied to an area with stone columns to induce consolidation, a stress concentration occurs due to the change in the stiffness between the stone columns and the surrounding soil. (See Figure 16.33b.) The stress concentration factor is defined as

$$n' = \frac{\sigma_s'}{\sigma_c'} \tag{16.53}$$

where

σ_s' = effective stress in the stone column
σ_c' = effective stress in the subgrade soil

The relationships for σ_s' and σ_c' are

$$\sigma_s' = \sigma'\left[\frac{n'}{1 + (n' - 1)a_s}\right] = \mu_s\sigma' \tag{16.54}$$

and

$$\sigma_c' = \sigma'\left[\frac{1}{1 + (n' - 1)a_s}\right] = \mu_c\sigma' \tag{16.55}$$

where

σ' = average effective vertical stress
μ_s, μ_c = stress concentration coefficients

The improvement in the soil owing to the stone columns may be expressed as

$$\frac{S_{e(t)}}{S_e} = \mu_c \tag{16.56}$$

where

$S_{e(t)}$ = settlement of the treated soil
S_e = total settlement of the untreated soil

Load-Bearing Capacity of Stone Columns

When the length L' of the stone column is less than about $3D$ and a foundation is constructed over it, failure occurs by plunging similar to short piles in soft to medium-stiff clays. For longer columns sufficient to prevent plunging, the load carrying capacity is governed by the ultimate radial confining pressure and the shear strength of the surrounding matrix soil. In those cases, failure at ultimate load occurs by bulging, as shown in Figure 16.34. Mitchell (1981) proposed that the ultimate bearing capacity (q_u) of a stone column can be given as

$$q_u = c_u N_p \tag{16.57}$$

where c_u = undrained shear strength of clay
N_p = bearing capacity factor

Mitchell (1981) recommended that

$$N_p \approx 25 \tag{16.58}$$

Based on several field case studies, Stuedlein and Holtz (2013) recommended that

$$N_p = \exp(-0.0096c_u + 3.5) \tag{16.59}$$

where c_u is in kN/m^2.

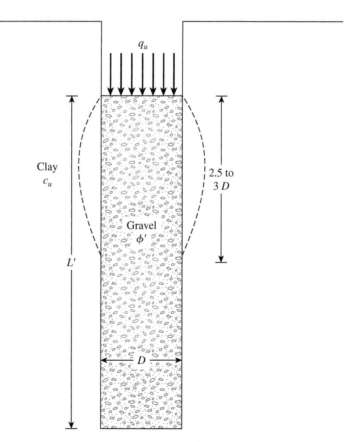

Figure 16.34 Bearing capacity of stone column

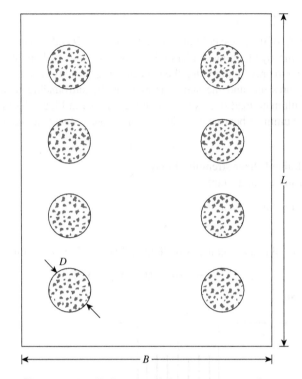

Figure 16.35 Shallow foundation over a group of stone columns

If a foundation is constructed measuring $B \times L$ in plan over a group of stone columns, as shown in Figure 16.35, the ultimate bearing capacity q_u can be expressed as (Stuedlein and Holtz, 2013)

$$q_u = N_p c_u \, a_s + N_c c_u \, (1 - a_s) F_{cs} F_{cd} \qquad (16.60)$$

where N_p is expressed by Eq. (16.59)

$N_c = 5.14$

F_{cs} and F_{cd} = shape and depth factors (see Table 4.6)

Then

$$F_{cs} = 1 + 0.2\frac{B}{L} \qquad (16.61)$$

and

$$F_{cd} = 1 + 0.2\frac{D_f}{B} \qquad (16.62)$$

where D_f = depth of the foundation.

Example 16.7

Consider a foundation 4 m $\times$ 2 m in plan constructed over a group of stone columns in a square pattern in soft clay. Given

Stone columns: $D = 0.4$ m
Area ratio, $a_s = 0.3$
$L' = 4.8$ m
Clay: $c_u = 36$ KN/m^2
Foundation: $D_f = 0.75$ m

Estimate the ultimate load Q_u for the foundation.

Solution

From Eq. (16.60),

$$Q_u = N_p c_u a_s + N_c c_u (1 - a_s) F_{cs} F_{cd}$$

From Eq. (16.59),

$$N_p = \exp(-0.0096 c_u + 3.5) = \exp[(-0.0096)(36) + 3.5] = 23.44$$

$$F_{cs} = 1 + 0.2\left(\frac{B}{L}\right) = 1 + 0.2\left(\frac{2}{4}\right) = 1.1$$

$$F_{cd} = 1 + 0.2\left(\frac{D_f}{B}\right) = 1 + 0.2\left(\frac{0.75}{2}\right) = 1.075$$

and

$$q_u = (23.44)(36)(0.3) + (5.14)(36)(1 - 0.3)(1.1)(1.075) = 406.31 \text{ kN/m}^2$$

Thus, the ultimate load is

$$Q_u = q_u BL = (406.31)(2)(4) = \mathbf{3250.48 \text{ kN}}$$ ∎

16.15 Sand Compaction Piles

Sand compaction piles are similar to stone columns, and they can be used in marginal sites to improve stability, control liquefaction, and reduce the settlement of various structures. Built in soft clay, these piles can significantly accelerate the pore water pressure-dissipation process and hence the time for consolidation.

Sand piles were first constructed in Japan between 1930 and 1950 (Ichimoto, 1981). Large-diameter compacted sand columns were constructed in 1955, using the Compozer technique (Aboshi et al., 1979). The Vibro-Compozer method of sand pile construction was developed by Murayama in Japan in 1958 (Murayama, 1962).

Sand compaction piles are constructed by driving a hollow mandrel with its bottom closed during driving. On partial withdrawal of the mandrel, the bottom doors open. Sand is poured from the top of the mandrel and is compacted in steps by applying air pressure as the mandrel is withdrawn. The piles are usually 0.46 to 0.76 m (1.5 to 2.5 ft) in diameter and are placed at about 1.5 to 3 m (5 to 10 ft) center to center. The pattern of layout of sand

Figure 16.36 Construction of sand compaction pile in Yokohama, Japan Harbor (*Courtesy of E. C. Shin, University of Incheon, Korea*)

compaction piles is the same as for stone columns. Figure 16.36 shows the construction of sand compaction piles in the harbor of Yokohama, Japan.

Basore and Boitano (1969) reported a case history on the densification of a granular subsoil having a thickness of about 9 m (30 ft) at the Treasure Island Naval Station in San Francisco, California, using sand compaction piles. The sand piles had diameters of 356 mm (14 in.). Figure 16.37a shows the layout of the sand piles. The spacing, S', between the piles was varied. The standard penetration resistance, N_{60}, before and after the construction of piles are shown in Figure 16.37b (see location of SPT test in Figure 16.37a). From this figure, it can be seen that the effect of densification at any given depth decreases with the increase in S' (or S'/D). These tests show that when S'/D exceeds about 4 to 5, the effect of densification is practically negligible.

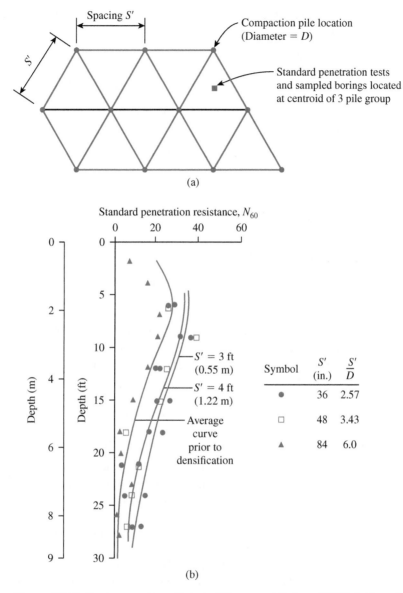

Figure 16.37 Sand compaction pile test of Basore and Boitano (1969): (a) layout of the compaction piles; (b) standard penetration resistance variation with depth and S'

16.16 Dynamic Compaction

Dynamic compaction is a technique that is beginning to gain popularity in the United States for densification of granular soil deposits. The process primarily involves dropping a heavy weight repeatedly on the ground at regular intervals. The weight of the hammer used varies from 8 to 35 metric tons, and the height of the hammer drop varies between 7.5 and 30.5 m

($\simeq$ 25 and 100 ft). The stress waves generated by the hammer drops help in the densification. The degree of compaction achieved depends on

- The weight of the hammer
- The height of the drop
- The spacing of the locations at which the hammer is dropped

Leonards et al. (1980) suggested that the significant depth of influence for compaction is approximately

$$DI \simeq \tfrac{1}{2}\sqrt{W_H h} \tag{16.63a}$$

where

DI = significant depth of densification (m)
W_H = dropping weight (metric ton)
h = height of drop (m)

In English units, Eq. (16.63a) becomes

$$DI = 0.61\sqrt{W_H h} \tag{16.63b}$$

where DI and h are in ft and W_H is in kip.

Partos et al. (1989) provided several case histories of site improvement that used dynamic compaction. In 1992, Poran and Rodriguez suggested a rational method for conducting dynamic compaction for granular soils in the field. According to their method, for a hammer of width D having a weight W_H and a drop h, the approximate shape of the densified area will be of the type shown in Figure 16.38 (i.e., a semiprolate spheroid). Note that in this figure b = DI. Figure 16.39 gives the design chart for a/D and b/D versus $NW_H h/Ab$ (D = width of the hammer if not circular in cross section; A = area of cross section of the hammer; N = number of required hammer drops). The method uses the following steps:

Step 1. Determine the required significant depth of densification, DI(=b).
Step 2. Determine the hammer weight (W_H), height of drop (h), dimensions of the cross section, and thus the area A and the width D.
Step 3. Determine DI/D = b/D.
Step 4. Use Figure 16.39 and determine the magnitude of $NW_H h/Ab$ for the value of b/D obtained in Step 3.
Step 5. Since the magnitudes of W_H, h, A, and b are known (or assumed) from Step 2, the number of hammer drops can be estimated from the value of $NW_H h/Ab$ obtained from Step 4.

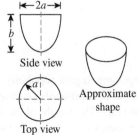

Figure 16.38 Approximate shape of the densified area due to dynamic compaction [Poran, C. J. and Rodriguez, J. A. (1992). "Design of Dynamic Compaction," *Canadian Geotechnical Journal*, Vol. 29, No. 5, pp. 796–802.]

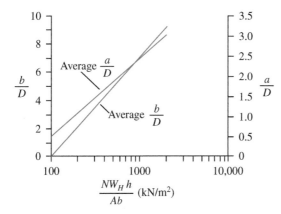

Figure 16.39 Plot of a/D and b/D versus $NW_H h/Ab$ (Based on Poran and Rodriguez, 1992) (Poran, C. J. and Rodriguez, J. A. (1992). "Design of Dynamic Compaction," *Canadian Geotechnical Journal*, Vol. 29, No. 5, pp. 796–802.)

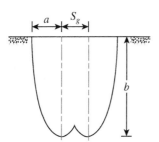

Figure 16.40 Approximate grid spacing for dynamic compaction

Step 6. With known values of $NW_H h/Ab$, determine a/D and thus a from Figure 16.39.

Step 7. The grid spacing, S_g, for dynamic compaction may now be assumed to be equal to or somewhat less than a. (See Figure 16.40.)

16.17 Jet Grouting

Jet grouting is a soil stabilization process whereby cement slurry in injected into soil at a high velocity to form a soil–concrete matrix. Conceptually, the process of jet grouting was first developed in the 1960s. Most of the research work after that was conducted in Japan (Ohta and Shibazaki, 1982). The technique was introduced into Europe in the late 1970s, whereas the process was first used in the United States in the early 1980s (Welsh, Rubright, and Coomber, 1986).

Three basic systems of jet grouting have been developed—single, double, and triple rod systems. In all cases, hydraulic rotary drilling is used to reach the design depth at which the soil has to be stabilized. Figure 16.41a shows the *single rod system* in which a cement slurry is injected at a high velocity to form a soil–cement matrix. In the *double rod system* (Figure 16.41b), the cement slurry is injected at a high velocity sheathed in a cone of air at an equally high velocity to erode and mix the soil well. The *triple rod system* (Figure 16.41c) uses high-pressure water shielded in a cone of air to erode the soil. The void created in this process is then filled with a pre-engineering cement slurry.

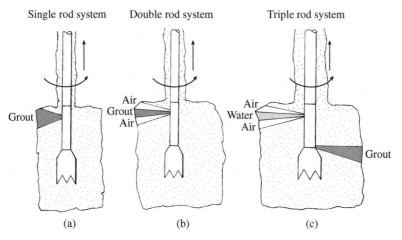

Single rod system   Double rod system   Triple rod system

(a)   (b)   (c)

Figure 16.41 Jet grouting

The effectiveness of the jet grouting is very much influenced by the nature of erodibility of soil. Gravelly soil and clean sand are highly erodible, whereas highly plastic clays are difficult to erode. A summary of the range of parameters generally encountered for the three systems above follows (Welsh and Burke, 1991; Burke, 2004):

Single Rod System:

A. Grout slurry

    Pressure . . . . 0.4–0.7 MN/m^2
    Volume . . . . 100–300 l/min
    Specific gravity 1.25–1.6
    Number of nozzles 1–6

B. Lift

    Step height. . . . . 5–600 mm
    Step time . . . . 4–30 sec

C. Rotation . . . . 7–20 rpm
D. Stabilized soil column diameter

    Soft clay. . . . . 0.4–0.9 m
    Silt . . . . 0.6–1.1 m
    Sand . . . . 0.8–1.2 m

Double Rod System:

A. Grout slurry

    Pressure . . . . 0.3–0.7 MN/m^2
    Volume . . . . 100 –600 l/min
    Specific gravity 1.25–1.8
    Number of nozzles 1–2

B. Air

    Pressure . . . . 700–1500 kN/m^2
    Volume . . . . 8–30 m^3/min

C. Lift

    Step height. . . . . 25–400 mm
    Step time . . . . 4–30 sec

D. Rotation . . . . 7–15 rpm
E. Stabilized soil column diameter

    Soft clay. . . . . 0.9–1.8 m
    Silt . . . . 0.9–1.8 m
    Sand. . . . . 1.2–2.1 m

Triple Rod System:

A. Grout slurry

    Pressure . . . . 700 kN/m^2–1 MN/m^2
    Volume . . . . 120–200 l/min
    Specific gravity 1.5–2.0
    Number of nozzles 1–3

B. Air

    Pressure . . . . 700–1500 kN/m^2
    Volume . . . . 4–15 m^3/min

C. Water

    Pressure . . . . 0.3–0.4 MN/m^2
    Volume . . . . 80–200 l/min

D. Lift

    Step height. . . . . 20–50 mm
    Step time . . . . 4–20 sec

E. Rotation . . . . 7–15 rpm
F. Stabilized soil column diameter

    Soft clay. . . . . 0.9–1.2 m
    Silt . . . . 0.9–1.4 m
    Sand. . . . . 0.9–2.5 m

16.18   Deep Mixing

Deep mixing method (*DMM*) refers to a ground modification technology in which soft, compressible, or other unstable soils are treated *in-situ* for safely and economically improving the physical and mechanical properties of the natural soil to meet the design requirements of various geotechnical applications. The treated soil generally has higher strength, lower compressibility, and lower permeability than the native soil. The technology involves mechanically blending *in-situ* soils with cementitious binder materials that are injected into the soil either in a dry form called *dry mixing* or in a slurry form called *wet mixing* through hollow rotating mixing shafts that are often mounted as multi-axis augers equipped with mixing paddles and cutting tools. Treated soils may be constructed as columns in various grid patterns or as overlapping columns to create soil mix walls. The columns are typically 0.6 to 1.5 m in diameter and may extend up to 40 m in depth (FHWA, 2000).

At the present time, *deep mixing* is used more as a generic name to describe the concept of deep soil mixing by using mechanical rotating shafts—as opposed to jet grouting, which uses hydraulically powered high pressure jets to achieve similar objectives. Depending on the characteristics of the mixing equipment, binder materials, treatment patterns, geographic locations, and the specialty contractors implementing the technique, a variety of acronyms and/or trade names are used globally to refer to the general concept of deep mixing.

Brief History of DMM

Various deep mixing methods evolved throughout the second half of the 20th century, primarily in Japan, the Scandinavian countries, and the United States. A chronological history of these developments and their applications is summarized by FHWA (2000). A brief review of those developments is given here.

DMM was first introduced in 1954 by Intrusion Prepakt Co. (USA) in the form of a single-auger *mixed in place (MIP)* piling technique. The MIP technique was used in Japan by the Seiko Kogyo Company of Osaka for excavation support and groundwater control (1961–early part of 1970s). Japan played pioneering roles in the development of several well-known deep mixing techniques. In 1972, the Seiko Kogyo Company developed the *soil mixed wall (SMW)* method for retaining walls by using overlapping multiple auger technique for the first time. The Port and Harbor Research Institute of Japan developed the *deep lime mixing (DLM)* methods through extensive laboratory and field research (1967–1977); the *cement deep mixing (CDM)* method using fluid cement grout in offshore soft marine soils (1975–1977); and other similar methods, such as *deep chemical mixing (DCM)*, *deep cement continuous mixing (DCCM)*, and *deep mixing improvement by cement stabilization (DEMIC)* over the following five years. The Public Works Research Institute of Japan developed the *dry jet mixing (DJM)* method using dry powdered cement (1976–1980). In 1979, the Tenox Company developed the *soil cement column (SCC)* system in Japan. The *spreadable wing (SWING)* method of deep mixing was developed in Japan in 1984, followed by various jet-assisted methods (1986–1991). In 1992, Fudo Company and Chemical Grout Company of Japan developed the *jet and churning system management (JACSMAN)* that combined the procedures of mechanical mixing (core of the column) and cross jet mixing (outside zone of the column).

The first major DMM development in Scandinavia was the *Swedish lime column* method to treat soft clays under embankments (1967 by Kjeld Paus, Linden–Alimak AB, in cooperation with Swedish Geotechnical Institute, Euroc AB, and BPA Byggproduktion AB).

The first commercial applications were in excavation support, embankment stabilization, and shallow foundations near Stockholm in 1975. The *lime cement column* method was first used commercially in Finland and Norway in the mid-1980s. Outside Scandinavia, the first European developments appear to be in France with the introduction of a compacted soil-cement mix called *"Colmix"* by Bachy Company in 1987 (constructed by reverse rotation of multiple augers during withdrawal), and in the UK at around the same time with the single-auger deep-mixing system developed by Cementation Ltd. In 1991, the City of Helsinki, Finland, and contractor YIT introduced block stabilization of soft clays to a depth of 5 m. In 1995, Finnish researchers introduced new binders such as slag and pulverized fly ash in addition to cement.

Major developments related to DMM in the United States include: (a) introduction of deep soil mixing (DSM) in 1987; (b) shallow soil mixing (SSM) in 1988 (both by Geo-Con, Inc.); (c) inclusion of DMM in the US EPA Superfund Innovative Technology Evaluation Program for *in-situ* stabilization of contaminated soils (1989); and (d) first full-scale demonstration of VERTwall DMM concept in Texas by Geo-Con, Inc. (1998).

DMM Treatment Patterns

Deep mixing techniques can be used to produce a wide range of patterns in the treated soil structure. The selected pattern depends on the construction location (land or marine), the purpose of the DMM applications, and the characteristics and capabilities of the method used. The treatment patterns can be single element (column), rows of overlapping elements (walls or panels), grids or lattices, or blocks.

Dry and Wet Mixing Methods

Deep mixing is carried out using either the dry method or the wet method. Dry mixing is possible when the natural moisture content of the *in-situ* soil is quite high, so the cement hydration reaction can take place for strength development. Deep deposits of organic and peat soils (with high water content) can be effectively stabilized with the dry method. The column diameter is typically 0.6 to 0.8 m with the depth of treatment reaching up to 25 m. Release of dry binder and the soil mixing occur during the withdrawal of the mixing rod, where the rotational direction is reversed compared to the direction during penetration. The binder dosage is maintained as desired by controlling the air pressure and the amount of binder during construction.

Wet mixing is more appropriate when the natural water content is low. Soft clays, silts, and fine sands are suitable for this method. The binder is introduced in a slurry form through a nozzle placed generally at the end of the auger. The specialized mixing tool contains transverse beams and can move vertically along the column length to achieve homogeneous mixing. The composition and the amount of slurry can be controlled to achieve design specifications. The column diameters are typically 0.4 to 2.4 m, depending on the application. Steel reinforcements can be inserted into the soft columns to improve bending resistance.

Cement injected in the wet method is typically in the range of 100 to 500 kg/m^3 of untreated soil (Bruce and Bruce, 2003). In the dry method, this range is 100 to 300 kg/m^3, provided the natural moisture content is in the range of 40 to 200%.

Problems

16.1 A sandy soil has maximum and minimum dry unit weights of 18.08 kN/m^3 and 14.46 kN/m^3, respectively, and a dry unit weight of compaction in the field of 16.35 kN/m^3. Estimate the following:
a. The relative compaction in the field
b. The relative density in the field

16.2 A silty clay soil has a plasticity index (PI) of 14. Estimate the optimum moisture content and the maximum dry unit weight of the soil when compacted using the procedure of:
a. Standard Proctor test
b. Modified Proctor test
Use Eqs. (16.11) and (16.12).

16.3 Redo Problem 16.2 using Eqs. (16.9) and (16.10). Given plastic limit (PL) = 18

16.4 The following are given for a natural soil deposit:
Moist unit weight, γ = 17.8 kN/m^3
Moisture content, w = 14%
G_s = 2.7
This soil is to be excavated and transported to a construction site for use in a compacted fill. If the specification calls for the soil to be compacted at least to a dry unit weight of 18.4 kN/m^3 at the same moisture content of 14%, how many cubic meters of soil from the excavation site are needed to produce 20,000 m^3 of compacted fill?

16.5 A proposed embankment fill required 8000 m^3 of compacted soil. The void ratio of the compacted fill is specified to be 0.6. Four available borrow pits are shown below along with the void ratios of the soil and the cost per cubic meter for moving the soil to the proposed construction site.

Borrow pit	Void ratio	Cost ($/m³)
A	0.82	9
B	0.91	7
C	0.95	8
D	0.75	11

Make the necessary calculations to select the pit from which the soil should be brought to minimize the cost. Assume G_s to be the same for all borrow-pit soil.

16.6 For a vibroflotation work, the backfill to be used has the following characteristics:
D_{50} = 2 mm
D_{20} = 0.7 mm
D_{10} = 0.65 mm
Determine the suitability number of the backfill. How would you rate the material?

16.7 Repeat Problem 16.6 with the following:
D_{50} = 3.2 mm
D_{20} = 0.91 mm
D_{10} = 0.72 mm

16.8 Refer to Figure 16.16. For a large fill operation, the average permanent load $[\Delta\sigma'_{(p)}]$ on the clay layer will increase by about 75 kN/m². The average effective overburden pressure on the clay layer before the fill operation is 110 kN/m². For the clay layer, which is normally consolidated and drained at top and bottom, given: $H_c = 8$ m, $C_c = 0.27$, $e_o = 1.02$, $C_v = 0.52$ m²/month. Determine the following:
   a. The primary consolidation settlement of the clay layer caused by the addition of the permanent load $\Delta\sigma'_{(p)}$
   b. The time required for 80% of primary consolidation settlement under the additional permanent load only
   c. The temporary surcharge, $\Delta\sigma'_{(f)}$, that will be required to eliminate the entire primary consolidation settlement in 12 months by the precompression technique

16.9 Repeat Problem 16.8 with the following: $\Delta\sigma'_{(p)} = 58$ kN/m², average effective overburden pressure on the clay layer $= 72$ kN/m², $H_c = 5$ m, $C_c = 0.3$, $e_o = 1.0$, and $C_v = 0.1$ cm²/min.

16.10 The diagram of a sand drain is shown in Figures 16.21 and 16.22. Given: $r_w = 0.25$ m, $r_s = 0.35$ m, $d_e = 4.5$ m, $C_v = C_{vr} = 0.3$ m²/month, $k_h/k_s = 2$, and $H_c = 9$ m. Determine:
   a. The degree of consolidation for the clay layer caused only by the sand drains after six months of surcharge application
   b. The degree of consolidation for the clay layer that is caused by the combination of vertical drainage (drained on top and bottom) and radial drainage after six months of the application of surcharge.
   Assume that the surcharge is applied instantaneously.

16.11 A 10-ft-thick clay layer is drained at the top and bottom. Its characteristics are $C_{vr} = C_v$ (for vertical drainage) $= 0.042$ ft²/day, $r_w = 8$ in., and $d_e = 6$ ft. Estimate the degree of consolidation of the clay layer caused by the combination of vertical and radial drainage at $t = 0.2, 0.4, 0.8$, and 1 year. Assume that the surcharge is applied instantaneously, and there is no smear.

16.12 For a sand drain project (Figure 16.20), the following are given:

   Clay: Normally consolidated
   $H_c = 5.5$ m (one-way drainage)
   $C_c = 0.3$
   $e_o = 0.76$
   $C_v = 0.015$ m²/day
   Effective overburden pressure at the middle of clay
   layer $= 80$ kN/m²
   Sand drain: $r_w = 0.07$ m
   $r_w = r_s$
   $d_e = 2.5$ m
   $C_v = C_{vr}$

   A surcharge is applied as shown in Figure P16.12. Calculate the degree of consolidation and the consolidation settlement 50 days after the beginning of the surcharge application.

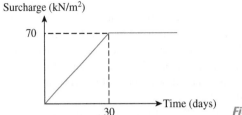

Surcharge (kN/m²)

70

30 Time (days)

Figure P.16.12

References

ABOSHI, H., ICHIMOTO, E., and HARADA, K. (1979). "The Compozer—a Method to Improve Characteristics of Soft Clay by Inclusion of Large Diameter Sand Column," *Proceedings, International Conference on Soil Reinforcement, Reinforced Earth and Other Techniques,* Vol. 1, Paris, pp. 211–216.

AMERICAN SOCIETY FOR TESTING and MATERIALS (2007). *Annual Book of Standards,* Vol. 04.08, West Conshohocken, PA.

BACHUS, R. C. and BARKSDALE, R. D. (1989). "Design Methodology for Foundations on Stone Columns," *Proceedings, Foundation Engineering: Current Principles and Practices* American Society of Civil Engineers, Vol. 1, pp. 244–257.

BARRON, R. A. (1948). "Consolidation of Fine-Grained Soils by Drain Wells," *Transactions,* American Society of Civil Engineers, Vol. 113, pp. 718–754.

BASORE, C. E. and BOITANO, J. D. (1969). "Sand Densification by Piles and Vibroflotation," *Journal of the Soil Mechanics and Foundations Division,* American Society of Civil Engineers, Vol. 95, No. SM6, pp. 1303–1323.

BROWN, R. E. (1977). "Vibroflotation Compaction of Cohesionless Soils," *Journal of the Geotechnical Engineering Division,* American Society of Civil Engineers, Vol. 103, No. GT12, pp. 1437–1451.

BRUCE, D. A. and BRUCE, E. C. (2003). "The Practitioner's Guide to Deep Mixing," *Grouting and Ground Treatment, Proceedings of the Third International Conference,* New Orleans, LA, pp. 474–488.

BURKE, G. K. (2004). "Jet Grouting Systems: Advantages and Disadvantages," *Proceedings, GeoSupport 2004: Drilled Shafts, Micropiling, Deep Mixing, Remedial Methods, and Special Foundation Systems,* American Society of Civil Engineers, pp. 875–886.

D'APPOLONIA, D. J., WHITMAN, R. V., and D'APPOLONIA, E. (1969). "Sand Compaction with Vibratory Rollers," *Journal of the Soil Mechanics and Foundations Division,* American Society of Civil Engineers, Vol. 95, No. SM1, pp. 263–284.

FHWA (2000). "An Introduction to the Deep Soil Mixing Methods as used in Geotechnical Applications," FHWA-RD-99-138, *Federal Highway Administration,* US Department of Transportation, p. 135.

GURTUG, Y. and SRIDHARAN, A. (2004). "Compaction Behaviour and Prediction of Its Characteristics of Fine Grained Soils with Particular Reference to Compaction Energy," *Soils and Foundations,* Vol. 44, No. 5, pp. 27–36.

ICHIMOTO, A. (1981). "Construction and Design of Sand Compaction Piles," *Soil Improvement, General Civil Engineering Laboratory* (in Japanese), Vol. 5. pp. 37–45.

JOHNSON, S. J. (1970a). "Precompression for Improving Foundation Soils," *Journal of the Soil Mechanics and Foundations Division,* American Society of Civil Engineers. Vol. 96, No. SM1, pp. 114–144.

JOHNSON, S. J. (1970b). "Foundation Precompression with Vertical Sand Drains," *Journal of the Soil Mechanics and Foundations Division,* American Society of Civil Engineers. Vol. 96, No. SM1, pp. 145–175.

LEONARDS, G. A., CUTTER, W. A., and HOLTZ, R. D. (1980). "Dynamic Compaction of Granular Soils," *Journal of Geotechnical Engineering Division,* ASCE, Vol. 96, No. GT1, pp. 73–110.

MATTEO, L. D., BIGOTTI, F., and RICCO, R. (2009). "Best-Fit Model to Estimate Proctor Properties of Compacted Soil," *Journal of Geotechnical and Geoenvironmental Engineering,* American Society of Civl Engineers, Vol. 135, No. 7, pp. 992–996.

MITCHELL, J. K. (1970). "In-Place Treatment of Foundation Soils," *Journal of the Soil Mechanics and Foundations Division,* American Society of Civil Engineers, Vol. 96, No. SM1, pp. 73–110.

MITCHELL, J. K. (1981). "Soil Improvement—State-of-the-Art Report," *Proceedings, 10th International Conference on Soil Mechanics and Foundation Engineering,* Stockholm, Sweden, Vol. 4, pp. 506–565.

MITCHELL, J. K. and FREITAG, D. R. (1959). "A Review and Evaluation of Soil–Cement Pavements," *Journal of the Soil Mechanics and Foundations Division,* American Society of Civil Engineers, Vol. 85, No. SM6, pp. 49–73.

MITCHELL, J. K. and HUBER, T. R. (1985). "Performance of a Stone Column Foundation," *Journal of Geotechnical Engineering,* American Society of Civil Engineers, Vol. 111, No. GT2, pp. 205–223.

MURAYAMA, S. (1962). "An Analysis of Vibro-Compozer Method on Cohesive Soils," *Construction in Mechanization* (in Japanese), No. 150, pp. 10–15.

OHTA, S. and SHIBAZAKI, M. (1982). "A Unique Underpinning of Soil Specification Utilizing Super-High Pressure Liquid Jet," *Proceedings, Conference on Grouting in Geotechnical Engineering,* New Orleans, Louisiana.

OLSON, R. E. (1977). "Consolidation under Time-Dependent Loading," *Journal of Geotechnical Engineering Division,* ASCE, Vol. 102, No. GT1, pp. 55–60.

OMAR, M., ABDALLAH, S., BASMA, A., and BARAKAT, S. (2003). "Compaction Characteristics of Granular Soils in the United Arab Emirates," *Geotechnical and Geological Engineering,* Vol. 21, No. 3, pp. 283–295.

OSMAN, S., TOGROL, E., and KAYADELEN, C. (2008). "Estimating Compaction Behavior of Fine-Grained Soils Based on Compaction Energy," *Canadian Geotechnical Journal,* Vol. 45, No. 6, pp. 877–887.

OTHMAN, M. A. and LUETTICH, S. M. (1994). "Compaction Control Criteria for Clay Hydraulic Barriers," *Transportation Research Record,* No. 1462, National Research Council, Washington, D.C., pp. 28–35.

PARTOS, A., WELSH, J. P., KAZANIWSKY, P. W., and SANDER, E. (1989). "Case Histories of Shallow Foundation on Improved Soil," *Proceedings, Foundation Engineering: Current Principles and Practices,* American Society of Civil Engineers, Vol. 1, pp. 313–327.

PATRA, C. R., SIVAKUGAN, N., DAS, B. M., and ROUT, S. K. (2010). "Correlations of Relative Density of Clean Sand with Median Grain Size and Compaction Energy," *International Journal of Geotechnical Engineering,* Vol. 4, No. 2, pp. 195–203.

PORAN, C. J. and RODRIGUEZ, J. A. (1992). "Design of Dynamic Compaction," *Canadian Geotechnical Journal,* Vol. 2, No. 5, pp. 796–802.

SHIBUYA, S. and HANH, L. T. (2001). "Estimating Undrained Shear Strength of Soft Clay Ground Improved by Preloading with PVD—Case History in Bangkok," *Soils and Foundations,* Vol. 41, No. 4, pp. 95–101.

STEUDLEIN, A. W. and HOLTZ, R. D. (2013). "Bearing Capacity of Spread Footings on Aggregate Pier Reinforced Clay," *Journal of Geotechnical and Geoenvironmental Engineering,* American Society of Civil Engineers, Vol. 139, No. 1, pp. 49–58.

THOMPSON, M. R. (1967). *Bulletin 492, Factors Influencing the Plasticity and Strength of Lime-Soil Mixtures*, Engineering Experiment Station, University of Illinois.

THOMPSON, M. R. (1966). "Shear Strength and Elastic Properties of Lime-Soil Mixtures," *Highway Research Record 139*, National Research Council, Washington, D.C., pp. 1–14.

TRANSPORTATION RESEARCH BOARD (1987). *Lime Stabilization: Reactions, Properties, Design and Construction*, National Research Council, Washington, D.C.

TULLOCK, W. S., II, HUDSON, W. R., and KENNEDY, T. W. (1970). *Evaluation and Prediction of the Tensile Properties of Lime-Treated Materials*, Research Report 98-5, Center for Highway Research, University of Texas, Austin, Texas.

WELSH, J. P. and BURKE, G. K. (1991). "Jet Grouting–Uses for Soil Improvement," *Proceedings, Geotechnical Engineering Congress*, American Society of Civil Engineers, Vol. 1, pp. 334–345.

WELSH, J. P., RUBRIGHT, R. M., and COOMBER, D. B. (1986). "Jet Grouting for support of Structures," presented at the Spring Convention of the American Society of Civil Engineers, Seattle, Washington.

YEUNG, A. T. (1997). "Design Curves for Prefabricated Vertical Drains," *Journal of Geotechnical and Geoenvironmental Engineering,* Vol. 123, No. 8, pp. 755–759.

APPENDIX A
Reinforced Concrete Design of Shallow Foundations

A.1 Fundamentals of Reinforced Concrete Design

At the present time, most reinforced concrete designs are based on the recommendations of the building code prepared by the American Concrete Institute—that is, ACI 318-11. The basis for this code is the *ultimate strength design* or *strength design*. Some of the fundamental recommendations of the code are briefly summarized in the following sections.

Load Factors

According to ACI Code Section 9.2, depending on the type, the ultimate load-carrying capacity of a structural member should be one of the following:

$$U = 1.4D \tag{A.1a}$$
$$U = 1.2D + 1.6L + 0.5(L_r \text{ or } S \text{ or } R) \tag{A.1b}$$
$$U = 1.2D + 1.6(L_r \text{ or } S \text{ or } R) + (1.0L \text{ or } 0.5W) \tag{A.1c}$$
$$U = 1.2D + 1.0W + 1.0L + 0.5(L_r \text{ or } S \text{ or } R) \tag{A.1d}$$
$$U = 1.2D + 1.0E + 1.0L + 0.2S \tag{A.1e}$$
$$U = 0.9D + 1.0W \tag{A.1f}$$

or

$$U = 0.9D + 1.0E \tag{A.1g}$$

where

U = ultimate load-carrying capacity of a member
D = dead loads
E = effects of earthquake
L = live loads
L_r = roof live loads
R = rain load
S = snow load
W = wind load

Strength Reduction Factor

The design strength provided by a structural member is equal to the nominal strength times a strength reduction factor, ϕ, or

$$\text{Design strength} = \phi \, (\text{nominal strength})$$

The reduction factor, ϕ, takes into account the inaccuracies in the design assumptions, changes in property or strength of the construction materials, and so on. Following are some of the recommended values of ϕ (ACI Code Section 9.3):

Condition	Value of ϕ
a. Axial tension; flexure with or without axial tension	0.9
b. Shear or torsion	0.75
c. Axial compression with spiral reinforcement	0.75
d. Axial compression without spiral reinforcement	0.65
e. Bearing on concrete	0.65
f. Flexure in plain concrete	0.65

Design Concepts for a Rectangular Section in Bending

Figure A.1a shows a section of a concrete beam having a width b and a depth h. The assumed stress distribution across the section at ultimate load is shown in Figure A.1b. The following notations have been used in this figure:

f'_c = compressive strength of concrete at 28 days
A_s = area of steel tension reinforcement
f_y = yield stress of reinforcement in tension
d = effective depth
l = location of the neutral axis measured from the top of the compression face
$a = \beta l$

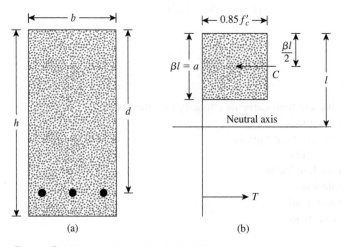

(a) (b)

Figure A.1 Rectangular section in heading

$\beta = 0.85$ for f_c' of 28 MN/m² (4000 lb/in.²) of less and decreases at the rate of 0.05 for every 7 MN/m² (1000 lb/in.²) increase of f_c'. However, it cannot be less than 0.65 in any case (ACI Code Section 10.2.7).

From the principles of statics, for the section

$$\Sigma \text{ compressive force, } C = \Sigma \text{ tensile force, } T$$

Thus,

$$0.85 f_c' ab = A_s f_y$$

or

$$a = \frac{A_s f_y}{0.85 f_c' b} \tag{A.2}$$

Also, for the beam section, the nominal ultimate moment can be given as

$$M_n = A_s f_y\left(d - \frac{a}{2}\right) \tag{A.3}$$

where M_n = theoretical ultimate moment.
The design ultimate moment, M_u, can be given as

$$M_n = A_s f_y\left(d - \frac{a}{2}\right) \tag{A.4}$$

Combining Eqs. (A.2) and (A.4)

$$M_u = \phi A_s f_y\left[d - \left(\frac{1}{2}\right)\frac{A_s f_y}{0.85 f_c' b}\right] = \phi A_s f_y\left(d - \frac{0.59 A_s f_y}{f_c' b}\right) \tag{A.5}$$

The steel percentage is defined by the equation

$$s = \frac{A_s}{bd} \tag{A.6}$$

In a balanced beam, failure would occur by sudden simultaneous yielding of tensile steel and crushing of concrete. The balanced percentage of steel (for Young's modulus) of steel, $E_s = 200$ MN/m²) can be given as

$$s_b = \frac{0.85 f_c'}{f_y}(\beta)\left(\frac{600}{600 + f_y}\right) \tag{A.7a}$$

where f_c' and f_y are in MN/m².
In conventional English units (with $E_s = 29 \times 10^6$ lb/in.²)

$$s_b = \frac{0.85 f_c'}{f_y}(\beta)\left(\frac{87,000}{87,000 + f_y}\right) \tag{A.7b}$$

where f_c' and f_y and in lb/in.²

To avoid sudden failure without warning, ACI Code Section 10.3.5 recommends that the maximum steel percentage (s_{max}) should be limited to a net tensile strain (ϵ_t) of 0.004. For all practical purposes,

$$s_{max} \approx 0.75\, s_b \qquad (A.8)$$

The nominal or theoretical shear strength of a section, V_n, can be given as

$$V_n = V_c + V_s \qquad (A.9)$$

where V_c = nominal shear strength of concrete
V_s = nominal shear strength of reinforcement

The permissible shear strength, V_u, can be given by

$$V_u = \phi V_n = \phi(V_c + V_s) \qquad (A.10)$$

The values of V_c can be given by the following equations (ACI Code Sections 11.2 and 11.11).

$$V_c = 0.17\lambda \sqrt{f_c'}\, bd \quad \text{(for member subjected to shear and flexure)} \qquad (A.11a)$$

and

$$V_c = 0.33\lambda \sqrt{f_c'}\, bd \quad \text{(for member subjected to diagonal tension)} \qquad (A.11b)$$

where f_c' is in MN/m², V_c is in MN, b and d are in m, and $\lambda = 1$ for normal weight concrete. In conventional English units, Eqs. (A.11a) and (A.11b) take the following form:

$$V_c = 2\lambda \sqrt{f_c'}\, bd \qquad (A.12a)$$
$$V_c = 4\lambda \sqrt{f_c'}\, bd \qquad (A.12b)$$

where V_c is in lb, f_c' is in lb/in.², and b and d are in inches.
Note that

$$v_c = \frac{V_c}{bd} \qquad (A.13)$$

where v_c is the shear stress.
Now, combining Eqs. (A.11a), and (A.13), one obtains

$$\text{Permissible shear stress} = v_u = \frac{V_u}{bd} = 0.17\phi\lambda\sqrt{f_c'} \qquad (A.14a)$$

Similarly, from Eqs. (A.11b), and (A.13),

$$v_u = 0.33\lambda\phi\sqrt{f_c'} \qquad (A.14b)$$

A.2 Reinforcing Bars

The nominal sizes of reinforcing bars commonly used in the United States are given in Table A.1.

Table A.1 Nominal Sizes of Reinforcing Bars Used in the United States

Bar No.	Diameter		Area of cross section	
	(mm)	**(in.)**	**(mm²)**	**(in.²)**
3	9.52	0.375	71	0.11
4	12.70	0.500	129	0.20
5	15.88	0.625	200	0.31
6	19.05	0.750	284	0.44
7	22.22	0.875	387	0.60
8	25.40	1.000	510	0.79
9	28.65	1.128	645	1.00
10	32.26	1.270	819	1.27
11	35.81	1.410	1006	1.56
14	43.00	1.693	1452	2.25
18	57.33	2.257	2580	4.00

The details regarding standard metric bars used in Canada are as follows:

Bar number	Diameter, mm	Area, mm²
10	11.3	100
15	16.0	200
20	19.5	300
25	25.2	500
30	29.9	700
35	35.7	1000
45	43.7	1500
55	56.4	2500

Reinforcing-bar sizes in the metric system have been recommended by UNESCO (1971). (Bars in Europe will be specified to comply with the standard EN 100080).

Bar diameter, mm	Area, mm²
6	28
8	50
10	79
12	113
14	154
16	201
18	254
20	314
22	380
25	491
30	707
32	804
40	1256
50	1963
60	2827

This appendix uses the standard bar diameters recommended by UNESCO.

A.3 Development Length

The development length, L_d, is the length of embedment required to develop the yield stress in the tension reinforcement for a section in flexure. ACI Code Section 12.2 lists the basic development lengths for tension reinforcement.

A.4 Design Example of a Continuous Wall Foundation

Let it be required to design a load-bearing wall with the following data:

Dead load = D = 43.8 kN/m
Live load = L = 17.5 kN/m
Gross allowable bearing capacity of soil = 94.9 kN/m^2
Depth of the top of foundation from the ground surface = 1.2 m
f_y = 413.7 MN/m^2
f_c' = 20.68 MN/m^2
Unit weight of soil = γ = 17.27 kN/m^3
Unit weight of concrete = γ_c = 22.97 kN/m^3

General Considerations

For this design, assume the foundation thickness to be 0.3 m. Refer to ACI Code Section 7.7.1, which recommends a minimum cover of 76 mm over steel reinforcement, and assume that the steel bars to be used are 12 mm in diameter (Figure A.2a). Thus,

$$d = 300 - 76 - \frac{12}{2} = 218 \text{ mm}$$

Also,

Weight of the foundation = $(0.3)\gamma_c$ = $(0.3)(22.97)$ = 6.89 kN/m^2

Weight of soil above the foundation = $(1.2)\gamma$ = $(1.2)(17.27)$
$= 20.72$ kN/m^2

So, the net allowable soil bearing capacity is

$$q_{net(all)} = 94.9 - 6.89 - 20.72 = 67.29 \text{ kN/m}^2$$

Hence, the required width of foundation is

$$B = \frac{D + L}{q_{net(all)}} = \frac{43.8 + 17.5}{67.29} = 0.91 \text{ m}$$

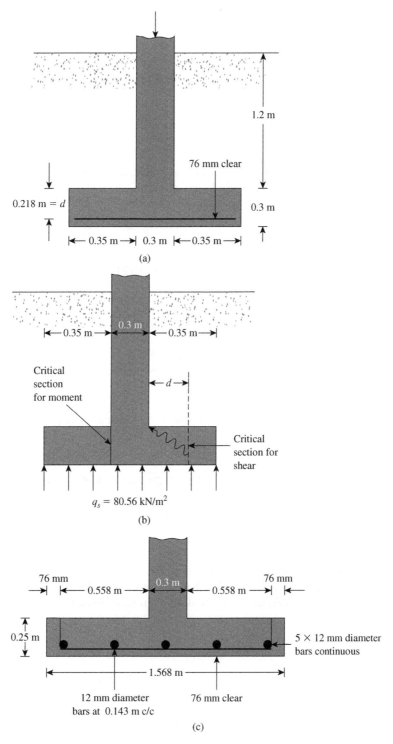

Figure A.2 Continuous wall foundation

So, assume $B = 1$ m.

According to ACI Code Section 9.2,

$$U = 1.2D + 1.6L = (1.2)(43.8) + (1.6)(17.5) = 80.56 \text{ kN/m}$$

Converting the net allowable soil pressure to an ultimate (factored) value,

$$q_s = \frac{U}{(B)(1)} = \frac{80.56}{(1)(1)} = 80.56 \text{ kN/m}^2$$

Investigation of Shear Strength of the Foundation

The critical section for shear occurs at a distance d from the face of the wall (ACI Code Section 11.11.3), as shown in Figure A.2b. So, shear at critical section

$$V_u = (0.35 - d)q_s = (0.35 - 0.218)(80.56) = 10.63 \text{ kN/m}$$

From Eq. (A.11a) with $\lambda = 1$,

$$V_c = 0.17\sqrt{f_c'}\,bd = 0.17\sqrt{20.68}\,(1)(0.218) = 0.1685\,\text{MN/m} \approx 168 \text{ kN/m}$$

Also,

$$\phi V_c = (0.75)(168) = 126 \text{ kN/m} > V_u = 10.63 \text{ kN/m—O.K.}$$

(*Note:* $\phi = 0.75$ for shear—ACI Code Section 9.3.2.3.)

Because $V_u < \phi V_c$, the total thickness of the foundations could be reduced to 250 mm. So, the modified

$$d = 250 - 76 - \frac{12}{2} = 168\,\text{mm} > 152\,\text{mm} = d_{min} \text{ (ACI Code Section 15.7)}$$

Neglecting the small difference in footing weight, if $d = 168$ mm,

$$\phi V_c = (0.75)(0.17)\sqrt{20.68}\,(1)(0.168) = 0.0974 \text{ MN}$$

$$= 97.4 \text{ kN} > V_u\text{—O.K.}$$

Flexural Reinforcement

For steel reinforcement, factored moment at the face of the wall has to be determined (ACI Code Section 15.4.2). The bending of the foundation will be in one direction only. So, according to Figure A.2b, the design ultimate moment

$$M_u = \frac{q_s l^2}{2}$$

$$l = 0.35 \text{ m}$$

So,

$$M_u = \frac{(80.56)(0.35)^2}{2} = 4.93 \text{ kN-m/m}$$

From Eqs. (A.2) and (A.3),

$$M_n = A_s f_y \left(d - \frac{a}{2} \right)$$

$$a = \frac{A_s f_y}{0.85 f'_c b} = \frac{(A_s)(413.7)}{(0.85)(20.68)(1)} = 23.5351 A_s$$

Thus,

$$M_n = (A_s)(413.7)\left(0.168 - \frac{23.5351}{2} A_s \right)$$

or

$$M_n(\text{MN-m/m}) = 69.5 A_s - 4868.24 A_s^2$$

Again, from Eq. (A.4)

$$M_u \leq \phi M_n$$

where $\phi = 0.9$.

Thus,

$$4.93 \times 10^{-3}(\text{MN-m/m}) = 0.9\,(69.5 A_s - 4868.24 A_s^2)$$

Solving for A_s, one gets

$$A_{s(1)} = 0.0128\ \text{m}^2;\ A_{s(2)} = 0.0001\ \text{m}^2$$

Hence, steel percentage with $A_{s(1)}$ is

$$s_1 = \frac{A_{s(1)}}{bd} = \frac{0.0128}{(1)(0.168)} = 0.0762$$

Similarly, steel percentage with $A_{s(2)}$ is

$$s_2 = \frac{A_{s(2)}}{bd} = \frac{0.0001}{(1)(0.168)} = 0.0006 < s_{\min} = 0.0018\,(\text{ACI Code Section 7.12.2.1})$$

The maximum steel percentage that can be provided is given in Eqs. (A.7a) and (A.8). Thus,

$$s_{\max} = (0.75)(0.85)\frac{f'_c}{f_y}\beta\left(\frac{600}{600 + f_y} \right)$$

Note that $\beta = 0.85$. Substituting the proper values of β, f'_c, and f_y in the preceding equation, one obtains

$$s_{\max} = 0.016$$

Note that $s_1 = 0.0762 > s_{\max} = 0.016$. So use $s = s_{\min} = 0.0018$. So,

$$A_s = (s_{\min})(b)(d) = (0.0018)(1)(0.168) = 0.000302\ \text{m}^2 = 302\ \text{mm}^2$$

Use 12-mm diameter bars @ 350 mm c/c. Hence,

$$A_s(\text{provided}) = \frac{1000}{350}\left(\frac{\pi}{4} \right)(12)^2 = 323\ \text{mm}^2$$

Development Length of Reinforcement Bars (L_d)

According to ACI Code Section 12.2, the minimum development length L_d for 12 mm diameter bars is about 558 mm (approximately equivalent to No. 4 U.S. bar). Assuming a 76-mm cover to be on both sides of the footing, the minimum footing width should be $[2(558 + 76) + 300]$ mm $= 1568$ mm $= 1.568$ m. Hence, the revised calculations are

$$q_s = \frac{U}{(B)(1)} = \frac{80.56}{1.568} = 51.38 \, \text{kN/m}^2$$

$$M_u = \frac{q_s l^2}{2} = \frac{1}{2}(51.38)(0.558 + 0.076)^2$$

$$= 10.326 \, \text{kN} \cdot \text{m/m} = 10.326 \times 10^{-3} \, \text{MN} \cdot \text{m/m}$$

$$a = \frac{A_s f_y}{0.85 f_c' b} = \frac{A_s(413.7)}{(0.85)(20.68)(1.568)} = 15.01 A_s$$

$$M_n = A_s f_y \left(d - \frac{a}{2}\right) = A_s(413.7)\left(0.168 - \frac{15.01 A_s}{2}\right)$$

$$\phi M_n \geqslant M_u$$

$$10.326 \times 10^{-3} = 0.9 A_s (413.7)\left(0.168 - \frac{15.01 A_s}{2}\right)$$

and

$$A_s = 0.00016 \, \text{m}^2$$

The steel percentage is $s = \dfrac{A_s}{bd} = \dfrac{0.00016}{(1.568)(0.25)} < 0.0018.$

(*Note:* Use gross area when $s_{min} = 0.0018$ is used.)

Use $A_s = (0.0018)(1.568)(0.25) = 0.000706 \, \text{m}^2 = 706 \, \text{mm}^2$. Provide 7×12 mm bars ($A_s = 565 \, \text{mm}^2$).

Minimum reinforcement should be furnished in the long direction to offset shrinkage and temperature effects (ACI Code Section 7.12.). So,

$$A_s = (0.0018)(b)(d) = (0.0018)\,[(0.558 + 0.076)(2) + 0.3](0.168)$$
$$= 0.000474 \, \text{m}^2 = 474 \, \text{mm}^2$$

Provide 5×12 mm bars ($A_s = 565 \, \text{mm}^2$).

The final design sketch is shown in Figure A.2c.

A.5 Design Example of a Square Foundation for a Column

Figure A.3a shows a square column foundation with the following conditions:

Live load $= L = 675$ kN
Dead load $= D = 1125$ kN

Allowable gross soil-bearing capacity = q_{all} = 145 kN/m²
Column size = 0.5 m × 0.5 m
f'_c = 20.68 MN/m²
f_y = 413.7 MN/m²

Let it be required to design the column foundation.

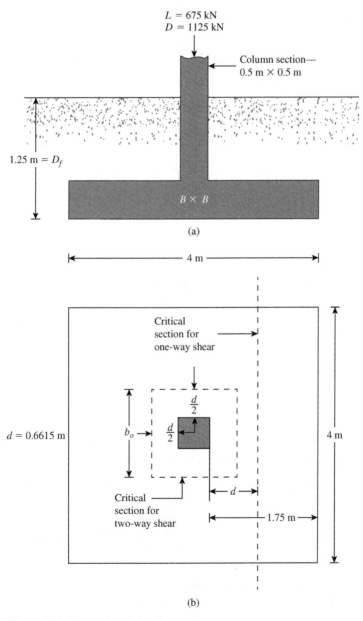

(a)

(b)

Figure A.3 Square foundation for a column

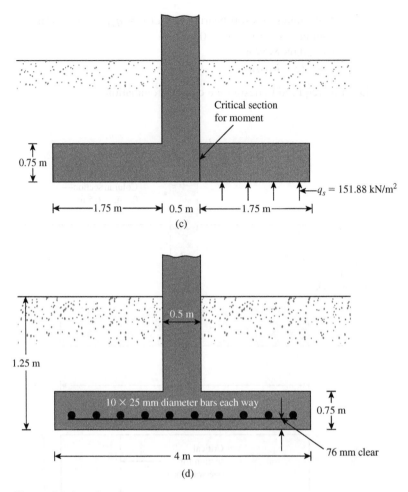

Figure A.3 (*continued*)

General Considerations

Let the average unit weight of concrete and soil above the base of the foundation be 21.97 kN/m³. So, the net allowable soil-bearing capacity

$$q_{\text{all(net)}} = 145 - (D_f)(21.97) = 145 - (1.25)(21.97) = 117.54 \text{ kN/m}^2$$

Hence, the required foundation area is

$$A = B^2 = \frac{D + L}{q_{\text{all(net)}}} = \frac{675 + 1125}{117.54} = 15.31 \text{ m}^2$$

Use a foundation with dimensions (*B*) of 4 m × 4 m.

The factored load for the foundation is

$$U = 1.2D + 1.6L = (1.2)(1125) + (1.6)(675) = 2430 \text{ kN}$$

Hence, the factored soil pressure is

$$q_s = \frac{U}{B^2} = \frac{2430}{16} = 151.88 \, \text{kN/m}^2$$

Assume the thickness of the foundation to be equal to 0.75 m. With a clear cover of 76 m over the steel bars and an assumed bar diameter of 25 mm, we have

$$d = 0.75 - 0.076 - \frac{0.025}{2} = 0.6615 \, \text{m}$$

Check for Shear

As we have seen in Section A.4, V_u should be equal to or less than ϕV_c. For one-way shear [with $\lambda = 1$ in Eq. (A.11a)],

$$V_u \leq \phi(0.17)\sqrt{f_c'}\, bd$$

The critical section for one-way shear is located at a distance d from the edge of the column (ACI Code Section 11.1.3) as shown in Figure A.3b. So

$$V_u = q_s \times \text{critical area} = (151.88)(4)(1.75 - 0.6615) = 661.3 \, \text{kN}$$

Also (with $\lambda = 1$),

$$\phi V_c = (0.75)(0.17)(\sqrt{20.68})(4)(0.6615)(1000) = 1534.2 \, \text{kN}$$

So,

$$V_u = 661.3 \, \text{kN} \leq \phi V_c = 1534.2 \, \text{kN—O.K.}$$

For two-way shear, the critical section is located at a distance of $d/2$ from the edge of the column (ACI Code Section 11.11.1.2). This is shown in Figure A.3b. For this case, [with $\lambda = 1$ in Eq. (A.11b)]

$$\phi V_c = \phi(0.33)\sqrt{f_c'}\, b_o d$$

The term b_o is the perimeter of the critical section for two-way shear. Or for this design,

$$b_o = 4[0.5 + 2(d/2)] = 4[0.5 + 2(0.3308)] = 4.65 \, \text{m}$$

Hence,

$$\phi V_c = (0.75)(0.33)(\sqrt{20.68})(4.65)(0.6615) = 3.462 \, \text{MN} = 3462 \, \text{kN}$$

Also,

$$V_u = (q_s)(\text{critical area})$$
$$\text{Critical area} = (4 \times 4) - (0.5 + 0.6615)^2 = 14.65 \, \text{m}^2$$

So,

$$V_u = (151.88)(14.65) = 2225.18 \, \text{kN}$$
$$V_u = 2225.18 \, \text{kN} < \phi V_c = 3462 \, \text{kN—O.K.}$$

The assumed depth of foundation is more than adequate.

Flexural Reinforcement

According to Figure A.3c, the moment at critical section (ACI Code Section 15.4.2) is

$$M_u = (q_s B)\left(\frac{1.75}{2}\right)^2 = \frac{[(151.88)(4)](1.75)^2}{2} = 930.27 \text{ kN-m}$$

From Eq. (A.2),

$$a = \frac{A_s f_v}{0.85 f_o' b} \qquad (Note: b = B)$$

or

$$A_s = \frac{0.85 f_c' B a}{f_y} = \frac{(0.85)(20.68)(4) a}{413.7} = 0.17 a$$

From Eq. (A.4),

$$M_u \leq \phi A_s f_v\left(d - \frac{a}{2}\right)$$

With $\phi = 0.9$ and $A_s = 0.17a$,

$$M_u = 930.27 = (0.9)(0.17a)(413700)\left(0.6615 - \frac{a}{2}\right)$$

Solution of the preceding equation given $a = 0.0226$ m. Hence,

$$A_s = 0.17a = (0.17)(0.0226) = 0.0038 \text{ m}^2$$

The percentage of steel is

$$s = \frac{A_s}{bd} = \frac{A_s}{Bd} = \frac{0.0038}{(4)(0.6615)} = 0.0015 < s_{min}$$
$$= 0.0018 \text{ (ACI Code Section 7.12)}$$

So,

$$A_{s(min)} = (0.0018)(B)(d) = (0.0018)(4)(0.6615)$$
$$= 0.004762 \text{ m}^2 = 47.62 \text{ cm}^2$$

Provide 10×25-mm diameter bars each way $[A_s = (4.91)(10) = 49.1 \text{ cm}^2]$.

Check for Development Length (L_d)

From ACI Code Section 12.2.2, for 25 mm diameter bars, $L_d \approx 1338$ mm. Actual L_d provided is $(4 - 0.5/2) - 0.076$ (cover) $= 1.674$ m > 1338 mm—O.K.

Check for Bearing Strength

ACI Code Section 10.14 indicates that the bearing strength should be at least $0.85 \phi f_c' A_1 \sqrt{A_2/A_1}$ with a limit of $\sqrt{A_2/A_1} \leq 2$. For this problem, $\sqrt{A_2/A_1} = \sqrt{(4 \times 4)/(0.5 \times 0.5)} = 8$. So, use $\sqrt{A_2/A_1} = 2$. Also $\phi = 0.7$. Hence, the design bearing strength = $(0.85)(0.65)(20.68)(0.5 \times 0.5)(2) = 5.713$ MN $= 5713$ kN. However, the factored column load $U = 2430$ kN < 5713 kN—O.K.

The final design section is shown in Figure A.3d.

A.6 Design Example of a Rectangular Foundation for a Column

This section describes the design of a rectangular foundation to support a column having dimensions of 0.4 m $\times$ 0.4 m in cross section. Other details are as follows:

Dead load $= D = 290$ kN
Live load $= L = 110$ kN
Depth from the ground surface to the top of the foundation $= 1.2$ m
Allowable gross soil-bearing capacity $= 120$ kN/m^2
Maximum width of foundation $= B = 1.5$ m
$f_y = 413.7$ MN/m^2
$f_c' = 20.68$ MN/m^2
Unit weight of soil $= \gamma = 17.27$ kN/m^3
Unit weight of concrete $= \gamma_c = 22.97$ kN/m^3

General Considerations

For this design, let us assume a foundation thickness of 0.45 m (Figure A.4a). The weight of foundation/m^2 = 0.45 γ_c = (0.45) (22.97) = 10.34 kN/m^2, and the weight of soil above the foundation/m^2 = (1.2)γ = (1.2) (17.27) = 20.72 kN/m^2. Hence, the net allowable soil-bearing capacity [$q_{net(all)}$] = 120 − 10.34 − 20.72 = 88.94 kN/m^2.

The required area of the foundation = $(D + L)/q_{net(all)}$ = (290 + 110)/88.94 = 4.5 m^2. Hence, the length of the foundation is 4.5 m^2/B = 4.5/1.5 = 3 m.

The factored column load = 1.2D + 1.6L = 1.2(290) + 1.6(110) = 524 kN.

The factored soil-bearing capacity, q_s = factored load/foundation area = 524/4.5 = 116.44 kN/m^2.

Shear Strength of Foundation

Assume that the steel bars to be used have a diameter of 16 mm. So, the effective depth $d = 450 - 76 - 16/2 = 366$ mm. (Note that the assumed clear cover is 76 mm.)

Figure A.4a shows the critical section for one-way shear (ACI Code Section 11.11.1.1). According to this figure

$$V_u = \left(1.5 - \frac{0.4}{2} - 0.366\right)Bq_s = (0.934)(1.5)(116.44) = 163.13 \text{ kN}$$

The nominal shear capacity of concrete for one-way beam action [with $\lambda = 1$ in Eq. (11.a)]

$$V_c = 0.17\sqrt{f_o'}Bd = 0.17(\sqrt{20.68})(1.5)(0.366) = 0.4244 \text{ MN} = 424.4 \text{ kN}$$

Now

$$V_u = 163.13 \leq \phi V_c = (0.75)(424.4) = 318.3 \text{ kN—O.K.}$$

The critical section for two-way shear is also shown in Figure A.4a. This is based on the recommendations given by ACI Code Section 11.11.1.2. For this section

$$V_u = q_s[(1.5)(3) - 0.766^2] = 455.66 \text{ kN}$$

The nominal shear capacity of the foundation can be given as (ACI Code Section 11.11.2)

$$V_c = v_c b_o d = 0.33\lambda \sqrt{f_c'} b_o d$$

where b_o = perimeter of the critical section

or

$$V_c = (0.33)(1)(\sqrt{20.68})(4 \times 0.766)(0.366) = 1.683 \text{ MN}$$

So, for two-way shear condition

$$V_u = 455.66 \text{ kN} < \phi V_c = (0.75)(1683) = 1262.25 \text{ kN}$$

Therefore, the section is adequate.

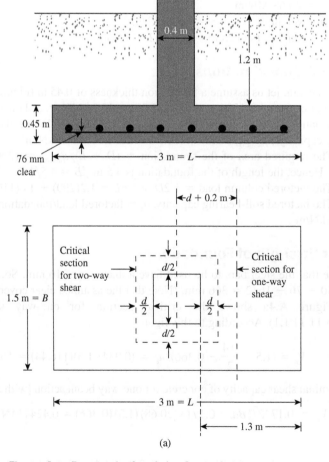

Figure A.4 Rectangular foundation for a column

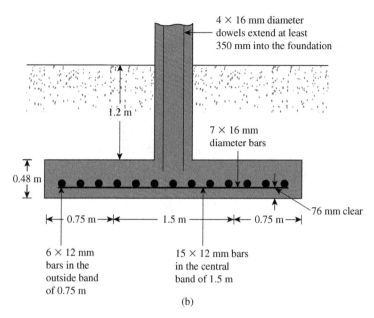

4 × 16 mm diameter dowels extend at least 350 mm into the foundation

1.2 m

7 × 16 mm diameter bars

0.48 m

76 mm clear

0.75 m ── 1.5 m ── 0.75 m

6 × 12 mm bars in the outside band of 0.75 m

15 × 12 mm bars in the central band of 1.5 m

(b)

Figure A.4 (continued)

Check for Bearing Capacity of Concrete Column at the interface with Foundation

According to ACI Code Section 10.14.1, the bearing strength is equal to 0.85 $\phi f_c' A_1$ ($\phi = 0.65$). For this problem, $U = 524$ kN $<$ bearing strength $= (0.85)(0.65)(20.68)(0.4)^2 = 1.828$ MN.

So, a minimum area of dowels should be provided across the interface of the column and the foundation (ACI Code Section 15.8.2). Based on ACI Code Section 15.8.2.1

$$\text{Minimum area of steel} = (0.005) \text{ (area of column)}$$
$$= (0.005) (400^2) = 800 \text{ mm}^2$$

So use 4 × 16-mm diameter bars as dowels.

The minimum required length of development (L_d) of dowels in the foundation is $(0.24 f_y d_b)/\lambda \sqrt{f_c'}$, but not less than $0.043 f_y d_b$ (ACI Code Section 12.3.2). So,

$$L_d = \frac{0.24 f_y d_b}{\lambda \sqrt{f_c'}} = \frac{(0.24)(413.7)(16)}{(1)(\sqrt{20.68})} = 349.33 \text{ mm}$$

Also,

$$L_d = 0.043 \, f_y d_b = (0.043)(413.7)(16) = 284.6 \text{ mm}$$

Hence, $L_d = 349.33$-mm controls.

Available depth for the dowels (Figure A.4a) is $450 - 76 - 16 - 16 = 342$ mm. Since hooks cannot be used, the foundation depth must be increased. Let the new depth be equal to 480 mm to accommodate the required $L_d = 349.33$ mm. Hence, the new value of d is equal to $480 - 76 - 16 - 16 = 372$ mm.

Flexural Reinforcement in the Long Direction

According to Figure A.4a, the design moment about the column face is

$$M_u = \frac{(q_s B)1.3^2}{2} = \frac{(116.44)(1.5)(1.3)^2}{2} = 147.59 \text{ kN-m}$$

From Eq. (A.2),

$$a = \frac{A_s f_y}{0.85 f_c' b} = \frac{(A_s)(413.7)}{(0.85)(20.68)(1.5)} = 15.69 A_s$$

Again, from Eq. (A.4),

$$M_u = \phi M_n = \phi A_s f_y \left(d - \frac{a}{2} \right)$$

or

$$147.59 = (0.9)(A_s)(413.7 \times 10^3) \left[0.396 - \frac{15.69}{2}(A_s) \right]$$

$$147.59 = 147{,}444.7\, A_s - 2{,}920{,}928 A_s^2$$

(*Note: d* = 0.396 m, assuming that these bars are placed as the bottom layer.)

The solution of the preceding equation gives

$$A_s = 0.00102 \text{ m}^2 \left[\text{that is, steel percentage} = \frac{A_s}{Bd} = \frac{0.00102}{(1.5)(0.396)} = 0.0017 \right]$$

Also, from ACI Code Section 7.12.2, $s_{min} = 0.0018$. Hence, provide 7 × 16-mm diameter bars (A_s provided is 0.001407 m²).

Flexural Reinforcement in the Short Direction

According to Figure A.4a, the moment at the face of the column is

$$M_u = \frac{(q_s L)(0.55)^2}{2} = \frac{(116.44)(3)(0.55)^2}{2} = 52.83 \text{ kN-m}$$

From Eq. (A.2),

$$a = \frac{A_s f_y}{0.85 f_c' b} = \frac{(A_s)(413.7)}{(0.85)(20.68)(3)} = 7.845 A_s$$

From Eq. (A.4),

$$M_u = \phi A_s f_y \left(d - \frac{a}{2} \right)$$

or

$$52.83 = (0.9)(A_s)(413.7 \times 10^3) \left[0.380 - \frac{7.845}{2}(A_s) \right]$$

(*Note: d* = 480 − 76 − 16 − $\frac{16}{2}$ = 380 mm for short bars in the upper layer.)

The solution of the preceding equation gives

$$A_s = 0.0004 \text{ m}^2 \qquad (\text{thus } s < s_{min})$$

So, use $s = s_{min}$, or

$$A_s = s_{min}\, bd = (0.0018)(3)(0.48) \approx 0.0026 \text{ m}^2$$

(*Note* : Use gross area when $s_{min} = 0.0018$ is used.)

Use 13 × 16-mm diameter bars.

According to ACI Code Section 12.2, the development length L_d for 16 mm diameter bars is about 693 mm. For such a case, the footing width needs to be $[2(0.693 + 0.076) + 0.4] = 1.938$ m. Since the footing width is limited to 1.5 m, we should use 12-mm diameter bars.

So, use 23 × 12 mm diameter bars.

Final Design Sketch

According to ACI Code Section 15.4.4, a portion of the reinforcement in the short direction shall be distributed uniformly over a bandwidth equal to the smallest dimension of the foundation. The remainder of the reinforcement should be distributed uniformly outside the central band of the foundation. The reinforcement in the central band can be given to the equal to $2/(\beta_c + 1)$ (where $\beta_c = L/B$). For this problem, $\beta_c = 2$. Hence, 2/3 of the reinforcing bars (that is, 15 bars) should be placed in the center band of the foundation. The remaining bars should be placed outside the central band. However, one needs to check the steel percentage in the outside band, or

$$s = \frac{A_s}{bd} = \frac{(2)(113 \text{ mm}^2)}{\left(\dfrac{3000 - 1500}{2}\right)(380)} = 0.00079 < s_{min} = 0.0018$$

So, use $A_s = (s_{min})(b)(d) = (0.0018)(750)(480) = 648 \text{ mm}^2$. Hence, 6 × 12-mm diameter bars on each side of the central band will be sufficient.

The final design sketch is shown in Figure A.4b.

References

American Concrete Institute (2011). *ACI Standard—Building Code Requirements for Reinforced Concrete, ACI 318-11*, Farmington Hills, Michigan.

UNESCO (1971). *Reinforced Concrete: An International Manual,* Butterworth, London.